The Green Cathedral

THE GREEN CATHEDRAL

Sustainable Development of Amazonia

Juan de Onis

A TWENTIETH CENTURY FUND BOOK

New York Oxford
OXFORD UNIVERSITY PRESS
1992

Oxford University Press

Oxford New York Toronto
Delhi Bombay Calcutta Madras Karachi
Kuala Lumpur Singapore Hong Kong Tokyo
Nairobi Dar es Salaam Cape Town
Melbourne Auckland

and associated companies in
Berlin Ibadan

Published by Oxford University Press, Inc.
200 Madison Avenue, New York, NY 10016

Library of Congress Cataloging-in-Publication Data

de Onis, Juan, 1927–
The Green Cathedral: sustainable development of Amazonia/Juan de Onis
p. cm.
"A Twentieth Century Fund book."
Includes bibliographical references and index.
ISBN 0-19-507460-2
1. Man—Influence on nature—Amazon River Region. 2. Economic development—Environmental aspects. Land settlement—Environmental aspects—Amazon River Region. 4. Amazon River Region—Economic conditions.
I. Title.
GF532.A4D4 1992 333.73′0981′1—dc20 91-36783

9 8 7 6 5 4 3 2 1
Printed in the United States of America
on acid-free paper

For my grandchildren
and yours

The Twentieth Century Fund is a research foundation undertaking timely analyses of economic, political, and social issues. Not-for-profit and nonpartisan, the Fund was founded in 1919 and endowed by Edward A. Filene.

Foreword

The clash of interests that is likely to determine the fate of the Amazonian rain forest has created one of the most intense human and public policy problems facing the world today. The dispute is environmentally significant because of the scale and potential global consequences of the region's transformation. It also symbolizes the differing priorities of the developed and developing countries. Much is at stake, and despite the rate of deforestation in the region, the outcome is still in doubt.

In the pages that follow, Juan de Onis, former long-time correspondent for *The New York Times*, looks past the headlines on this subject to present a clear discussion of ways to approach the goal of sustainable development for the people of the region. He explores the stark and uncompromising perspective of First World environmentalists, as well as the desire of Amazonians to partake of the fruits of development. The issues involved, of course, are considerably more complicated than a simple choice between commercialization and preservation.

De Onis expertly describes the approaches offered by the parties in conflict, as well as those seeking a middle way. In his work he also is explicitly prescriptive, discussing, for example, the essential roles of zoning, education, and financial assistance in the efforts to make a constructive transition at least a reasonable possibility. Achieving sustainable growth and a reasonable ecological balance in the Amazon region seems an overwhelmingly difficult task; but this book is the sort of informed and reliable guide that makes the mission at least comprehensible.

The Twentieth Century Fund is pleased to have supported de Onis's extensive research and analysis into this truly global public policy issue. In the past, the Fund has supported examinations of the conflicts between First and Third World nations, and analyses touching on the issue of areas that are claimed by many to be part of the common heritage of man, such as Philip Quigg's *A Pole Apart*, a Fund book that examined the issues surrounding the multinational Antarctic Treaty. De Onis continues that tradition. We thank him for his significant contribution to a wider understanding of a major question of our times.

February 1992

Richard C. Leone, President
The Twentieth Century Fund

Acknowledgments

My gratitude for personal assistance that made this work possible goes first to Murray Rossant, the late director of the Twentieth Century Fund, who set the course. Thanks go also to the Fund for its financial support and to the members of the staff who provided encouragement and editorial guidance, particularly Beverly Goldberg. Invaluable help in gathering materials came from Heloise Fontes in Rio de Janeiro, a resourceful researcher, and Diana Page in Washington, who was very generous with her time and knowledge. I am much indebted, for their hospitality, to Wim and Leticia Groeneveld and Rosemarie Gannon in Porto Velho, David and Marta Luz Atkinson in Brasília, Roger and Beverley Moeller in Paragominas, and the late Ludovico da Riva in Alta Floresta. Norman Gall in São Paulo provided helpful Amazonian contacts, and numerous other colleagues gave me valuable insights, particularly Ricardo Arnt. Many persons quoted in this book gave their views in personal communications with a spirit of collaboration that I value and have tried to reciprocate, but all judgments are entirely my responsibility. Finally, my thanks go to my wife, Manuela, for her patience and good cheer during the long separations required in the preparation of this book.

Preface

Amazonia has the seductive fascination of a riddle. This is an enigmatic world, and the more one seeks answers the more one is enmeshed by coils of contradictions. Enormous forests spring from soils that are so thin that agriculture can barely be sustained. Countless species flourish in the diverse habitats of the world's largest freshwater system, but only a handful of humanity has ever lived there. The dominant myths of the indigenous cultures celebrate the powers of nature in totemic creatures like the boa constrictor, the jaguar, and the alligator; yet this world is so fragile that the environment crumbles and the spirits become impotent at the first contact with modern society.

Nonetheless, Amazonia has a religious power, even for modern man. The primeval forest architecture of Amazonia can inspire an awe that Charles Darwin described when he first encountered the tropical forests of Brazil in 1832. "Wonder, astonishment & sublime devotion, fill & elevate the mind," wrote the great naturalist in his journal on the historic voyage to South America of *HMS Beagle.* Many visitors, before and after Darwin, have experienced the same mystical feeling in the presence of nature's creative power in Amazonia. Euclides da Cunha, one of Brazil's visionary writers, went to the Amazon in 1905 on a diplomatic mission to demarcate an unexplored border between Brazil and Peru. He discovered in that wilderness what he called the "unfinished last page of Genesis."

In an Amazonian forest, man is dwarfed by the trunks of immense hardwood trees that rise from a floor where narrow footpaths wind through a litter of leaves, hanging vines, damp ferns, and rotting logs. These stout columns support a dense canopy that hums overhead with the sounds of insects and bird life at heights of 150 feet or more. Like a stained glass dome, the canopy filters and diffuses the intense solar energy it receives in luminous shafts that dapple the shadowed galleries below with cool light. This humid tropical system, which harbors the most varied collection of living organisms in the world, so amazes scientists that they speak of it in images, like poets. Edward O. Wilson, Harvard University's dean of comparative zoology, has called the canopy of the tropical forests, and the millions of species that live there, "the last great unexplored environment of the planet."

The wonders of biodiversity are part of the Amazonian riddle, but that is not what drew me to study this region. My interest centers on the behavior of the human species in Amazonia. It grew out of observing the struggles of millions of men and women in search of new horizons on the frontier and the political and social dynamics of this historic occupation.

Some scientists speak of events in Amazonia today as if they were seeing the first page of the Apocalypse, not Genesis. They observe with horror the intrusion of powerful forces, driven by age-old ambitions of possession and profit and armed with modern technologies that unquestionably threaten life systems developed by nature over eons. But Amazonia is not a sacred grove reserved for worshippers of the wild, naturalists, or even the indigenous peoples who are studied by ethnographers. This region has been disturbed by forces of history and economics that have brought humanity into such proximity and intercourse that no part of the globe is free of intrusion. The Amazonian riddle is how to convert this disorderly occupation into a human society that can achieve environmentally sustainable economic development.

* * *

When I first encountered the Amazonian setting in 1963 as a foreign correspondent for *The New York Times* in Brazil, the new inland capital, Brasília, installed on the central plateau, made the Amazon basin seem no longer remote. The highway leading north from Brasília to Belém, the gateway port of the Amazon, was still a hazardous truck ride through quagmires of red mud in the rainy season, but the link to the interior had been created.

It took six days to drive from Goiás to Belém on a flatbed truck carrying a load of metal sheets. Joel, a durable driver from São Paulo, sped through the dry savannas of central Brazil, passing rangelands with thin cattle browsing on sparse, twisted trees. As we descended east of the Araguaia river, the red road entered a new world of increasingly dense forests. We loaded diesel fuel from steel drums, drank warm beer, and had an occasional hot water shave and shower at truck-stop towns, like Imperatriz and Paragominas, that are now substantial cities. Along the highway one saw frequent small land clearings where settlers had planted their first plots of rice, manioc, and corn among charred tree stumps. The surrounding rainforest seemed so immense that one could not imagine its being laid waste.

Three years later I had my second look at the new Amazon. The setting had already changed. I visited a vast ranch that was being cleared west of the Araguaia on the border between southern Pará and Mato Grosso. This had been the land of the Xavantes and Karajas, great warriors, who had a reserve on the Bananal, a huge island in the Araguaia. The island was being used by the new ranchers to ford their *zebu* cattle on foot into the virgin territory.

These were not small settlers. My trip had been arranged through a wealthy São Paulo businessman and coffee planter who had a stake in the new ranch. I flew in on a twin-engine plane, carrying supplies to the base camp where thousands of acres of forest were being hacked down by gangs of tough farmhands contracted along the Belém–Brasília highway. In the dry season, the cut forest was burned and the cleared land was seeded to provide pasture for the cattle. There was no road to truck out lumber, so there wasn't even a sawmill for the huge fallen trunks.

The work gangs cleared thousands of hectares of virgin forest in each dry season, and when put to the torch the fire-fronts stretched for miles. But when the rains came, the seeded grass called *coloniao* (panicum maximum) grew higher than the cows, and everyone connected with the ranch projects seemed sure that the Araguaia would become the great new cattle center of Brazil.

Two decades later, ranchers on the Araguaia no longer considered southern Pará a promised land. Many painful lessons had been learned. The nutrients in the poor soils of the region were soon exhausted; insect pests and weeds destroyed the pastures; the weight loss of cattle in the dry season was so severe that four-year-old animals, ranging on open pastures, produced 100 kilos of dressed meat, half that of a São Paulo steer. There were also growing social problems. Ranchers confronted an influx of peasants who were ready to fight for a piece of land, and the Indian communities had become more militant and better organized in defense of their tribal lands.

One of the most famous ranches, Suiá Missu, which once covered 600,000 hectares, has changed hands three times and is now owned by a subsidiary of ENI, the Italian state oil company, which is trying to solve the problems through a pasture research effort. The Rio Cristalino ranch opened by Volkswagen has been sold to a Paraná soybean farmer who is planting rubber trees. Answers are coming hard, and the most difficult problems are those of the tens of thousands of small farmers who have little access to credit, technical assistance, or marketing systems.

To my inexpert eye, the later problems were not apparent in 1965, when opening the Amazon seemed like a tropical version of the American march to the West. Who could argue that Brazil, with its dreams of nationhood, should do otherwise?

My second trip also provided my first direct contact with Amazonian Indians in tribal cultural conditions. From Conceição do Araguaia, a small river town with an airport, I flew with an American missionary pilot to a Kayapo Indian settlement at Baú, deep in the Amazon jungle beyond the Xingú river. This was the real frontier, I thought. Here were Indians, with crimson and black markings on their naked skins and stone discs lodged in their lower lips, with choke collars made from jaguar teeth and mother-of-pearl shells from a river mollusc. A group of Indian men returned from a nearby river bearing handwoven baskets filled with fish stunned by juices of native plants. It was December 31, and I remember spending that New Year's Eve watching nubile maidens and muscular braves perform a ritual dance. It had nothing to do with the Gregorian calendar, but with other mysteries.

What I also discovered at the Brazilian National Indian Foundation outpost at Baú was how vulnerable this supposedly protected culture was to outside influences. One member of the staff, a young anthropologist who had studied at the University of Pernambuco in Recife, said he had come to that remote Indian village mainly to escape political repression. In those days, the military intelligence services were arresting student leaders and purging universities of opponents of the military regime that had overthrown President João Goulart in 1964. The Fourth Army in Recife was particularly tough.

During a long evening of quiet talk in the jungle, the anthropologist said he was there under an assumed name, protected by friends and in touch with the underground political movements in Recife and São Paulo. I never saw him again, but years later, far from Brazil, I was reminded of this young radical after reading press reports of a guerrilla movement that was operating in the Araguaia–Tocantins region. There were only a few hundred guerrillas, but it took a special army command with thousands of troops trained in jungle warfare to eradicate the insurgents.

This experience influenced Brazilian military thinking on the "geopolitical" importance of the Amazon region. The economic development and territorial occupation of the eastern Amazon had been a strategic concept for many years. The insurgency underscored the need for settlements in the region to provide protection for major public investments in transportation, energy, and mining that were being installed in Amazonia. These "megaprojects" catalyzed the occupation that has unleashed such profound changes in the region.

My early Amazonian experiences also included trips made in Peru between 1965 and 1967 with Fernando Belaunde Terry, an urbane architect who had been elected president of Peru in 1963. He had studied Incan road building and believed that a system of highways along the Andean piedmont, girdling the low Amazonian forests, would make fertile valleys in the eastern region accessible for settlement of peasants from the Andean highlands who were pressing for agrarian reform. Belaunde saw himself as an argonaut of a new civilization in the Amazon that would stretch on an arc along the Andean piedmont from northern Bolivia to Venezuela, with interfluvial connections to the Paraguay–Paraná river system in the south and the Orinoco river in the north.

In practice, the peasants Belaunde wanted to settle have been the forerunners of the Maoist Shining Path guerrillas and the cocaine paste producers that have turned Peru's upper Amazon valleys into combat zones and extended a chain of drug operations through the western Amazon from Bolivia to Colombia, with serious damage to the environment, as well as to law and order.

Another factor leading to the occupation of Amazonia was oil. By 1960, oil had been found in Colombia along the Putumayo river, an Amazon tributary, by a Texaco–Gulf consortium. This discovery was followed by similar finds in Ecuador. Peru was the next to hit eastern jungle oil northwest of Iquitos. I visited these oil fields, where roads and pipelines built over the Andes to Pacific ports were followed by settlers, peddlers, and prostitutes, who hung their hammocks in the trees.

Amazon oil finds had lethal political consequences. Belaunde had financed his road-building projects with international loans, and he intended to pay for his Amazon program with the new oil money. But when the oil began to flow, Belaunde was overthrown by Peruvian military nationalists, who had other priorities. The jungle highway system was largely abandoned, and money went into state mining, fishing, and industrial enterprises and the purchase of French Mirage fighters and Soviet tanks. The arms race was joined by Ecuador, where the military also took power soon after Amazon oil was discovered.

In Brazil, the expansion into Amazonia continued with new highways, hydroelectric power dams, and mining camps springing up all over the Amazon basin. A five-day trip in 1984 from Cuiabá, capital of Mato Grosso, to Porto Velho, the capital of Brazil's new western state of Rondônia, found the frontier erupting with new settlers. They took on hardships with the resolve of the pioneer who does not look back. The display of machine power taming some of nature's toughest terrains made public heroes of engineers, like Bernardo Sayao, builder of the Belém–Brasília, who was killed by a falling tree.

All of this had the hallmarks of other modern sagas of colonizing the wilderness, but I had begun to see that there were dimensions that made Amazonia different. For political, social, ecological, and cultural reasons, comparisons with the American West or the Australian outback or the virgin lands of Siberia were not transferable to the tropical habitat, although in some respects, such as mining, there are similarities.

By the mid-1970s there was a growing outcry from ecologists and anthropologists over the reckless environmental costs being incurred in Amazonian development projects. Whenever I returned from my treks to the interior, my desk in Rio was piled high with mail from international environmental organizations, scientific committees, and church groups containing documented condemnations of ecological destruction, violation of indigenous cultures, and rural violence against peasant settlers. Amazonian affairs, far from being remote, had become burning issues for militant groups in the United States and western Europe, and for academic and scientific communities both abroad and within the Amazonian countries. The Roman Catholic Church, in a revival of its missionary traditions, was particularly impassioned over the violence being done to indigenous peoples.

It was clear that Amazonia had become a primary scenario for humanity's new concerns over protection of the environment. The future of tropical forests, along with that of the oceans, the atmosphere, Antarctica, and other areas of the "global commons," has given rise to new dimensions in international relations. Treaties have been negotiated that seek to reduce industrial pollution producing "acid rain" and the use of chemical products that appear to be depleting the ozone layer that filters out incoming ultraviolet rays. International conventions on reducing emissions of "greenhouse" gases and for protection of biodiversity are under negotiation.

An important role in this "global habitat diplomacy" is played by international development banks that lend billions of dollars to Third World countries seeking economic growth. These banks depend for their money on the legislatures of industrial countries which have been put under severe pressure by environmentalist movements to make lending to tropical countries subject to "green" conditionalities that protect the environment and tribal peoples. This adds a new factor to international tensions over the Third World's $1 trillion foreign debt because it brings First World pressures for environmental good conduct on countries that are struggling to develop their resource-based economies. Will the creditors convert these debts into financial instruments that can pay for protection of tropical environments that are of global concern? Is there really a common future that can serve as a bond between the rich North and the increasingly numerous poor of the South?

These are questions of ecological politics that are critical for the future of Amazonia. This immense domain in the heartland of South America offers mankind the best chance it may have to develop a major tropical habitat in a life-sustaining way. If, on the contrary, the political and social forces at work in Amazonia produce an ecological debacle, the whole world will pay a price that will only be measured fully after the damage is done. There is an ethical imperative for cooperation between the countries that are sovereign in Amazonia and the industrial nations that, as consumers of raw materials and sources of capital, can influence policies on growth and environmental protection beyond their borders. The common objective must be the discovery of new and better patterns of economic development that combine efficient use of the region's abundant resources with the stability of its unique ecological systems. This is the intricate challenge of the new Amazon frontier, which is more in the mind than in the trees.

Contents

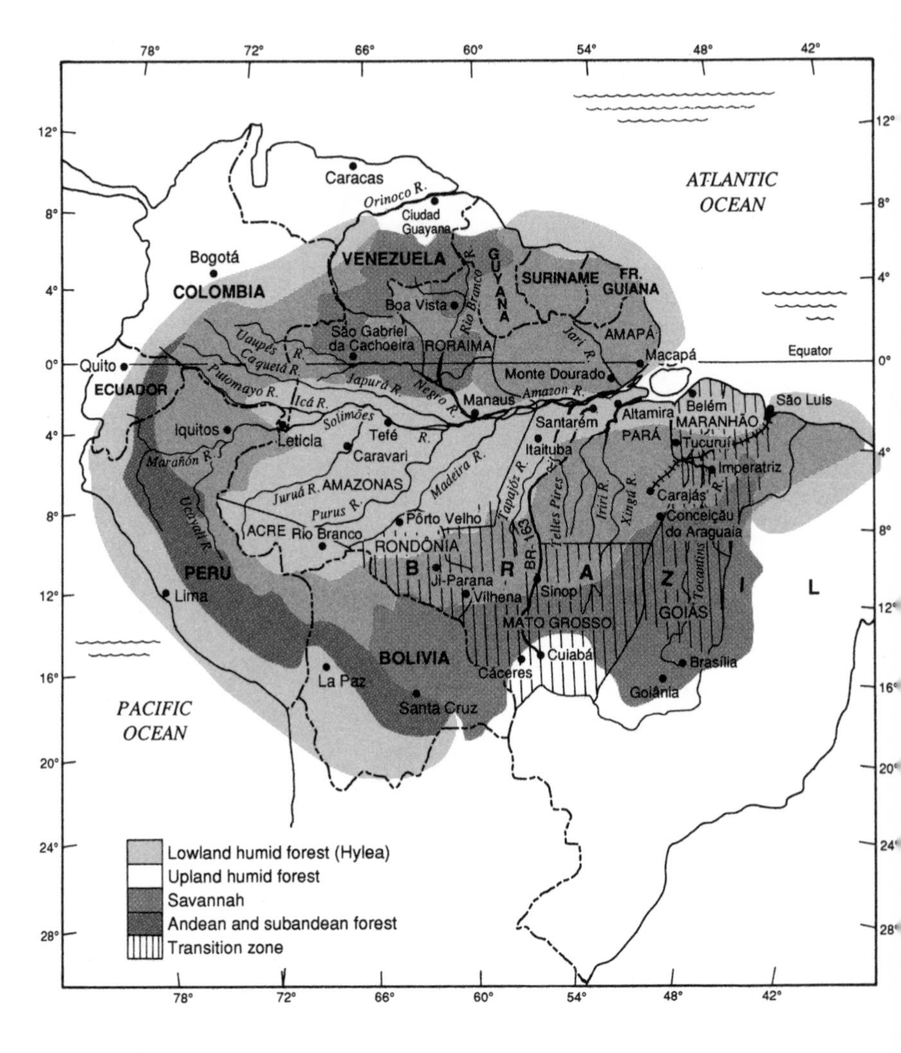

The hydrographic basin of Amazonia, which covers about 7 million square kilometers and contains a great diversity of biological "lifezones," from the dense, closed-canopy forest (*hylea*) to a transition zone of open forests, woody brush, palm stands, and savannahs.

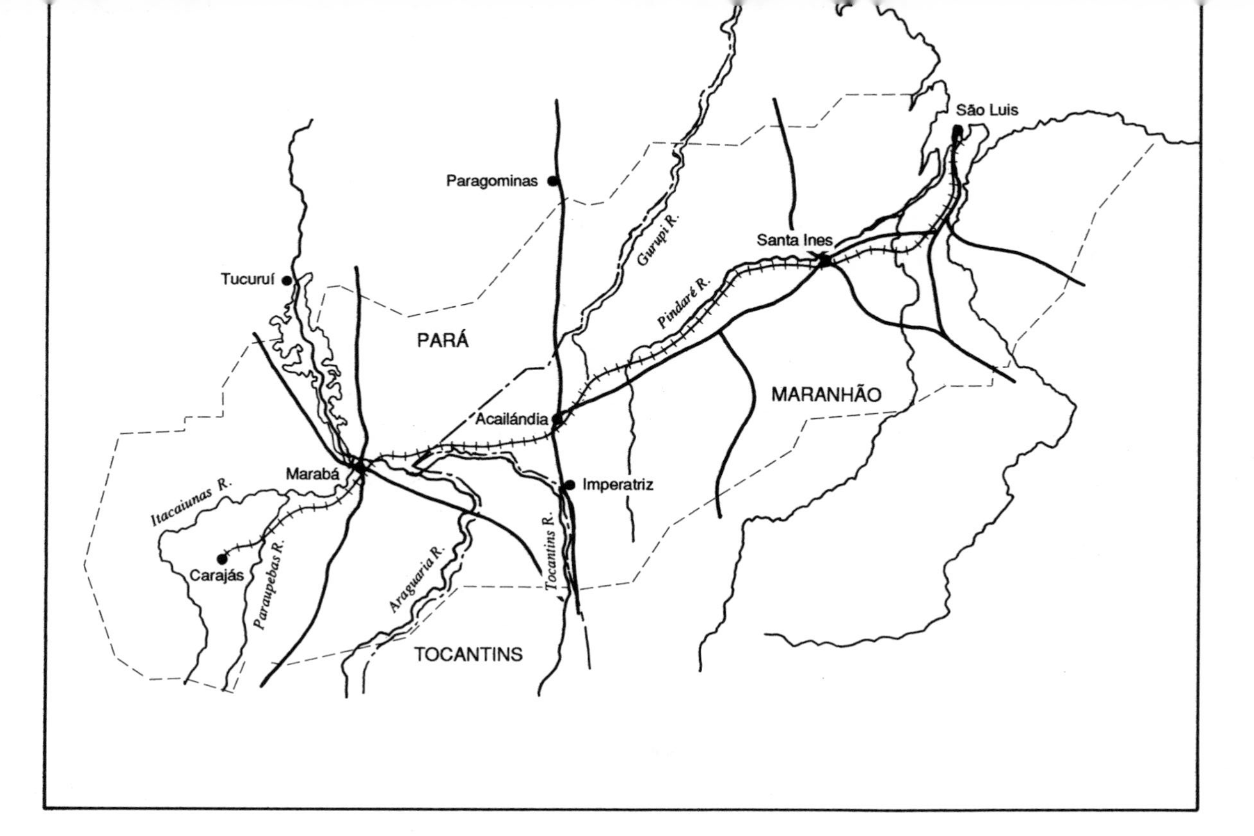

The Carajás corridor, combining highway, railroad, and power projects, is the most important route into Amazonia of the modern era. The railroad from the mines of Carajás to the Atlantic was a key component of a grandiose development scheme for the region.

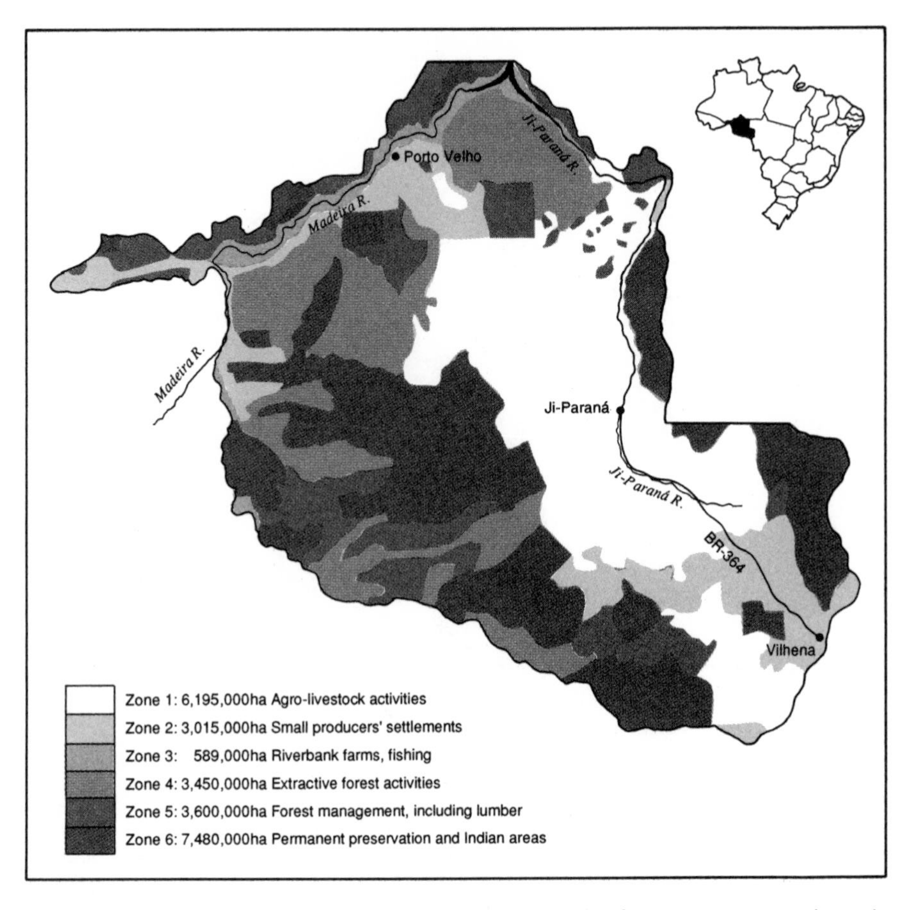

The Rondônia Agro-Livestock and Forestry Plan was the first major agro-ecological zoning blueprint developed for the Amazon region. The plan divided Rondônia into six zones, with forest conservation zones (4, 5, and 6) constituting 60 percent of the total area.

PART I

There is no "natural" ecology. Man has changed everything in nature. . . . There are no resources, only human resourcefulness.

René Jules Dubos

ONE

Introduction

All observation must be for or against some view if it is to be of any service.
Charles Darwin

Amazonia has acquired prominence in the eyes of the world with an aura of cataclysm. The vast tropical forest system of the Amazon basin that occupies the heartland of South America is undergoing severe changes. The cause for alarm is a recent human occupation, unprecedented in scale, mobility, and destruction of the natural environment. Scientists issue dire warnings that ecological blunders in this unusually sensitive quarter of the biosphere can bring unwanted global environmental changes, such as climate warming, and massive extinction of species. From Washington to Moscow, political leaders, "green" militants, rock singers, environmental evangelists, and protectors of primitive cultures and threatened species chant in chorus: "Save the Amazon!"

International response to Amazonian developments has been targeted by environmental activists and mobilized by the world media. "Torching the Amazon: Can the rain forest be saved?" asks the cover of *Time* magazine. *The Economist*, in a dirge titled "The Vanishing Jungle," warns of the disappearance of virtually all tropical forests by the year 2000. A television series, "A Decade of Destruction," exposed millions of living room viewers to a vivid chronicle of the violations of the primeval Amazon forests and its native Indians. "Can Man Save This Fragile Earth?" was the gloomy theme chosen by *National Geographic* magazine for its centennial edition.[1]

These are valid questions when considering the contemporary Amazon region. They bring a sense of urgency and a plea for action to discussions, long confined to scientific circles, on the disturbing events in Amazonia. This region had been considered an isolated backwater, remote from any international conflict or perceived crisis. This is no longer the case. Amazonia figures high on the "global common future" agenda as a region for special international cooperation. Brazil, the host for the United Nations Conference on Environment and Development (UNCED) in 1992, has been chosen as the proving ground for an ecological rescue program to save the world's rainforests.

The environment of Amazonia is threatened because a new breed of native settlers and exploiters of natural resources entered this habitat without a thought to the sustainability of their behavior. This has happened rapidly, with little time to adapt; more than half of the 16 million people now living in Amazonia have migrated there in the past three decades. The main attraction was cheap land that was made accessible by new roads, airports, and telecommunications. Large ranchers and homesteaders have reduced some 500,000 square kilometers of forest to ashes. Fortune seekers are plundering the rivers of gold and the forests of timber. Mining companies have built railroads to remote jungle sites where the earth is blasted and clawed open by ravenous machines. Gigantic hydroelectric dams impound the waters of wild rivers and the forest canopy trembles as helicopters airlift oil rigs to remote locations where geologists drill.

The world—including the people who live in Amazonia—has little doubt about the extent of the environmental damages associated with economic activities in the region. Satellite images provide accurate measurements of enormous deforested areas. Scientists sample riverbeds dangerously polluted by mercury used in gold extraction. Meteorological stations record erratic changes in rainfall and atmospheric gases. Health teams combat epidemic increases in malaria among settlers and the devastating effects of communicable diseases on Indian communities invaded by gold miners.

These are some of the ecological and social costs that economic expansion has inflicted on the world's greatest tropical forest. This occupation has been the consequence of public policies that conform to national objectives of the countries of the region. The opening up of the heartland of South America is an event of considerable historic importance for seven countries and an overseas territory of France that share the region. Many of the participants are proud actors in a genuine frontier saga. Yet, economic benefits have not lived up to the expectations of the planners, and the ecological and social costs have been unexpectedly high.

This painful reality is accepted by most political leaders and public opinion in the Amazon countries. But what should be done about it? The settlement that has taken place cannot be reversed. What are the practical solutions that will provide a livelihood for these settlers? What priority should be given to preserving tropical forests for ecological reasons? These programs are costly, and the Amazonian countries face other pressing Third World environmental problems bred by poverty in their swollen cities and famine-stricken rural populations, as well as air and water pollution in their industrial centers. At the same time, they are struggling to create jobs, raise living standards, and generate export surpluses that will finance economic development. Are the wealthy industrial countries going to help them pay for ecological programs on a global scale?

This study subscribes to the view, supported by an abundant literature, that it is imperative that efforts be made to reduce the destruction of the tropical humid rainforests that girdle the earth's equator.[2] The reasons are as much sociopolitical as purely ecological. This belt of frost-free lands, abundant in water and year-round sunshine, is where two-thirds of mankind will be trying to make a living if the human population reaches 8 billion people, as is expected by the year 2125. Can the environments of the tropics, stripped of their forests, maintain the water and land fertility to produce the food, fuel, and fibers, or equivalent income, for 6 billion people? This is one of the life-or-death questions facing the coming century. The outcome will have a powerful impact on North–South political and economic relations.

In the case of Amazonia, there is more to this than the demographic alarm and the humanitarian compassion that are both awakened by the depressing spectacle of how the poorest half live and die in the Third World. Only a small fraction of the new tropical population will be in Amazonia. But they can produce a deforestation that will alter the ecology of a watershed that drains 5 percent of the world's land surface. In the opinion of some scientists, this could disrupt rainfall cycles in South America and produce global climate changes.

Amazonia's evergreen treetop canopy covers nearly 40 percent of the world's remaining tropical humid forests. The annual burning to clear land involves so much combustible biomass that scientists who study the chemistry of the global atmosphere consider the fiery deforestation of Amazonia a major source of emissions of carbon dioxide. Growing concentrations in the atmosphere of CO_2 and other "greenhouse" gases that trap earthly heat radiation could increase global temperatures, according to many scientists. Although climate modelers differ on the degree and speed of change, many predict severe climate alterations, with drastic consequences not only for the Amazon region, but for the entire world. This places Amazonia in the context of the global climate change debate, where the relative importance of biomass burning compared with fossil fuel emission has unleashed a North–South political controversy.[3]

Amazonia's potential role in global climate change is closely connected with loss of biodiversity in the world's greatest natural gene bank. Amazonia contains enormous areas that are untouched by the kind of population pressures that are destroying the tropical forest regions of southeast Asia. But in thinly settled Amazonia, humans are migrating into heretofore "empty" spaces that teem with a unique variety of other species. In some "hot spots," the diversity of plants, insects, and fish species is regarded as the highest in the world. Botanists have counted 297 species of trees coexisting in 1 hectare near Manaus. Loss of tree cover is presented by biologists as the cutting edge of a process of genetic extinction on a scale that has not been seen since the last ice age 15,000 years ago. They argue that as Amazonia goes, so will go mankind's largest repository of species. This is the "biodiversity for humanity" question that has made conservation of genetic resources the subject of another North–South controversy. Supposing that scientific researchers and genetic engineers create "miracle" seeds and pharmacological wonders from the genetic structures of Amazon species, who is going to receive the economic benefits?

These Amazonian environment issues undergo constant debate, often with alarmist overtones of "ecocide." Some of the more radical views have proven to be simply bad guesswork. One widely publicized forecast, based on mathematical projections, claimed in 1982 that all Amazonian primary forests would be gone in one human generation.[4] Time has shown that this is an implausible scenario. Most of Amazonia's rainforest ecosystems are still intact and capable of being protected. But there are sobering examples around the world of regional ecological disasters, such as the drought-stricken Sahel region of Africa, that serve as a warning of how fast things can go wrong if there are neglect and lack of timely action. So, drawing more attention to Amazonia, even with an alarmist edge to attract public attention, is better than omission. Better yet is more accurate monitoring.

With all things considered, this study finds the outlook for the tropical forests of Amazonia less alarming than others do, not because the dangers are unreal, but because governments and peoples in Amazonia are beginning to show that they can learn from past errors to deal in new and better ways with their natural environment. Amazonia is not ecologically ruined or economically crippled. There is time to bring about changes in human behavior that underlie the dangers to the ecosystems if a multinational North–South action plan is politically concerted and adequately financed.

The rate of deforestation in the Brazilian Amazon, after reaching an annual peak in 1987, dropped sharply, by as much as 60 percent in 1990, because people found forest burning economically less attractive than during early stages of frontier occupation. The governments of the Amazonian countries have responded to the international outcry for protection of tropical forest by strengthening environmental controls, increasing protected areas, and reversing tax and credit policies that led to deforestation. There is less resistance to foreign assistance for environmental enforcement and scientific research programs. Most importantly, there are signs on the Amazon frontier that cattle ranchers, large-scale miners, land developers, small farmers, lumbermen, and forest product extractors are beginning to change behaviors that have been the main causes of environmental damage.

Environmental reform in the Amazonian context is being put to a crucial test in Brazil, the world's principal tropical rainforest country. The First World's "Group of 7" leaders, at their summit meeting in London in June 1991, gave financial and scientific backing to a Brazilian Tropical Rainforest Conservation project that is supposed to make the Amazon the proving ground for sustainable management of tropical rainforests all over the world.[5] This unprecedented act of international cooperation on Amazonian problems presents an opportunity to reverse the environmental degradation of the region, symbolized by the flaming forests during the "decade of destruction," and bring forth sustainable economic growth from its vast natural resources.

This turn of events toward environmental responsibility is at variance with much received wisdom. Many contemporary studies of Amazonia have tended to record and denounce the environmental effects of the occupation as if predatory forms of economic exploitation were inevitable.[6] These studies present the interaction between humans on the Amazon frontier in terms of class warfare that is projected into the area by social inequality and economic exploitation in the national societies. This model establishes categories of evildoers and victims in a state of permanent conflict. Environmental and social violence go together and are usually blamed on "agents of capitalist expansion," on the "geopolitical strategies" of military planners during authoritarian periods, on landowners resisting agrarian reforms who force rootless peasants to destroy forests, or simply on "greed and plunder." More recently, foreign bankers have been added to the blacklist for allegedly forcing Latin American debtor nations to lumber their rainforests and pollute their rivers for timber and gold to pay their debts.

Rural violence on Amazonia's frontiers is endemic. The Land Pastoral Commission (CPT) of the Brazilian Roman Catholic bishops recorded 1,655 homicides in Amazonia's rural areas in the decade ending in 1990. The victims include priests, nuns, and lay workers involved in peasant organizations, as well as leaders and lawyers of rural unions, of which the most famous is Francisco "Chico" Mendes of the Acre rubbertappers. These conflicts attracted the attention of human rights organizations such as Amnesty International, which has given Amazon violence a high profile and put pressure on Brazilian and Peruvian authorities to investigate and prosecute cases in which killers, including police, have been identified. Most cases never get to court and impunity remains widespread.

The field research for this inquiry into human behaviors that underlie modern changes in the Amazon environment encountered abundant evidence of violence and lawlessness. But conflict analysis tends to overlook, or minimizes, experiences that show a potential for cooperation and social development. These moves toward sustainable growth are insufficiently reflected in the existing literature. The peasants and Indians have not lost every battle, and through better organization and political alliances they are forcing public authorities to deal more fairly with their frontier problems. Throughout the Amazon, from Ecuador and Bolivia to Colombia and Brazil, the Indian communities are acquiring rights to indigenous land and political influence that they never had before. Small peasant cooperatives and rural unions are a major force in many frontier townships, and wildcat miners are not pushed around lightly when it comes to mining claims.

Many examples are given in coming chapters that illustrate constructive changes in behaviors that are part of the "new frontier." Two cases, both in areas of well-documented violence in the past, will serve as an introduction to this basic theme.

The *fazenda* Bela Vista, 150 kilometers north of Conceição da Araguaia, is one of the many large ranches in the "Parrot's Beak," an area between the Tocantins and Araguaia rivers known for violence between ranchers and peasants. The zone was placed at one time under military control when a minor guerrilla insurgency emerged in 1971. Bela Vista was a project of a big São Paulo autoparts manufacturing company, Bardella, which was attracted to Amazonia by federal tax incentives that made colonization projects seem cheap. But after clearing 5,000 hectares of the fazenda's 17,500 hectares for cattle pasture, Bardella found it had to invest more money than it had planned and it abandoned the ranch in 1978. Bela Vista was promptly invaded by 400 peasant families. Bardella then sold the property to Jurandir Goncalves Siquiera, a big São Paulo cattle rancher. The conflict began when the new owner tried to repossess Bela Vista with a squad of hired gunmen under covert police protection.

Oity Faria Leite, a leader of the Bela Vista peasant resistance, is a former policeman from Minas Gerais who went to the Araguaia valley with five brothers looking for land. He got a lot at an official agrarian reform site, but the land was so poor and roads were so bad that it collapsed. The peasants in search of better land invaded Bela Vista peacefully, but when the new owner brought pressure for eviction, Oity organized a peasant militia to protect their land. In 1987, he was listed by Amnesty International as a "prisoner of conscience" when he was jailed for two months and tortured by police after a clash in which four peasants (including two children) and one gunman were killed. The Bela Vista peasants held their ground, and the National Institute for Agrarian Reform and Colonization (INCRA) expropriated the ranch in their favor in 1988.

With the land conflict settled, Oity saw that without capital or equipment the Bela Vista peasants would need outside help to make the land produce. Through a rural union lawyer who had defended him, Oity contacted a private center for Indian and peasant agriculture in Goiás, run by Wanderley de Castro, a rural sociologist, who helped Oity's group of 80 peasant families form an "agroecological association." With assistance from Rolf Wagner, a German "regenerative" agriculture technician from the Brazilian Gaia Foundation, the peasants adopted new agricultural technology that give them a basis for stable production without having to deforest more land.

The system adopted after demonstration trials is a succession of rice and corn in consortium with legumes, such as *guandu* and *mucuna* beans, that fix nitrogen. There is no soil tillage and special care is given to maintaining organic mulch on the ground for protection and fertility. Oity, who has a wispy blond beard and sad blue eyes, smiles when he talks about the annual yields obtained in the second year of cultivation on the same ground: 2,000 kilos of rice, 3,500 kilos of corn, and 3,000 kilos of *guandu*, which has a market as seed in southern Brazil. Those are good yields for a low-cost organic system. The association bought a rice huller and a truck, with private financing, and the peasants decided to diversify commercial production to include *inhami*, a starchy tuber. "*Inhami* has a good urban market, and this will be our best cash crop," said Oity, for whom the struggle is no longer for a piece of land but for a share of the market.[7]

Bela Vista's successful resistance is an example of what can be done by peasant agriculture in Amazonia, and it is not an isolated case. Jean Hebette, a former French priest and rural sociologist who is a leading Amazonian researcher at the University of Pará in Belém, has created a peasant agriculture research center at Marabá, on the Tocantins, that serves four big rural workers' unions representing 40,000 small holders or farm workers who aspire to land. "The peasant organizations in the Araguaia–Tocantins have demonstrated their ability to fight off land speculators who have tried to drive them from the land. . . . It is to be expected that as industrial capital becomes more important in the region, the state will be less dominated by the big landowners. Industry doesn't want land conflicts, it wants cheap labor," said Hebette, who is an advocate of family farming. He believes that small rural proprietors will also resist being reduced to the condition of a worker proletariate.[8]

Another example comes from the northeast of Mato Grosso, an area occupied by the biggest corporate ranches. One of these, the Companhia de Desenvolvimento do Araguaia (CODEARA), figures prominently in the literature on modern Amazonia as a symbol of occupation of Indian lands and eviction of old-time peasant settlers. CODEARA, a São Paulo agricultural investment company, took over virtually a whole township, 300 kilometers south of Conceição do Araguaia, with an initial claim in 1975 based on a state land grant of 350,000 hectares. The violence included an attempt to eradicate the old village of Santa Terezinha while forests were cut and burned by gangs of peons to open cattle ranges. The determined resistance of a French priest, Francisco Jentel, who was arrested and deported, and of his bishop, Dom Pedro Casaldaliga, drew international attention to CODEARA's methods, and under government pressure the company distributed land to about 100 local farmers and conceded the town a site of 517 hectares.[9]

CODEARA'S complete domination of the Santa Terezinha area still continued when Judith Lisansky, researching a Ph.D. at the University of Florida, arrived for a year's residence in the town in 1978. As she saw it then, "the cattle fazendas . . . relations with the local population could best be described as a 'cold war.'" Good land was inaccessible to small farmers. Lack of credit, technical support, or farm-to-market roads was forcing peasants settled by INCRA to sell their lots. The town was dying, and merchants were planning to abandon Santa Terezinha and relocate at Vila Rica, a new town to the northwest. It was a picture of increased land concentration in a few big ranches on a frontier "closed" to the peasants by capitalist agriculture.[10]

Yet, when Lisansky returned in 1987, after less than a decade, she was "overwhelmed by the signs of change apparent everywhere." There was electricity at Santa Terezinha and public lighting on improved streets lined with trees, a town water system, a new post office, a satellite reception dish for telephone and television, two banks, a new primary school, a municipal sawmill, a garbage truck, and a cluster of low-cost homes financed by the government housing bank. The population had grown from 1,930 in 1979 to 3,350 in 1987, or nearly 75 percent. This was a new Santa Terezinha, eager to compete with Vila Rica, which had become the region's main commercial center, with supermarkets and regular bus service to Cuiabá, the capital of Pará, and Belém. The great dream of Santa Terezinha was a road connection to the Belém–Brasília highway to the east that would win back the commercial lead taken by Vila Rica.

As for CODEARA, the company was employing over 400 full-time workers. Many new jobs had been created by a CODEARA rubber plantation with a million trees occupying what had been cattle pasture. A company bus service carried workers from the town to the ranch along with small settlers who had been allowed to open farms within the fazenda. Local people had access to a company hospital, and butchers in Santa Terezinha were supplied with cheap beef by CODEARA. More than half of CODEARA's land was being offered for sale to settlers under a private colonization project at prices one-tenth of the value of farmland then in southern Brazil. The emphasis of the cattle operation had shifted from burning forest to regenerating old pasture, and CODEARA had diversified its core area of 150,000 hectares into production of rubber, grains, and oil seeds, for which the company created a research and seed selection center. For the townspeople, Lisansky observed, "the general perception of the fazenda as the enemy was being replaced by a more benevolent and almost paternalistic image of a company providing a variety of benefits for the region and its inhabitants."[11]

The changes at Santa Terezinha and in many other parts of Amazonia suggest that the question of who benefits and who loses from public and private development investments is complex. The initial impact of modern economic organization on a backward area produces conflicts in which the weak are vulnerable. Nowhere in the colonization of Amazonia, whether the Caqueta of Colombia, the Yucumo of Bolivia, the Huallaga of Peru, or the Carajás corridor of Brazil, to name a few of the major settlement areas, is there an example in the literature of what could be called "socially equitable regional development" that could serve as a model. The most that can be learned from past experiences is the many mistakes that should be avoided in future settlement programs.

The record does provide many examples, however, of the ability of entrepreneurs to find solutions to problems that are basic for economic stability on the frontiers. Conventional conflict analysis often disregards the adaptability and creativity of a free-enterprise system. CODEARA's research program, which provides seeds and plants to farmers in the region, and choice of diversification, such as the shift from cattle to rubber, is the way solutions are found to economic uncertainties and technical doubts. This often involves an exchange with public research centers, but the risk of application of a new system is an adaptive act by enterprises that are seeking a foundation for sustainable economic production. When there are costly "externalities," like water pollution or damage to forest ecosystems, this has to be controlled; but initiative must not be stifled because successful ventures normally create positive "external" benefits in indirect jobs, taxes, social services, and community organization. It seems clear from the experience at Santa Terezinha that a major change came with the consolidation of the municipal township. This would not have happened without the economic basis provided by CODEARA.

North of Vila Rica there is another striking example of the foundational role of private enterprise in Amazonia. This is the Fazendas Reunidas S.A. (FRENOVA) Gameleira sugar alcohol distillery and the adjoining Cofresa colonization project. This is a settlement with a township nucleus of 3,000 people, including 70 families of farmers who came to Amazonia's tropical lands from the wheat-growing Passo Fundo region of Rio Grande do Sul. FRENOVA has a core area of 150,000 hectares that remain with the original owners of a project that began in 1963 as a 850,000 hectares land speculation in the virgin forests. "We acquired thousands of hectares for the price of a silk necktie for a land official," recalled José Augusto Leite de Medeiros, a prosperous Sao Paulo businessman.

FRENOVA and Medeiros are on the blacklist of the "progressive" Roman Catholic pastoral agents under Bishop Casaldaliga, who denounce working conditions of contract labor gangs that clear forest and till the fields on ranches in northern Mato Grosso as "slave labor." The minimum day wage and unhygenic living quarters received by 800 field hands at Gameleira are a crude from of subsistence. As a result of agitation for land reform, 90,000 hectares of FRENOVA's best land along the highway to Vila Rica were expropriated by INCRA in 1988. Three years later, only 25 squatters were on this land, and neither the government or the church had organized any settlement.

During this time, FRENOVA was selling small farms and ranches of between 1,000 and 5,000 hectares at about $125 per hectare to buyers coming into the relatively fertile area between the Araguaia river and the Serra do Roncador. Banana production was booming. A large rice mill at Cofresa was in operation, and contractors with road building equipment were busy opening roads within the large ranches, many of which were replanting original pastures in new varieties. The contrast between the initiative of the private developers and the immobile agrarian reform area was striking.

Medeiros was only a few years out of law school when he and some friends decided on the frontier venture. Medeiros had been a radical student leader at the university, but after a trip to the Soviet Union and China he decided that socialism was not the wave of the future for Brazil. Mato Grosso's frontier lands offered an attractive opportunity for investing money provided by his wealthy family, but Medeiros had no idea of how to use the land. It took him years of experimentation to learn. First, some forest was cleared for cattle; then improvements were made using federal tax incentives; and then most of the property was sold to raise capital for a sugar distillery with an innovative technology based not on heavy crushing equipment, but on low-pressure separation by water of sucrose from shredded cane. This process, invented by a German chemist, William Leibig, has attracted international attention.

The Gameleira distillery produces 600,000 liters a month of ethanol, a renewable, low-pollutant fuel consumed in Brazil by alcohol-driven cars and trucks. Gameleira has an irrigation system for 8,000 hectares of sugarcane on soils limed to reduce acidity and fertilized by a combination of chemicals and nutrients recovered from the extraction stills. This sophisticated agronomic system supplies alcohol for much of northeast Mato Grosso and neighboring areas of Goiás. With a big Pernambuco sugar refiner as a new partner, Medeiros said he planned to increase cane volume and shift part of the output to sugar because he can compete with sugar brought from São Paulo at high transport costs. Cattle ranching continues, with new pastures on already cleared land, but as a secondary activity to sugar and to the corn, rice, and banana production of the settlers.

Gameleira is another example of entrepreneurial adaptation to Amazon conditions creating an economic foundation for a stable community. Who benefits? Medeiros and his partners, of course; but also, in one degree or another, about 1,500 wage earners and settlers who have chosen to go to Mato Grosso to make a living, and all the suppliers of goods and services for this successful agro-industrial community. Is this "progress"? It is not the Fabian ideal, but CODEARA and Gameleira "work" in a market economy. They have permanence, which is essential for sustainable development.

This study is an empirical inquiry into the nature and the causes of human behaviors that underlie modern changes in the Amazon region. This is a complex experience produced by a contradictory mixture of modern and traditional factors operating in a natural environment that is still at an early stage of discovery and scientific study. As a historical experience, there are serious gaps in social, political, economic, and cultural information that make generalizations and conceptual assertions perilous. The full story is still to be told of the regional development process surrounding the Brasília–Belém highway or the migration that overwhelmed Rondonia.

Recognizing these limitations, this inquiry has been guided by only one thesis, namely, that close observation and analysis of human behaviors in Amazonia make it possible to identify ways in which the interaction of man and nature can be modified to achieve sustainable development, understood as economic growth with environmental stability. This modest aim rests on certain pragmatic assumptions that need to be stated, without theoretical pretensions, only as a framework for the personal observations and research information contained in this study:

Proposition 1. A geographic definition of Amazonia is necessary because this is a loose term that can be given wider or more restrictive meanings. This study defines Amazonia in the broad hydrographic sense as all the territory drained by rivers flowing into the Amazon system, from north and south of the equator, and from the eastern slopes of the snow-peaked Andes up to an altitude of about 1,500 meters, above which the life-zone or ecosystem ceases to be warm, evergreen rainforest.[12]

This Amazonia is a continuous land mass from the Atlantic to the Andes on the scale of Australia, with an area of about 7 million square kilometers, of which about 5.5 million is densely forested. The largest part stretches across the ancient Brazilian and Guyanese shields, which are eroded plateaus covered by interfluvial forests and open savannas at altitudes between 50 and 300 meters above sea level. These are the weather-beaten shoulders for the sedimentary plain, covering about 1.25 million square kilometers of annually flooded land, through which the Amazon trench carries the world's largest river to its outlet in the Atlantic ocean (see map in front matter).

In its larger geophysical dimensions, Amazonia is a unified natural resource system made up not only of the flora and fauna of the biome, but of the land, water, mineral, and energy resources that are the region's principal economic assets. This perspective extends from the deep geological formations underlying the river system to the atmospheric regions that contain the potent climate dynamics of the rainforest. Most of Amazonia is a warm, humid world where water cycles driven by equatorial solar energy maintain a permanent rainforest vegetation. Trees grow faster in Amazonia, and in greater variety, than anywhere in the world.

But within this broadly defined Amazonian unity there are differences in soils, rainfall, topography, and temperature that produce a mosaic of different life zones, or biomes, containing local species of flora and fauna. Therefore, Amazonia is not as homogeneous as it appears. The closed rainforests of the so-called *hylea* that dominate the major interfluvial plateaus are very different from the floodplain forests of the annually inundated riverbank areas, called *várzeas* in Brazil. The approaches to Amazonia from central Brazil constitute a transition area between the *hylea* and the so-called *cerrados*—flat shrub lands with open forests that have proved productive under mechanized farming. This transition area in the eastern and southern Brazilian Amazon is where most of the modern settlement has taken place, in a crescent from Pará through northern Mato Grosso to Rondonia. This arc is totally differently from the Rio Negro region or the western limits of the Amazon basin in Bolivia, Peru, and Ecuador. Consequently, the diversity of soils, forests, rivers, and geological formations in the ecological geography of Amazonia requires differentiated strategies for sustainable development or ecosystem management of each specific area.

Proposition 2. Land use zoning of Amazonia is an essential concept for the territorial organization, environmental protection, and efficient development of the region. Experience has shown that Amazonia is exposed to irrational forms of occupation if settlement is not based on agro-ecological-economic zoning. This is the scientific basis for differentiated strategies suitable to each ecosystem. Zoning is both a technical and a legal instrument that sets the ground rules for regional development and the best use of natural resources in Amazonia.

Detailed scientific study of Amazonia's different regions will take years, but that is not an impediment for the immediate establishment of protection areas for ecosystems that have already been identified as critical. There are strong scientific arguments for considering large Amazon rainforest more valuable for their ecological "services," as regulators of climate and hydrological systems, than as "economic assets" for their land or timber.[13] Given the importance assigned to the preservation of Amazonian ecosystems for their biodiversity, zoning could begin by exclusion of large areas of intact rainforest from any form of development until conservation units are defined and established. Priority should be given to the demarcation of Indian reserves and extractive forests because they are occupied by forest people whose low-intensity economic activities do not endanger their ecosystem—except when they sell their commercial timber and the gold in their rivers to outsiders, which happens all too often.

In Brazil, the total area assigned to conservation units, which was about 5 percent of the Amazon region in 1991, could easily cover 30 percent on the basis of studies already carried out. An international scientific workshop held in Manaus in 1990 prepared a detailed biodiversity map for the entire Amazon that covers the habitats of all the known species from butterflies and parrots to monkeys and water lilies. The priority areas were marked out by the leading naturalists on Amazonia, and enormous tracts of pristine forest were nominated as biodiversity havens.

The Amazonia biodiversity map provides a blueprint for action by governments, researchers, and ecologists. All that is needed is organized management for the area. But the token funding provided for biodiversity does not match the cataclysmic rhetoric of biological extinction. The World Bank and United Nations Environment Program (UNEP) provided a $5.5 million grant for the eight-nation Amazon Pact group to prepare a plan.

Private debt-for-nature swaps offered some money. But the economic dimensions of a biodiversity conservation effort on an Amazonian scale are far beyond these trifling sums or the current means of the Amazon countries. Maria Tereza de Padua, a former director of national parks in Brazil and president of a leading Brazilian conservation foundation, told a UNEP international biodiversity conference in Brasília in 1991 that it would cost $12 billion to set up the necessary conservation units in Brazil alone. No offers were heard from the First World delegates.

Therefore, zoning alone, without an economic foundation for environmental programs, is not enough. But zoning can encourage productive investments in Amazonia to build the foundation by reducing the uncertainties over regional development in Amazonia that inhibit investments. These doubts can only be removed by defining appropriate areas for resource development, as well as the preservation areas. International financial and scientific support will always be insufficient, and ultimately, the resources for environmental protection programs must come from the Amazonian countries themselves.

The Brazilian presentation to the Group of 7 of the Amazon conservation project spoke a plain political truth when it said that "efforts at environmental protection will only be successful if they are compatible with the development goals of the Brazilian people and with the attainment of reasonable standards of living for the residents of the region." But there is another economic truth that is equally plain: There is no reasonable prospect that the Amazonian governments will have the fiscal resources or the administrative abilities to conduct environmental programs without a major expansion of Amazonia's productive base and an active participation by the enterprise sector in environmental protection. As will be seen in Chapters 8 and 9, there is no societal solution for Amazonia's environmental stability without a solid economic foundation.

This has nothing to do with laissez-faire "business as usual" in Amazonia. Private developers, who had to be lured to the region initially by tax waivers and cheap credit, should now be required to pay a "green" royalty for access to Amazonian resources, such as minerals and timber. These revenues should be specifically directed to environmental programs. If the Amazonian governments had collected even 5 percent of the value of the gold and timber that have been plundered from the frontier, the environmental agencies and local governments would not be bankrupt and riddled with corruption. Administrative reform and adequate funding to staff environmental and taxation agencies are essential for the enforcement of zoning policies.

Proposition 3. The international environmental movement has played, and will continue to play, a decisive role in support of environmental activism in the Amazonian countries. Without the remarkable lobbying success of the "conservation community," made up of a wide variety of nongovernmental organizations (NGOs), the governments of both the

First World and the Amazonian countries would not have reached the agenda for environmental action that is now before them. The "green" campaigns have taken hold politically in the Amazonian countries and there has been an explosion in the number of domestic scientific groups, NGOs, and indigenous movements that keep environmental issues alive. Within the governments there are also ministers, presidential advisers, legislators, and state governors who have taken up the environmental banners. These are important allies of the international groups that are working for the development and implementation of practical solutions for the environmental problems of Amazonia.

But "ecological politics" and implementation of environmental programs are not necessarily the same thing. A militant vigilance by international and local NGOs over the environmental policies and programs of governments and the multilateral development banks will remain a major task. The same can be said for indigenous rights groups influencing Indian policies. However, the amount of international financial support for Amazonian "advocacy" groups, which rose sharply in the 1980s, subsequently declined. The emphasis in private foreign aid is shifting from the environment as a "cause" to support for grassroots projects that can help peasants, Indians, rubbertappers, and other forest communities make a better living. Hopefully, this will mean less money spent on airfares and hotels for international symposiums that constantly repeat the same themes, and more support for practitioners in Amazonia of community development, agroforestry, and applied research in tropical agriculture.

The decade of the 1990s will probably see a modest increase in the flow of governmental resources for environmental ends, although far less than Brazil and other governments expect. The wealthy members of the Group of 7 are keeping a tight hold on their pursestrings. Despite the easing of East–West "cold war" tensions, military spending cuts do not seem likely to provide a "green" dividend beyond the modest $1.5 billion Global Environment Facility, a new World Bank "window." Even if there were greater generosity, however, it is unclear that the Amazonian countries could absorb a large increase in funds without significant domestic reforms.

For instance, the great stumbling block for the World Bank in considering a $167 million loan project for natural resource management and environmental protection in the Brazilian state of Rondônia was the bankrupt condition of the state government and the woeful deterioration of the state administration. A $137 million World Bank loan to Brazil in 1990 to finance a national environmental protection plan was on the shelf for a year because the Brazilian Ministry of Economy could not find a noninflationary source for the local "counterpart" funds necessary to release the foreign portion. Brazil's financial disarray prevented disbursement of a German DM250 million grant for environmental projects in Brazilian Amazonia.

This points to the need for solid financial and administrative reforms in Amazonian countries as a condition for large-scale international loans or grants that require local counterpart funds. In the past, international loans for Amazonian "development" projects have too often financed inefficient state agencies, bloated bureaucracies, high-priced consultants, and pork barrel politicians. The World Bank's $450 million Polonoroeste loan to settle peasants along a paved highway built between Mato Grosso and Rondônia is a well-documented example.[14] An Interamerican Development Bank loan of $140 million to extend the highway to Acre also went on the rocks as Brazil repeatedly violated the terms of the contract.

In place of this, what is needed from donor governments and development banks is funding for small projects prepared with the participation of the people who are directly affected in the design, organization, and implementation. Funding of this kind would involve state and local governments with NGOs that have begun to emerge in Amazonia as consultants and partners with rural associations, unions, and indigenous communities. This is not the way multilateral bank officials or federal bureaucrats usually like to work, but in Amazonia it is the best way to educate the public in environmental and natural resource management. Any loan involving these areas should carry an educational component as a straight social grant from the lending bank, without cost to the borrower. The top-down technocratic way of the past not only has not worked well, but has failed to educate the people whose behavior in relation to their environment is what sustainable development is all about.

Proposition 4. The central concern of this inquiry is sustainable development, a concept that combines economic growth with environmental stability. This apparently simple proposition is slippery ground that should be entered with care, in Amazonia or anywhere else. It is not easy to define sustainability because, in practice, it is not a "natural" process. It is an artificial creation of human societies.

Sustainable development has raised great polemics among ecologists, economists, and entrepreneurs.[15] The Brundtland Commission report of 1988, widely circulated under the title "Our Common Future,"[16] adopted a definition of sustainability based on the altruistic principle that the consumption of one generation should not exhaust the stock of resources needed by the next generation to maintain the same standard of living. A group of 50 international business leaders, in a position paper prepared for the UNCED conference, presented a vision of sustainable development based on "open markets, ecologically realistic prices, improved property rights, efficiency and innovation," opening the way to "new opportunities, new markets, new ideas . . . and an increase in economic competitiveness."[17] This definition is criticized by a school of "hard ecology" that regards the pursuit of constant economic growth, particularly by Third World political leaders and multilateral lending institutions, such as the World Bank, as an "ideology" that is the enemy of a sustainable environment.[18]

The definition of this study is that sustainability is the result of a historic social learning process that is culturally transmitted through a society's internal value patterns and institutions. Sustainability is the result of limitations that are imposed by the society's value system on human behavior in relation to the natural surroundings, or external physical situation. These limitations are often ritualized in totems, taboos, and other symbolic information in a primitive society, where sustainability is equivalent to finding enough food for the survival of the tribe. This leads to cultural practices based on an intimate knowledge of the natural surroundings on which life depends, which is environmentally benign but insufficient to sustain a large, diversified population.

In a technological society, which manipulates natural forces, the information system for sustainability is far more complex. The biological concept of natural carrying capacity, which is an iron limit for primitive societies, is replaced by economic behaviors that produce goods and services for the expanded needs of a modern society. The limit to growth in this technological system is not resources, but levels of waste and pollution that undermine ecological stability, or the maintenance of a socially accepted environment. This creates tensions in the value system between economic growth and "quality of life" that must be resolved politically.

Each society has its own choices to make, and the ecological problems created by the industrial world (such as acid rain, nuclear waster, and fossil-fuel air pollution) are not Amazonia's problems. The problem that weighs on most of the people in this region is economic survival today, not extinction tomorrow, so there is a strong demand for economic growth. If the needs and wants of a society are not met, social protest can make government unsustainable, and without sustainable government there can be no long-term environmental policy, any more than there can be an economic development policy. Something like this breakdown is afflicting the Amazonian countries.

The slippery dichotomy of economic growth with environmental control is reflected in the derivation of the words *ecology* and *economy* from the same Greek root *oikos*, in the sense of household. Ecology implies a natural totality, a whole; economy is a division of the whole into parts for some specific use, as in fencing a field to pasture a selected herd of animals, not a free range for all herbivores. The rule of sustainable development is that the two must function in combination, not in opposition. Human husbandry and technical creativity can select and modify components of the natural habitat to serve man's economic needs, but not at the risk of destroying the human abode. There is a proper balance.

The anthropological record shows that man is a dominant species who creates his environment by selecting and manipulating other species to serve his needs and wants in a great variety of natural situations. This is a process that is guided by internal patterns of the social system (technology, beliefs, laws, work motivation) but that depends also on human interaction with an external situation (natural resources, climate, disease). The ecologically sustainable experiences are easy to identify: They last.

For example, the protofarmers of ancient Syria or Anatolia who switched from hunting and foraging to soil tillage and cropping of selected wild species of grains, along with the domestication of cattle and fowl, launched an agricultural revolution. Over centuries, this revolution advanced from the Black Sea to the Atlantic, profoundly altering human behavior from nomadic ways to permanent settlements.[19] What became of the forests where the Roman legions faced the barbarians on the Danube, in Germania, Gaul, and Britain? They are mostly gone, and in their place came colonies of settlers, wheat fields, vineyards, olive groves, dairy pastures, and hay meadows that provided the economic foundation for western civilization. Carried across the Atlantic, this technology converted forests from the Connecticut, the Hudson, the Ohio, and the Mississippi river valleys into farmland that produced the greatest industrial power in the world. That is a process of sustainable development that has bridged 10,000 years from its neolithic origins to the present and has been adapted worldwide to different environments.

It is argued, correctly, that Amazonia's climate, soils, species, and ecological relationships differ radically from the environmental conditions of the temperate northern regions. There are natural barriers to simple imitation in Amazonia of some agriculture and forestry practices that have been developed over centuries in the North. But it is a mistake to conclude from these obvious differences that there are insuperable environmental obstacles in Amazonia to highly productive agriculture and forestry adapted to specific local conditions, or to the development of a technological society. In this respect, with its remarkable variety of natural resources, Amazonia should be a high-tech laboratory in the tropical world for the discovery of how to make the most of its resources in a pattern of sustainable development. Freedom to experiment must be encouraged.

A pattern of sustainable development has not emerged on the Amazon frontiers for two principal reasons. The first is that the initial stage of the occupation did not produce an organized, stable, and integrated Amazonian economy. The second, which is partly a consequence of the first, is that stable social patterns have not developed and, as a consequence, social controls are virtually nonexistent. The social information "codes" for sustainability do not exist because they don't have an economic foundation.

On the surface, this shows up as a state of social disorder in which a turbulent occupation, marked by violent disputes over land, gold, timber, and drugs, has outrun the limited ability of national, state, and local authorities to maintain control. The lawlessness involves not only the predictable cafe shootouts and bank robberies, but sophisticated airborne commando raids on mining camps, international drug wars, and political assassinations of mayors, judges, and candidates for governorships or federal office. Juvenile prostitution is rampant in the Brazilian Amazon towns on a scale that indicates a massive breakdown of family cohesion and desperate economic need. It goes without saying that schools and health clinics are woefully inadequate for the education of a new generation and for the care of mothers and infants. The lack of social fabric and community pride are among the causes that make settlers in Amazonia so wasteful of their natural environment.

Human behaviors in Amazonia constantly overstep the limits required for sustainability in land use, in wildcat mining, in timber extraction, and in predatory hunting, fishing, and extraction of forest products. A powerful symbol of the despoiling of Amazonia is the plight of the indigenous Indian communities that have been displaced in many areas by the new settlers. Their traditional cultures have often been deeply wounded if they have not actually been killed by the invaders or fallen prey to diseases.[20] A more secure, stable existence for the remaining indian communities as protectors of forested areas is an important part of any strategy for environmentally stable development.

But this alone will not halt the destructive practices of settlers who do not put down roots on land that is suitable for farming. With the slowing of forest clearing for cattle ranges, shifting agriculture is the most important single cause of deforestion. This unstable situation can be attributed in part to survival needs; but it is also due to lack of experience in how to use Amazonia's land and forest resources productively. Costly mistakes in Amazonian agriculture have been made by ranchers and commercial farmers, who have paid a price in failed crops, degraded pastures, and economic returns that have fallen far short of expectations. For tens of thousands of small settlers on frontier land, the consequences are more serious. The rapid decline in fertility of land that is cleared turns the "settler" into a migrant who has to move on and cut more forest to survive.

The greatest single challenge facing Amazonian agricultural research and extension agents is to help small farmers develop technologies for the humid tropics that will make them stable landholders. The possibilities for agriculture in the soils and climates of Amazonia is much debated. A number of analysts have reached the conclusion that small producers of annual cereal crops are condemned to failure,[21] and soil limitations have been blamed for colonization busts from the Transamazon highway to the Chimore in eastern Bolivia. But numerous other studies on frontier agriculture that show that management, site and crop selection, and transportation are the key factors for success or failure.[22]

This inquiry tends toward the latter conclusion. The evidence of successful agriculture, tree plantations, and cattle raising in well-chosen, suitable locations in Amazonia suggests that some entrepreneurs, large and small, are learning from failures and making necessary adaptations. Settler success depends on finding the right combination of land use, forest products, and animal husbandry that is appropriate for the special features of each Amazon ecosystem.

Large corporate farms, like the *fazenda* Itamaraty in western Mato Grosso of Olacyr de Moraes, the so-called King of Soybeans, produce higher soybean yields than in southern Brazil thanks to years of research in species adaptation and agrochemical technology that corrects soil acidity and nutrient deficiencies. The example of this technology is spreading across the Amazon *cerrado* transition zone, where the main problem is the high cost of road transport. The solution is the Ferronorte railroad, a $1.4 billion project that Moraes has begun, to link Cuiabá and the eastern Amazon with São Paulo by rail by 1995.

The central economic problems of Amazonia will not be solved without a creative input from entrepreneurs who have a talent for discovering new ways to produce goods competitively and create markets for Amazonian products. The importance of private enterprise for the solution of the Amazonian problems has been neglected in discussions of the environmental issue because frontier economics have been equated with quick-buck speculation and outright plunder. That is a political judgment that makes entrepreneurs appear to be enemies of the environment and not, as they should be, a dynamic part of the solution.

The "new frontier" is not just an undertaking for a few economic magnates, however. It is a social enterprise that will require the experience, initiative, and talent for discovery of all the Amazonians. This is a creative challenge for Indians and scientists, ranchers and settlers, miners and foresters, union leaders and university professors, bankers and politicians. It is they who will have to create a society that is economically productive and ecologically stable. This is a work for social architects, not bulldozers; for nature craftsmen, not slash-and-burn gangs; for patience, not haste. It will be a slow process, like the construction of a great cathedral. A cathedral is a metaphor for man's quest for sustainability. If ecology is a new religion, as some say, what better place to build a "Green cathedral" than Amazonia, the greenest part of planet Earth?

TWO

The Worth of Amazonia

Amazonia was regarded, until recently, as a primitively backward and economically marginal part of the world. Suddenly a great human occupation has taken place. What has been discovered that makes Amazonia seem so valuable now to so many people?

In economic terms, the answer is quite a lot. Improved transportation made cheap land available in abundance for ranchers and subsistence farmers. Increased access and exploration turned up vast mineral, energy, and forest resources, which are worth billions of dollars. These raw material assets include iron ore, gold, timber, oil and gas, tin, bauxite, and the cheap hydroelectric energy to convert these materials into metals and industrialized products. With further exploration there is every likelihood of more to come. The development of these resources and the prospects they offer for Amazonian development are considered in detail in Part II.

An inventory of Amazonia's natural resource potential must be combined with the region's ecological "assets" when the objective is the design of a strategy for sustainable development. But beyond these abundant material riches, Amazonia's "worth" also depends on what people see it to be. This is a matter of subjective values and beliefs, which have undergone extraordinary changes over time. Each age of western civilization projected into Amazonia its own values, tastes, and fantasies. Today's age is no different.

Since the discovery of America, Amazonia has had a "magical" image of a land where the real and the fantastic were inseparable. The European name for the region was inspired by the sight of Indian women archers who seemed to an early chronicler to be the Amazons, or mythical female warriors of ancient Greek legend. The early Iberian explorers, a mixture of rapacious conquerors and quixotic adventurers, created the legend of El Dorado, the gilded man. This legend launched expeditions down unknown Amazon rivers in search of Manoa, a city said to be paved in gold. No such city was ever found, although it appears on maps until the eighteenth century. The fantasy of metallic riches became part of the Amazon heritage; but it was not until this century that the discovery of vast mineral wealth in Amazonia showed that the fantasy was not misplaced.

Another fantasy growing out of Old World religious and philosophical speculation attributed to Amazonia the condition of a virtual paradise on earth, peopled by natives innocent of original sin—a setting for utopias. The Spanish missionaries who chronicled the early exploration, perhaps carried away by the prospects of converting the multitudes of new tribes, painted an idealized picture of the new lands. As we know now, this attraction had tragic consequences for the natives.

Cristobal de Acuña, a Jesuit who traveled the length of the "River of the Amazons," asserted in a report to the king of Spain that it was the greatest river of the globe, surpassing the Ganges, the Euphrates, and the Nile "in the provinces it waters, the land it fertilizes, the people it feeds, and the oceans it increases."[1]

Such glowing prospects were invoked by the missionaries of the discovery as reasons for the kings of Spain and Portugal to support their organization of the Amazon territories in the service of God and empire. With no less enthusiam, Sir Walter Raleigh argued at the court of King James I for a royal charter to carry out a colonization project in Guiana. Raleigh asserted that this was a land that "for health, good air, pleasure and riches I am resolved cannot be equaled by any region east or west."[2]

Raleigh was regarded by his Elizabethan contemporaries as a learned man, knowledgeable of New World opportunities. He maintained that Manoa, the city of gold, existed and would make the fortune of Englishmen who invested in the Guiana venture. Raleigh was beheaded before this prospectus could be tested, but it led to the establishment of Britain's one colony on the South American mainland, now the independent republic of Guyana.

Over time, better knowledge of Amazonia was accumulated from the travel narratives and descriptive reports of explorers, missionaries, adventurers, colonizers, and students of nature. The coming of the Enlightenment to western Europe brought savants to Amazonia with new visions of the age of science. The world had to be measured and all things classified to extend the circle of learning. The geography and the tropical flora and fauna of Amazonia presented fascinating opportunities for these students of natural science.

The succession of notable observers of Amazonia included Charles–Marie La Condamine, Alexander von Humboldt, Karl Friedrich, Philippe von Martius, Johann Baptist von Spix, Alfred Wallace, Richard Spruce, Henry Bates, Louis Agassiz, and Karl Von Den Steinen, among others. Their learned treatises created the initial foundations for the biological, geological, and anthropological knowledge of the region and its native people.

These scientific missions were not without commercial importance, for this was science with practical applications for the expansion of world trade. The reports of the riches to be found in the tropical latitudes also had political effects because they stimulated colonial expansion by mercantile nations in Africa, Asia, and Latin America.

For Baron von Humboldt, the much-traveled Prussian naturalist and political thinker, Amazonia seemed to hold the promise of being the future "granary of the world." He circulated this enduring misconception without ever having set foot in the Brazilian Amazon, from which he was barred by suspicious Portuguese colonial authorities who gave orders that he be "arrested on sight."

Other nineteenth-century observers who did spend time in the Brazilian Amazon shared von Humboldt's faith in the potential of Amazonia for human settlement and profitable investment. Alfred Russell Wallace, an English naturalist who spent four years (1849–52) on the Amazon and Negro rivers, wrote that it was a "vulgar error" to suppose that the luxuriant vegetation of the tropics overpowered man's efforts in agriculture and husbandry.

> Just the reverse is the case: nature and the climate are nowhere more favorable to the labourer, and I fearlessly assert that here the "primeval" forest can be converted into rich pasture and meadow land, into cultivated fields, gardens and orchards, containing every variety of produce, with half the labour, and what is more importance, in less than half the time than would be required at home. . . .[3]

In the spirit of limitless progress through human enterprise and invention that animated the advances of science and technology during the nineteenth century, Amazonia was seen by most outside observers as a naturally magnificent but torpid resource base awaiting the uplifting hand of "civilization."

When Theodore Roosevelt crossed the great divide of South American waters, the Parecis plateau between Mato Grosso and today's Rondonia, the gaze of the former U.S. president and archtype liberal progressive turned from the wild marshes of the Paraguay watershed behind him to the even greater wilderness of the Amazon basin ahead. The vision he saw during his exploratory trip to Amazonia with Marshal Candido Rondon of Brazil in 1914 was of a western civilizing mission through which, he predicted, "the devil of evil wild nature" would be tamed by science and Christianity.

> Surely in the future this region will be the home of a healthy civilized population. It is good for cattle raising and the valleys are fit for agriculture. . . . when railroads are built into these interior portions of Mato Grosso the whole region will grow and thrive amazingly, and so will the railroads.[4]

Yet, despite the optimistic judgments of Jesuit missionaries, enlightened scientists, and progressive world leaders on the fertility of the Amazon basin and its potential for human settlement, these expectations were still unfulfilled after four centuries. Colonial enslavement and diseases brought from Europe and Africa had destroyed most of the native Indian labor force, undermining the extraction from the forest of exotic beverages, like wild cacao, spices, medicinal barks, gums, and cosmetic oils which provided the only exports. Total population had declined in the region, despite European immigration schemes promoted by Portugal before Brazil, and the Spanish colonies broke their Old World ties and became independent in the 1820s.

Amazonia was all but abandoned by the mid-nineteenth century when the industrial revolution handed the region an unexpected bonanza—rubber. Elastic balls of latex extracted by Indians from *hevea brasiliense* and other tropical trees had been known in Europe since the voyages of Columbus, and La Condamine displayed samples he had collected on his Amazon trip to his learned colleagues at the Academy of Science in Paris. The waterproof qualities of rubber had been used in foul-weather clothing and shoes. But it was not until the invention of rubber products for carriage wheels, railroad couplings, industrial tubing, telegraph lines, electric insulation, and, above all, the automotive and bicycle tire that rubber became a major commodity.

Natural rubber was a virtual monopoly for Brazil, Peru, Colombia, and Bolivia, where the principal sources grew profusely. Industrial demand sent prices soaring. Imports to Britain rose from 10,000 kilos in 1857 to 58,700 kilos in 1874, but this was just the start of an extraordinary expansion. Gathering rubber became a frenzy all over Amazonia, ultimately involving some 150,000 laborers, who were mostly migrants. In 1912, this army of indebted, malaria-ridden jungle workers had brought Brazil's production to over 31 million kilos, providing as much export revenues as coffee.[5]

With the rivers providing transport, the extractive system extended into the remotest corners of Amazonia, creating trading settlements throughout the interior. But wealth stayed in a few hands and was squandered in Belém, Manaus, and Iquitos, the ports from which rubber was shipped to Europe and United States. These urban centers, where pounds sterling and dollars were the common currency, lavishly imported the latest Parisian fashions, including the dancing girls, and installed foreign bank branches, telegraph lines, and opera houses built by Italian architects. Yellow fever was banished from Belém by sanitary engineering, and an American financier, Percy Farquahar, built a modern port there before his ill-fated venture to build a railroad around the rapids on the Madeira river.

Rubber brought Amazonia into the world trading system on an industrial scale, but the boom lasted only until World War I. To escape the Amazon monopoly, British and Dutch botanical researchers and colonial administrators opened rubber plantations in Malaysia, Singapore, Sumatra, and other southeast Asian locations. "In Brazil, there was no comparable effort to investigate the problems of rubber cultivation," commented Warren Dean in his authoritative environmental history of rubber.[6] By 1932, Brazil's rubber exports had fallen 85 percent, to 4,582,000 kilos, and Amazonia's economy was prostrate again.

When the rubber boom collapsed, Amazonia was unable to keep pace with the economic and social development of the more dynamic regions of the countries to which it belongs. The picture faded of a land of great resources and exceeding potential for mankind. Rather than a promise of wealth, national governments saw the region as an economic and social problem, a tropical Appalachia or Mezzogiorno.

Why the rush into Amazonia now? There is no single answer, but various factors have come together.

1. The pessimistic view of Amazonia underwent a remarkable change after World War II. Significant amounts of oil began to be found on the Amazon side of the Andes, first in Bolivia and Colombia, then in Ecuador and Peru. Geological fieldwork begun during the war, when resource scarcities were a major U.S. security concern, found huge manganese deposits in Amapá, Brazil, and later high-grade iron ore and bauxite in Pará, opening up unsuspected mining possibilities.

2. The backwardness of Amazonia in relation to more dynamic areas of the national economies could not be ignored for domestic political reasons. The apparent solution was integration of these inaccessible territories into the national economies through aggressive government programs in road building and planned settlement, as well as major national and foreign investments in energy, mining, and ranching. The new ventures were possible because the modern sectors of the semi-industrialized Amazonian countries provided capital, manpower, and technologies of transport, energy, and communications on a scale unknown in Amazonia before.

3. International concerns over resource scarcities were perceived in Amazonian countries to represent a potential danger to their control over their undeveloped regions. This widely held view was expressed by Eneas Salati, a Brazilian climatologist, who was appointed in 1990 as executive director of the Brazilian National Research Institute for Amazonia (INPA) in Manaus. Salati is a strong advocate of environmental protection and of scientific rationality in the management of Amazonia's resources, but he is not blind to "developmental" pressures. Putting Amazonia in a global context of population growth and resource demand, Salati wrote in 1982:

> If the present conditions of growth and development continue, the world population in the year 2000 will be approximately 6.5 billion people, or 50% more than now. Without considering other problems of survival related to health and industrialization, this number already indicates the pressing need to increase by 50 percent the production of food by the end of this century. . . . In addition to the problem of producing food, the world is experiencing a hunger for wood and paper, aggravated by the approaching end of the forest reserves of Asia and Central America. Therefore, the industrialized countries turn their eyes toward the last forest reserve of the world, estimated to contain in Brazilian Amazonia alone, a volume of 50 billion cubic meters of commercial wood.[7]

4. Another explanation that has been given wide currency claims that national capitalists began an internal "colonization" of Amazonia to expand the domestic market for industrial goods and to invest their accumulated

wealth in land speculation. This theory argues, with Marxist inspiration, that government programs in Amazonia are designed to favor investment by a "dominant class" of big investors from outside the region. Octavio Ianni, summing up a study of land conflicts in the frontier areas of Pará between wealthy ranchers (mainly from southern Brazil) and peasant squatters, put it this way:

> At first sight, the government seems to be inducing a new economy, a new society, by generating economic changes in the region. In practice, however, what is underway is an alliance between private enterprise and the state in which the principal result is rapid formation and accumulation of private capital, both national and foreign.[8]

This model of technologically advanced, capitalist development is seen by Ianni, and others who share this view,[9] to be closing the Amazonian frontier to the traditional residents, including indigenous people, and to millions of landless peasants who could be settled on the virgin soils if governments followed more egalitarian social policies. An extension of this theory to the role played by foreign capital is that the natural resources of Amazonia are being exploited to pay for the foreign debts accumulated by the governments of the Amazonian countries, with little regard for the environment or stable development of the region.

What all of these contemporary beliefs have in common, in contrast with the past, is that they are homegrown and have not been imposed from abroad. They have their origins in nationalist and developmentalist ideologies that are widespread in Latin America. The new version of the theory that Amazonia will be called upon to "serve humanity" as a source of food, minerals, and fibers, and the view that the frontier resources are chips in a high-stakes capitalist poker game, both assume that Amazonia's lands, minerals, energy, and forests are economically valuable and will be developed by someone, either domestic or foreign.

But Amazonia's long-awaited encounter with economic development has provoked an unexpected backlash. The economic-growth models that inspired "developmentalism" a generation ago have come under heavy attack in the affluent, industrialized countries that have entered their "postindustrial" age. Highly vocal environmental and scientific movements question whether the present system of industrial production can be extended to the Third World's burgeoning populations without provoking an ecological disaster on a global scale.

As Robert Goodland, a senior environmentalist at the World Bank, has put it:

> . . . we allow ourselves and our globe to be operated on a system of frontier economics created when the world was essentially empty of people; when resources were scarcely tapped; air and water were clean; land and forests abundant. This idyllic era has long since vanished, but the economic orthodoxy remains to be modernized to reflect today's world

> which is full (or overfull). Possibly the single most influential change toward sustainability is to revamp neoclassical economics to accomodate environmental concerns. . . .[10]

This perspective has had a substantial influence on North–South relations because of the developing countries' dependence for capital from the industrial countries. Lobbying by environmental activists has produced a more critical political attitude toward development lending. Multilateral development banks have been forced to impose tighter ecological conditionality on lending for development projects. This represents a new constraint on international financing for deeply indebted Third World nations that are already hard pressed to achieve what they regard as minimally acceptable rates of economic growth.

The environmental dimension produced serious friction between lending agencies and borrowers over deforestation and other ecological damages in Amazonia. But, in a short span of a decade, "green conditionality" led to significant changes in the way borrowing countries deal with environmental issues. Amazonian leaders have accepted the scientific arguments that buttress the world's environmental concerns, and they have joined in the international conferences on climate change, ozone depletion, biodiversity, and other global issues.

There was angry rejection, however, of more radical ecological arguments coming from sectors of the First World that seem to place Amazonia in the category of a permanent ecological reserve, an untouchable rainforest conserved for "humanity." A comment by President François Mitterand suggesting that the global stakes in Amazonia's environment require an international agreement limiting Amazonian national sovereignties produced outraged reactions in Brazil. "Ecological colonialism" was the term applied by Brazil's Minister of Justice Jarbas Passarinho, and a leading economic columnist wrote:

> The interest of the rich countries is to keep Amazonia in a state of hibernation, not development. They demand that Brazil keep the world's refrigeration system working nicely while they go on polluting and heating up the planet.[11]

The preservationist view of Amazonia reflects a psychological need among some people in the First World for a relationship between man and nature that is less threatening and more secure than that offered by technological society. From this perspective, in a world of acid rain, nuclear accidents, oil spills, and ozone holes, the survival of man is threatened by the very systems of production that are organized to satisfy the demands of growing masses of consumers. This is also a philosophical position on the relationship between man and nature that began with the German romantics, at the start of the industrial revolution, and reached darker nihilist forms in Martin Heidegger. The fixation on Amazonia now reflects this yearning for nature untouched by artificial human change.[12]

If this is the case, there is a clash of cultures. The Amazonians feel less somber about the future of humanity, and they do not see themselves on the threshold of a world of exhausted resources. Their existential problems are immediate—hunger, housing, and health. They believe that the primary "worth" of Amazonia lies in what the resources of the region can do for the people who make up their community of nations. They want international cooperation in the form of capital, technology, science, and markets to develop their economies. They are tired of lectures based on experiences that they have not lived.

The question then becomes: Can the resources of the tropical forests be exploited in a rational way that contributes to economic growth and, at the same time, sustains Amazonian ecosystems on which the future of the region and its people depend?

Fernando Collor de Mello, the dynamic young president of Brazil, made environmental protection a top domestic and foreign policy priority on taking office in 1990. He launched a program that reduced deforestation in Amazonia by enforcing penalties for illegal clearing and eliminating tax incentives for land speculation. He committed Brazil to an international program for preservation of the Amazon rainforest and took measures to create major indigenous reserves in controversial areas, such as the Yanomami and Kayapo territories.

But Collor's main ambition was to put Brazil into the economic ranks of the First World. He was out to show that economic development of Amazonia can be compatible with preservation of the region's ecosystems. Announcing a program to reforest thousands of devastated hectares of Amazonia along the Carajás railway with commercial species, Collor said:

> In environmental questions, Brazil no longer rides in the wake of international denunciations. We don't proclaim our innocence or invoke falacious nationalist arguments. We have left the bench of the accused and are in the vanguard of international ecological initiatives. The question is not to establish who is more or less to blame, but to find forms of international cooperation that will avoid new disasters in the future. Our basic directive is that Amazonia should be and will be preserved and that Amazonia and Brazil should be and will be developed in perfect harmony with the protection of our ecosystem.[13]

This theory of development with environmental protection is the new belief—or fantasy—for the occupation of Amazonia. To be successful, the Amazonian governments will have to assume the responsibility for designing policies and creating administrative structures that protect entire ecosystems. The creation of protection areas, national parks, biological reserves, national forests, and indigenous areas that protect ecosystems are decisions for national governments because such land use allocations usually go beyond the jurisdiction and competence of state or local authorities. National governments are also the natural negotiators for international environmental cooperation, which in the case of ecosystem protection, such as reforestation and forest management, can involve billions of dollars.

The ecosystem approach calls for some changes in conventional economic criteria for valuation of resources. Amazonia's economic production is based on natural resources that include water, land, minerals, energy, biomass, and genetic reserves. Some of these are renewable, depending on how they are exploited, and some necessarily are exhausted over time. In either case, these economic uses produce environmental alterations, but these can be costed and regulated through the legal mechanism of environmental impact assessments. At this level, the active participation of local communities in environmental protection is critical (see chapter 9).

But a true inventory of Amazonia's resources should recognize that beyond the economic use value, there is a social value in ecosystems that provide conditions for the region's enormously diversified forms of life. Although these systems cannot be quantified in the way natural resources usually are measured, sustainability of the ecosystems must be the foundation for any rational strategy to develop the economic potential of Amazonia. Damages to the Amazonian ecosystems as a result of resource development must be counted as a social cost, and in some cases an unacceptably high one. These are public-interest decisions that can't be left to the marketplace.

Water will serve as an example. As is well known, the Amazon river system represents the world's largest source of fresh water, 15 percent of the total carried by all the rivers on the globe. The Amazon pours into the ocean four times as much water as the Congo and ten times more than the Mississippi. Through its various mouths on the Atlantic, the Amazon delivers a year-round average water outflow of 165,000 cubic meters per second. That is enough water to supply the annual consumption of ten New York cities.

The uses of water for agriculture, energy, fisheries, transportation, and sanitation in Amazonia are basic to the economy. But conventional economic measures of the value of water resources in Amazonia don't provide an adequate accounting of the total importance of water to the ecological sustainability of the region. "The circulation of water . . . is the essential factor that controls life in Amazonia, and there is nothing in the region, alive or dead, that does not testify to this fact," wrote Swiss zoologist Hans Bluntschl in 1912.[14]

Amazonia's water supply is constantly replenished by clouds that are carried into the region from the Atlantic ocean by the prevailing westward flow of the trade winds in the southern hemisphere. In their passage across the mainland, and particularly when they encounter the Andean barrier, these clouds unload their water contents, which contributes to the heavy rainfall. The ecosystem is not just a passive recipient in this water cycle. The vegetation that covers the region is constantly recycling water back to the atmosphere through evapotranspiration.

The total amount of rainfall in the Amazon basin is 12 x 10^{12} cubic meters of water a year and the outflow through the rivers is only about 5.5 x 10^{12} cubic meters. The surface outflow of water from Amazonia, while the highest in the world, is therefore only the part that the ecosystem releases from a much larger volume. The rest of the rain must return to the atmosphere in the form of vapor.[15] This explains the permanent high humidity of the area, with all that this implies for the myriad forms of life that exist in Amazonia.

The water cycle is responsible for the mild temperatures in the region. At any given time there are more than a 100 billion tons of water suspended over the region. This humid blanket is an important factor in the Amazonian energy balance because the vapor absorbs infrared radiation from the earth surface, and this makes for a moderate range of daily temperature variations. The daily high and low are in a range of 28° to 19°, compared with differences in the deserts, in the same latitudes as Amazonia, of from 40° by day to below freezing at night. Among other things, this year-round temperature range makes it possible to grow up to three crops a year on the same ground.

This system is driven by solar energy, which has been calculated at the high level of 700 calories per square centimeter per day in central Amazonia. Half is consumed in the evaporation of water through the plants. Another part of the solar energy drives the process of photosynthesis which creates phytomass, or plant material. In Amazonia, forest vegetation is created at a far faster rate than in temperate forests. In one year, the Amazonian "green machine" creates up to 90 tons of phytomass per hectare, compared with 20 to 30 tons a year in a temperate forest.

Looking at Amazonia in terms of the ecosystems opens up new economic horizons. The vigor of photosynthetic conversion is potentially one of the most valuable resource characteristics of Amazonia, although the practical means of using this aptitude for creation of phytomass have not yet been worked out for agriculture, forestry, or energy production. In the future, the relative advantage of Amazonia's ecosystems for growth of wood, and therefore as a carbon "sink" withdrawing CO_2 from the atmosphere, may be assigned a greater market value than the return that could be obtained by using land for nonforestry production.[16]

The disruptive effects of severe deforestation on the water cycle, and therefore on Amazonia's climate and ecosystems, are self-evident. The elements of the ecosystem are all interdependent, and the water balance that keeps Amazonia green depends on the equilibrium of the forest. If large areas of the Amazon are converted into pasture or annual crops, with removal of the forest cover, there will be a change of climate "with a more defined prolonged dry season, a deficit of water in the soil and larger temperature variations,"[17] Salati predicted. The results for the forest will be severe. "The plants that were initially selected and developed as a function of the initial conditions of the evolving ecosystem are integral and fundamental parts today of the equilibrium that has been established . . . [at] the present level of rainfall," Salati maintained.[18]

Another area of research and speculation is in the possible effects of a major disturbance in the Amazonian water and atmospheric gas cycles on global climate. One of the scientists who have been measuring this is Luiz Carlos Molion, a senior weather researcher at the Brazilian National Space Research Institute (INPE) at São José dos Campos who is now director of an Amazonian research consortium near Manaus. Molion led the Brazilian team that conducted a series of experiments in Amazonia between 1985 and 1988 with the U.S. National Aeronautics and Space Agency (NASA) to study the influence of Amazonia on the chemical make up of the global atmosphere, including the formation of gases contributing to the "greenhouse effect."

In a paper prepared in 1987, Molion asserted that one of the biggest dangers of Amazonian deforestation, and consequent disruption of the water cycle, would be a reduction in heat transfers from Amazonia to colder temperate regions that receive less solar radiation. This theory is based on evidence that clouds formed by vapor arising out of the Amazonian ecosystem transfer heat outside the region through atmospheric circulation. Molion wrote:

> If these tropical regions receive less heat, they will become relatively colder with consequent reduction in the growing season for plants. This would reduce the production of grains, because the major producers today are outside the tropics. In addition, the reduction of temperatures in the polar regions would cause the ice line to move toward lower latitudes, which would tend to accelerate the begining of a new glacial era.[19]

Molion's hypothesis is untested, beyond mathematical models, but so are other climate change theories. If Molion's view should prove accurate, the effect would be the opposite of the expected warming of temperate and polar regions from the "greenhouse effect" caused by gases that trap radiant heat from the earth in the atmosphere. Only time and further research will tell.

In any case, whether Amazonia contributes to warming or cooling the world, or both, the link between man-made disruptions in Amazonia's water cycle, primarily from deforestation, and climate alterations in South America and in the world are now under careful scientific observation. The INPA in Manaus is coordinating this research with strong international cooperation. Only with better understanding of how Amazonia's basic ecosystems function will it be possible to say what water is "worth" and how much should be invested in protecting this resource. Even in Amazonia, water can't just be taken for granted.

The close connection that is believed to exist between the water cycles and the removal of tropical forests is not only a threat that has to be controlled, but an opportunity. The burning of Amazonian forests, in addition to contributing to "greenhouse" gas levels, removes tree stands that consume atmospheric CO_2 in the process of growth through photosynthesis. A growing rainforest provides an efficient system for converting atmospheric carbon into cellulose making cellulose fiber, the main element of wood. On a very large scale, this natural system may be a cheap way to remove CO_2 produced in the northern industrial countries, which are the main sources of fossil fuel gases that have doubled atmospheric CO_2 during the past 100 years.

In their natural state, Amazonia's primary tropical forests can absorb nine kilos of carbon per hectare daily, according to measurements made by NASA/INPE researchers. In the whole region, this would would remove a net 6.3 million tons of carbon a day. But new growth forest would accelerate the rate of absorption, so replanting of deforested areas, or regeneration of forest in depleted areas, would create what is known as a "carbon sink." Just how much the net carbon removal would be is a matter under debate among researchers. According to Roger Sedjo, a forester at Resources for the Future in Washington, D.C., global CO_2 levels could be stabilized, reducing "greenhouse" risks for decades, by tree plantations covering 4,650,000 square kilometers.[20] That figure happens to coincide with almost the whole of the Brazilian Amazon, where the amount of deforestated land theoretically available for reforestation is 400,000 square kilometers.

Any sizable amount of reforestation produces an ecological dividend, and this factor can be added to the economic benefits that can be obtained from plantations in tropical countries. Reforestation helps maintains the water cycle in an ecosystem and reduces the dangers from soil erosion. But plantations are expensive and have to be paid for by providing a raw material for pulp industries or as energy wood.

In Brazil's Amazon rainforest conservation project there is a proposal that solves the financing problem by combining the environmental and the commercial goals. Companhia Vale do Rio Doce, which developed the Carajás iron mine, built a railroad that runs through a devastated area of degraded cattle ranches. This land is cheap, and if it is reforested it can be a source of thousands of jobs and numerous forest products. For both commercial and ecological reasons, Vale do Rio Doce has launched an international forest management project in the Carajás corridor, which forsees two groups of paper companies. Each would plant 150,000 hectares in pulp species, such as eucalyptus, on degraded land and would manage native forests in preservation areas under their control.

Another ecosystem "product" that has great undeveloped potential in Amazonia is tourism. The intact rainforest, the majestic rivers, the tropical wildlife, the regional cuisines and native handicrafts, the marketplaces and nightlife are all natural or cultural expressions of an imposing habitat.

Amazonia can still offer, in many undisturbed locations, the visions of the primeval sylvan world described by Darwin, Bates, and da Cunha.

The preservation of this natural world can be helped by a style of tourism that is sensitive to the ecological values. In Brazil and Peru, private investors have developed hotels and transport infrastructure that provide controlled access to this world for "nature" tours at a treetop lodge in the rainforest canopy near Manaus, on the Telles Pires river in northern Mato Grosso, and in the Manu national park of Peru. But these small ventures just scratch the surface, and international investors are being encouraged to develop proposals. "The development of tourism in Amazonia offers a unique opportunity for opening ourselves to the world through an activity that contributes to economic growth and preservation of natural resources," said Egberto Batista, Brazil's presidential secretary for regional development. But there is nothing in Amazonia that can be compared as a tourist attraction with the wild animal reserves of east Africa or with the "outward bound" appeal of the Himalayas. Without a trained scientist or native guide, the Amazonian forests and rivers can be monotonous and, in most places, uncomfortable. Tourism will require a great organizing effort before this becomes a major economic activity.[21]

The most futuristic vision of the value of Amazonia comes from biological scientists who see the region as the world's most important genetic reserve. Amazonia's ecosystems are the home for more than 300 mammals, 2,000 species of fish, at least 60,000 plants, and 2 million species of insects and microscopic forms of life. A few hectares of Amazon forest hold more species of plants and insects than are found in all of Europe. The variety is due to the efficient use of resources by interdependent species in ecosystems that are natural laboratories for the production of complex organic chemical molecules. These forms of life, and the ecosystems in which they coexist, are said to be Amazonia's greatest wealth.[22]

The preservation of Amazonia for its biodiversity has been argued powerfully by eloquent advocates. Peter Raven, director of the Missouri Botanical Garden, rails against the destruction of Amazon rainforests in speeches that ring with the moral fervor of a Cotton Mather.[23] Harvard University's Edward O. Wilson, a world authority on social insects, compares the removal of the rainforest canopy to create pasture to an act of barbarian ignorance, "like burning a Renaissance painting to cook dinner."[24] Les Kaufman, scientific director of the New England Aquarium, began to get a sense of the economic value of nature in the Bronx, where he supplied his high school's scientific laboratory with snakes from his private zoo. As an ecologist specializing in coral reefs, he sees ecosystems as far more valuable for their genetic creativity than for whatever possible yields can be obtained from "stupid and wasteful" alternative use of rainforests for lumbering, cattle, or agriculture.

> The major commodity in the world's rain forests lies dormant and largely unrecognized: It is information. Sequestered in the countless plants and

> insects of the forest is a vast chemical arsenal. To the organisms these chemicals are defensive and offensive weapons, but to people they offer medicines and chemical conveniences of great potential. The importance of this chemical warehouse lies not only in the substances themselves but also in the fact that for each there already exists a genetically coded blueprint. It is such a blueprint, extracted and placed in bacteria or yeast cells, that can allow us to produce huge quantities of any desired substance. The genes of rain forest organisms also contain the secrets needed to create animals and crop plants that can live effectively under moist tropical conditions. . . . biotechnology (has) made natural diversity too valuable to let it be wasted on narrow-minded, one-shot business ventures. This is one resource that can be protected only by keeping it absolutely intact.[25]

The weight of the scientific arguments for conservation of biodiversity is overwhelming, but the practical possibilities for biotechnology and genetic products lack an economic foundation in Amazonia. Historically, this region has played an important role in the international exchange of cultivars and natural products. Amazonia is the source of rubber and cacao, which were succesfully transplanted elswhere for plantations on an industrial scale. The bark of an Amazon tree provided quinine, which was synthesized by laboratories abroad as a medicinal product. The powerful poison used by Amazon Indian hunters, curare, which produces a fatal respiratory paralysis in large doses, is the basis for medicines used to treat heart disorders. But all the biotechnology uses of organic substances from the Amazon have been developed abroad. The current production of all of the native plant products extracted from the Amazon rainforests (excluding coca leaf) does not equal the value of the basic food crops grown in the region—rice and manioc.

Researchers at the New York Botanical Garden's Institute of Economic Botany, such as Charles T. Power and Douglas Daly, following in the footsteps of Ghillean T. Prance, have made great collections of Amazonian plant species with potential for medicinal uses or as new products that could be marketed beyond the local markets. These include the *açaí* palm fruit used as an ice cream flavor in cities from Manaus to Belém or the delicious tart juice of the wild *camucamu* of the Peruvian Ucayali river valley. Darell Posey, an American ethnobotanist, studied the forest knowledge and agriculture of the Kayapo Indians for a decade and identified 50 kinds of bees known to the Kayapo for their honey, insect control, and pollenization functions. But this knowledge has not yet been turned into any new product involving biotechnology.

One of the reasons is that genetic engineering is a very expensive high-tech activity that is concentrated in a few First World government laboratories, like the U.S. National Institutes of Health, and large corporate research centers. Feeble efforts to develop a national capacity for biotechnology in Amazonian countries have been made in Brazil, both at universities and by a few private corporations, such as Agroceres, a major seed company, and Souza Cruz, a subsidiary of British Tobacco Company. But a combination of fiscal crisis—restricting funds for scientific research—and lack of clear legal definition of property rights for private developers reduced investment, and some projects were abandoned.[26]

International assistance for the Amazonian countries in the biotechnology area has been insignificant. Since the Carter administration, the U.S. government has been officially pronouncing the loss of biological diversity to be a matter for international action. Congress passed the International Environment Protection Act of 1983, under which the president is authorized to help countries protect habitats and develop sound wildlife and plant conservation programs. Congress added Section 119 to the Foreign Assistance Act, making USAID the lead agency for furthering the biological diversity strategy in developing countries. An interagency task force asserted in 1985 that "threats to biological diversity warrant global concern . . . [because] this is the deep and largely unexplored pool in which new foods, fibres, fuels, chemicals, medicines, pharmaceuticals, herbicides, insecticides and raw materials for industry will discovered." But in 1991, the amount of USAID money for environmental projects in Amazonia did not exceed $15 million.

While it was neglected financially, biodiversity became a new source of North–South friction as negotiations developed over an international convention on protection of biodiversity. The main issue revolved around the benefits to be assigned host countries for species that were converted into biotechnology products. Given the difficulties surrounding this new area of international cooperation, the most that could be expected in the middle term was agreement on protection of specific Amazon conservation areas, scientifically selected for their importance as biological-ecological reserves, with substantial financial support from the First World and lots of diplomatic tact to avoid charges of "ecological colonialism."

THREE

The Owners of Amazonia

Amazonia's prodigious natural environment seems at first sight to be a historical void, an extravagant ecology that has yet to be shaped by human culture into a social economy. Arnold J. Toynbee, British historian of civilizations, skipped over the Amazon during his visits to South America because, he said, there was nothing "historic" to be seen there.[1] Indeed, judged by monuments like the Incan ruins of Peru, the greatest river system in the world can show nothing comparable to the ancient civilizations of the Tigris–Euphrates, the Nile, or the Ganges.

Such comparisons are misleading. The voices of history can be heard when the standpoint is cultural anthropology and ecology. The native people of Amazonia were more advanced 500 years ago in the development of a genuinely original culture adapted to their environment than the population of the region is now. The modern history of Amazonia arises from the clash between autochthonous cultures and invading forms of political and social organization that imposed domination and exploitation on the natives. The transformations this brought about, including extensive interbreeding, created new social forms that are more advanced in many ways, but that have not yet achieved the compatability with their environment that was the case in the past.

The Amazon was discovered in 1539 by Francisco de Orellana, a Spanish captain, who descended from the snow-peaked volcanoes of Ecuador, navigated 2,500 miles through unknown rivers, and emerged at the mouth of the Amazon two years later. Returning to Spain, he reported having seen white-skinned female warriors, whom he called Amazons. These figures from Greek mythology inspired the name by which the waterway became known in Europe. The name, alien to the Indian tongues, was the first act of Amazon colonization.

The written account of the Orellana odyssey by Gaspar de Carvajal, a Spanish Jesuit who made the trip, and subsequent early chronicles pictured the rivers as heavily populated. Carvajal saw permanent settlements of well-fed, healthy natives who made a living from a combination of agriculture in the floodplains, fishing in the rivers, and hunting and foraging in the forest. These Indians had well-developed ceramics, carving and weaving skills, an extensive pharmacopeia that included poisons and hallucinatory drugs, and a mobility over great distances in dugout canoes that greatly impressed the Europeans.

A century later, when the river was reconnoitered upstream in 1638 by a Portuguese military expedition led by Captain Pedro Texeira, the chronicles still reported "a population so great on both shores of the river that sailing from three hours before sunrise until nightfall there was no end to the edifices nor a landfall that was not occupied by houses."[2]

Where these people came from is uncertain, and their past may be beyond archeological reconstruction. Some estimates of the numbers of the indigenous people at the time of discovery run into the millions, but evidence is scanty. The perishability of forest building materials and the absence of hard metal has impoverished the archeological record and obscured the level of evolution of these Indians who migrated throughout their great interior, with considerable knowledge of star movements.[3]

Contemporary research has unearthed abundant ceramic materials at ancient Indian agricultural sites at the mouth of the Tapajós river that correspond to early chroniclers' reports of dense Indian populations at these locations. There is further evidence in the discovery of identical ceramics in widely separated places that the Amazonian Indians, in their river migrations, extended their culture from the tropical basin westward into the upper valleys of the Andes and northward to the Atlantic coast of South America by way of the Orinoco river.[4]

The Indians of Amazonia had no real metallurgy. They marveled at European metal knives and axes, cutting tools for which they had only bone and shell in their sylvan technology. But what bedazzled the first explorers was the gold that the Indians gathered in the rivers and made into decorative objects.

This discovery proved to be the colonial curse that destroyed the Indians. A century after the Texeira expedition, when the French savant Charles de La Condamine repeated the journey from Ecuador to the Atlantic in 1743, he found only a handful of mission towns, where Indians lived under the tutelage of Spanish and Portuguese Jesuits, Franciscans, and Capuchines.

> The banks—wrote La Condamine in his chronicle—were peopled no longer than a century ago by a great number of nations, who withdrew into the inner parts of the continent as soon as they saw the Europeans. There are now to be seen but a small number of little towns. . . .[5]

The abandonment of the river for the freedom of the forest was the result of the pressures the colonizers placed on the Indians for labor and head taxes. At the junction of the Maranon and Chachapoyas rivers in Peru, La Condamine observed that gold was to be found in the riverbed.

The Indians there, he reported, "gather up exactly the quantity necessary for paying their tribute and capitation; and that only when they are greatly pressed to bring it; at any other time they would tread it under foot, rather than take the pains to gather and cleanse it."[6]

A few days downstream, at the confluence of the Santiago river, La Condamine found the Spanish population terrified of the head-shrinking Jibaro Indians, who had been "christianized" and forced to work in the goldmines until they revolted and fled to the forests.[7] The Jibaros are still feared; but this has been a terribly unequal contest in which the Indians, and the habitat, have lost.

The first European dicoverers, who were Renaissance men, often gifted in letters, like Amerigo Vespucio, perceived and admired the ecological stability of the native cultures. "They live in harmony with nature," wrote Vespucio after living among the Amazon natives during a voyage to the coast of Brazil in 1500. But he also put his finger on the weaknesses of their political and social system.

> They know nothing of the immortality of the soul, they have no private property among them, for everything is held in common; they have no boundaries to their kingdoms or provinces; they have no king; they obey nobody; each man is his own master.[8]

The lack of political organization of the Indian peoples, and the absence of any common intertribal language or religious unity, must be seen as the basic reasons why handfuls of Spaniards and Portuguese were able to impose their dominion over multitudes of natives. When the Europeans arrived, the Amazon valley was undergoing continuous intertribal warfare with frequent migrations by displaced groups. But the profound disruption of the native societies by the Europeans has a deeper explanation in the clash between two profoundly different cultures. Along with the steel of their weapons and armor, their horses and blunderbusses, their common language, military organization and religion, the Europeans brought a "logic" of conquest that would overwhelm those who had neither property, boundaries, nor king.

The European concept of man was radically different from that which inspired the magical, symbolic world of the Amazonian natives. In his penetrating study of the roots of European colonial expansion in America, Africa, and Oceania, Alfred W. Crosby speculated that the key to dominance was not only superior technology, but a vision of man and nature that gave sanction to "the direct control and exploitation of many species for the sake of one: Homo Sapiens."[9] In contrast, the "Amerindians placed a prohibitively high valuation on animals, considering them fellow creatures equal to or even superior to humans, not as potential servants."[10]

The Europeans ruptured the totemism and rituals by which the Amazonian Indians lived and the natural environment they worshiped. The use of gold, for instance, as a means of exchange and measure of wealth subverted the Indian view of gold as a sacred substance, a symbol of fertility, a link between man and the sun, and therefore an element in shamanistic rituals and a substance for exquisite artistic creations.[11]

This is the essential element of El Dorado, the king-priest whose body was covered with gold dust as he floated on a raft bearing gold objects that were cast into the waters of a sacred lake as a sacrifice to the gods. The potency of this myth launched the European adventurers on expeditions into unknown parts of South America in search of the lost gilded city of Manoa and the sacred lake.

One of the ironies of Amazonian history is that the cultural incompatibility between the Europeans and the natives undermined the effective development of the region for centuries. The pressure the colonizers exerted on the Indians to produce gold drove away the native labor force without which the Europeans could not make the Amazonian environment yield economic gains. The irony became tragic when contagious diseases introduced by the Europeans completed the dispersal of the Indians, for whom flight into the forest was the only escape from extinction.

The story of the Amazonian Indians has been told eloquently by a succession of anthropologists, from Claude Levi-Strauss[12] to John Hemming, whose scholarly work is a moving salute to the "vanished tribes."[13] There is no need to repeat that here, although the important role that Indians play in environnmental policy for Amazonia is returned to in Chapter 10.

For the history of Amazonia, the importance of the Indians needs to be underscored mainly as evidence that Amazonia was not, and is not, an empty quarter without a cultural past. It has been a habitat for millions of humans who have made their living in the forest and were able to maintain and reproduce their societies over centuries until they were overcome by invaders who came not only with stronger weapons but with a culture based on domination of nature. There is no turning back from this, but the Indian cultures continue to make a contribution to what Amazonian human society is, and what it may become.

Some scholars have pictured the attrition of the natives as genocide. The national societies of the Amazonian countries are all increasing their penetration of the interior, so the judgment refers not only to times past, but to current events. David Maybury-Lewis, a founder of Cultural Survival, a center for militant pro-Indian anthropologists in Cambridge, Massachusetts, says that a "second conquest" is under way in which the development programs of the Amazonian governments represent a threat of final extinction of the native Indians still left in the backlands.

> It is in this interior—the last refuge of the Indian peoples—that the second conquest is now taking place. It is powered by a worldwide quest for resources in which all the Amazonian nations find themselves engaged. The new conquest, unlike the first one, is not particularly interested in Indian labor. It is very interested in Indian land. The threat to the

> Indians is not this time one of slavery, but of expropriation of their lands and total destruction of their way of life, if not of their persons as well.[14]

The term *genocide*, implying moral responsibility for purposeful extinction of a people, is an oversimplification of the complex Amazonian cultural conflict. The Amazon territories today are characterized by a mixture of races, so, rather than a genocide, the past four centuries have seen the genetic integration—forced or voluntary—of most of the surviving Indians into a new mixed race. The autochthonous peoples live on in the demographics of Amazonia today.

Of the 16 million people in the region, a majority are mestizos, a mixed population of white, black, and Indian blood. The genetic presence of the Indians is evident in the oval eyes, long straight black hair, supple bodies, and facial features of people at all levels of society. Miscegenation has perpetuated many basic customs drawn from Indian culture, such as the preparation of food.

In the countryside, the basic daily fare is based on the same combination of mandioc flour and fish with local condiments and fruits that the Indians eat. Home construction is with split bamboo poles and palm thatch, sometimes on pilings to avoid floodwaters, just like the Indian villages. In riverfront cities, like Belém or Iquitos, canoes are as much a part of urban transport as buses or motorbikes. Clothing is a product of "civilized" society, but daily washing and bathing in the rivers is a practice that maintains an indigenous form of personal hygiene in which the body and hair are treated with oils and essences from Amazonian plants.

In addition to these Indian strands in the world of Amazonia, the Indians who retain a tribal identity are an active minority that is engaged in an ongoing political confrontation with the "modern" sectors over the exploitation of Amazonia. The desires of the "native" or Indian communities to preserve their cultural autonomy in lands that they consider theirs by ancestral right is one way by which they can compete for a share of the "modern" economic and social goods, which they do not reject.

The fact that the Indians have not disappeared, and that they maintain an interactive tension with the "modern" national societies of which they are now a part, makes the "Indian question" relevant for this study for three reasons.

1. The Amazonian Indian drama will continue while there remain tribal people who struggle to retain their cultural identity. Indian rights, including preservation of tribal lands, are a political issue in Amazonian countries, with important implications for development of land, mineral, oil, water and timber resources. The intercultural conflict bears on domestic policies dealing with Amazonian social integration, education, linguistics, ethnic pluralism, churchstate relations, and national security.
2. The cultural survival of primitive people is on the agenda of interna-

tional relations. It is a human rights issue. The scientific study of "savage" societies by cultural anthropologists in this century developed the view that Indian cultures represent something of "value" in human affairs. This has produced a movement by scientific, religious, and political action groups in industrial countries for protection of Indian rights, with growing influence on foreign-aid decision makers. The scientific and humanitarian pressures from abroad, combined with the activism of domestic "indigenist" groups, makes this a delicate foreign policy issue.

3. The Indian issue cannot be separated from strategies for the environmental protection of Amazonia. The Indians form part of this ecology, and just as they have been passive victims of the destruction of the environment, they can be active agents for its preservation. The creation of areas legally reserved for occupation by Indian communities in Amazonia is considered by some environmental activists as a way to preserve tropical forests and their valuable genetic resources.

* * *

The international political–diplomatic history of Amazonia provides another important element for this study because it is the foundation for the system of interstate relations among the nations that make up this region.

Amazonia is a territory under the jurisdiction of eight sovereign states that act autonomously in their respective portions of the region, although they increasingly recognize their common interests in economic development, environmental protection, and security. These regional interests are reflected in a Treaty for Amazon Cooperation, signed in 1978 by all the nations of this region, each with its own piece, large or small, of the Amazonian natural system. The signatories of the pact are Bolivia, Brazil, Colombia, Ecuador, Guyana, Peru, Surinam, and Venezuela.

It must be stated clearly that Amazonia is not, and cannot be placed, in the political and legal situation of a tropical Antarctica. The ice-covered southern polar continent was internationalized by a treaty in 1959 signed by 37 countries, most of which, including the United States and the Soviet Union, are in another hemisphere. The Antarctic treaty acknowledges claims to territorial sovereignty in the sixth continent by seven countries, including Argentina, Chile, and New Zealand (the nearest to the region), but the signatories have set up an international condominium to regulate the development of its economic resources and share the benefits among United Nations member countries. Mining exploration concessions can be granted under the treaty, but only with the consent of an international governing body, and extractive activity has been banned for 50 years to protect the environment, according to a treaty amendment adopted in 1991.

In Amazonia, there is no territory without a national owner and no shared rights to resources other than those that the states are prepared to concede under their sovereign laws. The borders are well defined, and there is only one significant unresolved territorial dispute, involving Peru and Ecuador, which fought a war in 1941 over access to the Amazon river. Ecuador signed a treaty ending the conflict, but the demarcation of the border remains a festering source of dispute because Ecuador, from where the Amazon river was discovered, is unreconciled to being deprived of direct access to the great river.

Contrary to what some global environmentalists assert, Amazonia and its forests do not belong to "humanity." Amazonia belongs to national societies, and they are responsible for what happens there, hopefully guided by good sense and a decent regard for world opinion. What other nations, and people in general, can demand legitimately is that these actions not damage the global climate and other areas of common interest, as defined by international conventions.

The present distribution of Amazonia is the result of five centuries of diplomatic maneuvering and occasional military confrontations that shaped national areas of influence even before accurate maps could be drawn. As a result of this process, Brazil is preponderant in the region by virtue of possessing two-thirds of the Amazonian territory, the Atlantic mouth of the river system, and relatively easy access to its Amazonian interior by way of navigable rivers.

This does not reduce the importance of Amazonia for the Andean countries—Peru, Bolivia, Ecuador, and Colombia. Some have the largest part of their national territories in Amazonia. The ways in which these countries manage their Amazonian resources and the influence their governments bring to bear on regional development can make significant contributions. Even the minor border areas of Venezuela, Guayana, and Surinam that are on the northern Amazonian flank make them participants in the region. But Brazil necessarily is the key political and economic factor in any regional approach to Amazonia.

The Iberian colonization of South America took, from the start, two distinctly different routes. The Spanish conquistadores came first, and they took the high road, advancing through the mountain areas of the Andes, which led to occupation of the Pacific coast. The Portuguese, at a slower pace, took the low road along the northeastern hump of Brazil, establishing their colonies on the Atlantic coast. For a long time, Amazonia was simply ignored.

When the kingdoms of Spain and Portugal were world leaders in the discovery of Africa, Asia, and America, they were also the pillars of the Roman Catholic world. The Portuguese navigators took a head start in finding the routes around Africa to India. The discovery by Columbus, in the service of the Spanish crown, of the western passage to the Antilles laid Spanish claim to the northern coast of South America. This competition created a potential conflict between the two powers over lands that were about to be discovered.

To head off this possibility, Pope Alexander VI issued his famous bull of

1493, a year after Columbus' first voyage, establishing a division of the spoils even before they were known. He drew a line along a longitude 100 leagues west of the Azores, with everything to the east for Portugal and all to the west for Spain. On paper, this meant Africa for the Portuguese and America for the Spaniards.

It was then that Portugal made an inspired move toward securing a claim to a new land that until then no European had laid eyes on, and that would be named Brazil when it was discovered six years later. In 1494, negotiators for Portugal and Castille met in a monastary at the town of Tordesillas, where the Duero river runs west toward its mouth on the Atlantic at Porto, and agreed by a treaty to shift the pope's line of demarcation westward to a longitude 370 leagues west of Cape Verde, on the African Atlantic coast.

Portugal's insistence on moving the line westward has raised speculation that Lisbon may have had secret knowledge of the New World that Spain lacked.[15] If not, it was an extraordinary stroke of intuition that changed the course of South American history. The Tordesillas line gave Portugal all the coastal lands of Brazil when these were discovered in 1500 by Pedro Alvares Cabral, leader of a Portuguese expedition that was bound for India.

Cabral's unforeseen landfall on the coast of what is today the state of Bahia virtually coincided with the discovery of the mouth of the Amazon by a Spanish expedition, in which Vespucio was the navigator. But for 40 years the Amazon basin remained, for Europeans, an unknown no-man's land until Orellana made his unplanned journey down the great river.

The news Orellana brought back to Spain of the river of the Amazons stirred up great plans. Emperor Charles V and his advisers evidently thought the mouth of the Amazon, as well as the interior, fell on Spain's side of the Tordesillas line. The Spanish crown invested in a return expedition by Orellana in 1549 with three ships and a large military contingent to claim the Amazon for Spain.

The expedition was destroyed by disease, shipwreck, and ultimately the death of Orellana. This failure discouraged Spanish interest in the Amazon, and for the next 50 years Spain's policy was to concentrate all its colonization efforts on the Pacific and the Caribbean. Peru's silver and Colombia's gold came to Spain in ship convoys, with a portage in Panama, by way of naval stations at Portobello, Cartagena, and Havana.

In this scheme, Spain's Amazon policy was to shut off any exploration that could open a "back door" for foreign powers to reach the Andean treasures of Spain's colonial empire by way of the rivers of Amazonia. For half a century the Amazon was unexplored, except for an expedition in quest of El Dorado by Spaniards Pedro de Ursua and Lope de Aguirre that ended in murder, mutiny, and madness.

Spain's abandonment of Amazonia set the stage for Portuguese/Brazilian moves that disregarded the Tordesillas line and began a sustained effort to bring all the eastward-running rivers of the Amazon under Lusitanian control. The success of this policy ultimately decided the question of who would have the upper hand in the region.

The first step in this direction was the military expedition led by Pedro Texeira, a Portuguese creole captain, who went up the Amazon from Belem in 1638 with 70 soldiers, 1,200 Indians rowing 47 canoes, and 2 Spanish priests who had floated down the river escaping from Indians in Ecuador. After eight months Texeira reached Quito, repeating in reverse Orellana's initial journey of discovery. On the return trip to Belém, Texeira took time to explore the interior rivers, including the Madeira and the Negro. This laid the foundation for Brazil's eventual Amazonian predominance.

At the time of Texeira's expedition, Spain and Portugal were still united in a kingdom that had been formed in 1580 to maintain a Roman Catholic power base on the Iberian peninsula against the rising Protestant forces in Europe. The Amazonian region, remote as it was, could not remain outside the competition this religious conflict unleashed. The French, who scoffed at the Tordesillas division of the New World, had set up a fortified colony at São Luis in Maranhão, a pre-Amazon territory in northern Brazil. The French had also sent trading and exploration expeditions up the Tocantins river. English and Dutch traders had gone as far inland as the Xingú river, setting up forts and tobacco plantations as well as occupying positions on the Guiana coast. There was even an Irish outpost.

News of these incursions revived the fears of the Spanish crown over the security of the inland areas, and orders were sent to push the invaders out of the Amazon. The military muscle for this reaction came not from Europe but from Brazilian colonists, many of whom were by then creoles with their own reasons for protecting their native turf.

A military expedition recruited in Pernambuco and supported by friendly Indians laid siege to the French settlement in Maranhão and recovered São Luis in 1615. A fort was raised at Belém in 1616, and this became the base of operations against the Dutch and English, whose forts and warships in the Amazon were reduced in a series of naval and land operations between 1616 and 1630. Texiera's name figures prominently in the reports on the combat with the English and Dutch.

Texeira's journey up the Amazon came as an unwelcome surprise to the Spanish authorities in their Andean mountain capital at Quito. The Portuguese were not enemies, but the route they had found could also be discovered by the Dutch and English, who were known to be trading with Amazon Indians from bases in the Guianas. The viceroy in Lima dispatched reports to Spain and ordered that Texeira be escorted on the return trip by a Spanish Jesuit, Cristobal de Acuña, who was to proceed from the mouth of the Amazon back to Spain and report to King Philip IV on the "new discovery" of the Amazon.

Acuña's account in 1641 of his ten-month descent of the Amazon with Texiera is one of the most lucid chronicles written in the age of discovery on life in Amazonia.[16] It is rich in details on the natives, the geography, and the natural exuberance. ("I measured with my own hands a cedar that was thirty handspreads in girth.")

Acuña was also an astute political observer. He reported in a "memorial" to the Council of Indies that during the return trip Texeira and his men, while nominally loyal subjects of Spain, had formally asserted a Portuguese claim to a vast wedge of Amazonia between the Madeira and Negro rivers, all of which now forms part of Brazilian territory. He bore witness that the Portuguese/Brazilians whom he had come to know had the intention "beyond any doubt" to extend their control from the mouth of the Amazon deep into the "nation of Peru," a development he saw as prejudicial to the interests of Spain.[17]

Acuña's enthusiasm over what he had seen ("in fertility the shores are Paradise") led him to urge Spain's immediate occupation of the Amazon, but starting from Quito, not Belém. Acuña argued that effective occupancy would provide a safer, shorter inland river route for the Andean riches bound for Spain than the trip through the corsair-infested Caribbean.

The Jesuit's warning either fell on deaf ears or came too late. In 1642, the union between the royal houses of Spain and Portugal was dissolved. Smoldering rivalries resumed. Texeira's expedition, which discovered large Indian populations and the existence of abundant gold, inspired new incursions. A string of Portuguese forts established control of key points on the middle Amazon, up the Madeira to the south and up the Negro and Branco rivers to the north. There was no matching effort in the Spanish colonies.

Outside of America, Portugal's fortunes were on the decline from the days when Portuguese navigators and soldiers, such as Vasco da Gama, created a world empire. This small maritime nation was displaced by the French, Dutch, and English in Asia, and its African colonies were threatened. But Portugal still had sources for slaves in Africa and a great colony to be developed in Brazil.

In 1755, an energetic statesman, Sebastião José Carvalho e Mello, Marquis de Pombal, came to power as the king's chief minister in Portugal. He was determined to restore Portugal's fortunes, and Brazil became the keystone for Portugal's overseas policy.

Pombal may have sensed that Portugal's failure to develop the huge Brazilian territory, including the Amazon, would be an invitation to other, stronger powers to step in. Spain had signed away the last vestiges of Tordesillas in the treaty of Madrid in 1750 which recognized the western limits of Brazil's territory as the Guapore and Madeira rivers. This provided some security for Portugal's goldmines and placers in Mato Grosso. But the Amazon could not be consolidated against outside threats without a military presence, a system of river communication, stable communities, and production of goods.

Pombal was a mercantilist who sought to make the colony a source of wealth for both the crown and the merchants, shipowners, and manufacturers of Portugal, who received a monopoly over trade with Brazil. For this system to yield returns and finance the cost of colonial administration, the colony had to produce tradable goods. The South provided abundant gold from the mines of Minas Gerais and Goiás. The plantations of Bahia and Pernambuco, with slave labor, produced exportable sugar, tobacco, and cotton.

Amazonia was a problem, however, because its wealth depended on extractive activities that were ecologically damaging. There were good markets in Europe for Amazon forest products such as cacao, the chocolate bean, which grew wild; spices such as cravo, a tree bark with clove scent, and *canela*, a cinnamon-like bark found in the western sub-Andean forests; oils from plants and turtles; and medicinal roots such as *salsa* (sarsaparilla).

But the extractors, using Indian labor, were foragers, not farmers. They made no effort to domesticate and cultivate the plants and fauna from which they drew their products.[18] It took over 1,000 turtle eggs to produce a pot of cooking oil, and the extraction was so predatory, says one student of the period, that it produced extinction of turtle populations and eradication of vast groves of *salsa* and *cravo*. As a result, production declined.

A far more serious ecological disturbance was affecting the economic exploitation of Amazonia when Pombal came to power. The European colonists and their African slaves had introduced smallpox, measles, tuberculosis, and other contagious diseases that were unknown in Amazonia and for which the Indians had neither immunity nor native remedies. Epidemics devastated whole Indian villages. The death toll undermined both the extractive and the plantation activities that depended on Indian labor. Another problem was malaria, apparently not endemic to the Amazon before the coming of Europeans, but which has flourished in the insect paradise that the tropical Amazon provides, with fatal or debilitating effects on both Europeans and Indians.

With characteristic vigor, Pombal launched the first of the "Amazon development plans" through which government administrators have attempted to achieve human settlement and stable production in Amazonia. Pombal's development plan called for stimulation of Portuguese private investment in Amazonia and expansion of trade and production, using the Indians as both laborers and consumers.

To this end, he persuaded King José I to issue two decrees that appeared to revolutionize the existing Amazon situation. One decree freed Indian slaves, restored to all Amazon Indians "the liberty of their persons, goods and commerce," and recognized Indian tribal lands as inviolable. The second decree struck at the power of the missionaries, who were replaced as civil administrators of Indian villages by a system of government-appointed "directors" who reported to the military governor. When the Jesuits resisted the innovations, Pombal had the king expel them from

Brazil in 1760, an example that was followed shortly thereafter in Spanish America by King Charles III of Spain.

Pombal created a General Company of Commerce of Greater Pará and Maranhão—modeled on the Dutch and British East India trading companies of the time—which was supposed to promote plantation agriculture in Amazonia. This scheme never enjoyed much success because of the high cost of slaves.

During the 27 years that Pombal remained in royal favor in Lisbon, the "Directorate" system created 61 towns, most of which remain until today active communities, and the creation of a new captaincy general in Manaus consolidated Brazil's control over the Negro river and its huge share of the western Amazon. The "Directorate" was a modest economic success, in terms of production of goods for export. Roberto Santos, in his *Economic History of Amazonia*, says exports from Belem to Lisbon of cacao, coffee, rice, and cotton (the main products) went from an average annual rate of 30,554 pounds sterling in 1756–70 to an annual 41,848 pounds sterling in 1780–84, and averaged 65,000 pounds sterling to the end of the century, with cacao increasing the most.

Yet, the growth in population of non-Indian settlers during this period was tiny for such a vast area—from 55,000 counted in a census in Para in 1773 to 93,000 in 1816.[19] There was also no improvement in the condition of the Indians, who continued to be exploited as "free" laborers and reduced in numbers by disease. Because of the insoluble labor problem, there were no real advances in plantation agriculture or industrialization of raw materials. Groups of agricultural settlers imported from Europe barely survived without Indian labor, and many drifted to the cities and towns. The century ended with no momentum toward economic development in Portuguese Amazonia. The situation in the Spanish-speaking countries was even less dynamic.

However, great events were at hand that would change Latin American and Amazonian history. The political earthquakes of the Napoleonic period in Europe reached the Iberian peninsula when French troops occupied Spain and Portugal. The French invasion of 1807 sent the Portuguese royal household across the Atlantic into a temporary exile in Brazil. When Napoleon was defeated by the Grand Alliance led by Britain, King Joao IV of Portugal was restored to his throne and returned to Lisbon; but his son and heir, Dom Pedro I, decided to remain in Brazil and in 1822 was proclaimed emperor of a new independent nation. This coincided with the independence from Spain won by the neighboring republics of Argentina, Paraguay, Peru, Colombia, and Venezuela.

The achievement of independent nationhood by the new South American countries did not affect the distribution of sovereignty over Amazonian space between Brazil and its neighbors. This was already an accomplished fact. But independence did heighten the assertion of national interests.

A nineteenth-century equestrian monument of Dom Pedro I in Rio de Janeiro's Tiradentes Square is a parable in bronze of Brazil's pride of possession of its vast interior, including the Amazon. The heroic bronze statue shows the emperor, in Napoleonic uniform and sword in hand, astride a steed that paws a rectangular base, each side of which represents the four great rivers of Brazil and the native peoples who lived there. One is the San Francisco river of the Northeast, with an Indian warrior holding a feathered spear sitting beside an anteater. The next is the Parana river of the Southeast Quarter, where an Indian couple is seated atop a *capivara* wild boar and an armadillo. Next comes the Madeira, of the Northwest, where an Indian brave, seated on a turtle, draws a bow and arrow. The quadrant is completed by the Amazon river, represented by another native couple in feathered headdress seated on a huge alligator.

Yet, the early stages of independence in Brazil brought a period of anarchy in Amazonia. First, there was resistance from pro-Portuguese loyalists who felt their commercial interests lay with the metropolis. "We have no relations with the south of Brazil, while our parents are in Portugal, which is the true market for our products, and toward which all communications are easier than they are to Rio de Janeiro, as if nature herself, with the currents and winds, had chosen to show us the union that is most in our interest," declared a junta of merchant-citizens of the North. Oliveira Lima, a distinguished Brazilian historian, says that the loyalist movement in Maranhão–Pará secretly harbored the intention of declaring a "free Amazonia in all its equatorial vigor."[20]

This separatist dream was disposed of by a small navy assembled by Dom Pedro under the command of Lord Cochrane, an adventurous British naval officer, who imposed a blockade on Bahia that ended resistance there and in Maranhão and maintained the union between Brazil and the northern Amazon states. But the evidence of loyalist sympathies among the northern merchants, the beneficiaries of the Pombal's policies, revealed a political cleavage in the the Amazonian states that would soon contribute to a social explosion with strong underlying racial antagonisms.

In 1830, Para was the scene of an uprising by mestizos, mulattoes, and Indians against their masters, who were predominantly Portuguese merchants and landowners. The movement, known as the *cabanagem* in Brazilian history, spread throughout the Amazonian towns and plantation. Rioters sacked ships, warehouses and the homes of merchants in Belem and other large towns, while Indians in the interior killed landowners and their families, burned down whole plantations, and destroyed herds of cattle before melting into the forest.

There was civil war between political factions for control of Belém, which changed hands several times, and anarchy for a decade. Much of what had been built up was destroyed before order was restored in Pará and the western Amazon. Nothing similar happened elsewhere in Brazil after independence, and the traumatic effects of the violence were still evident to travelers in the Amazon years later. Despite the atrocities, Brazilian nationalist historians regard the *cabanagem* as a creole assertion of "Amazonia for the Amazonians."

In the Spanish-speaking countries of the Amazon basin, the wars of independence and the subsequent power struggles over exercise of republican self-government took place far from Amazonia and changed little in the region. During the first half of the nineteenth century, the central governments were absorbed with problems of political organization and economic development that barely took note of the empty and inacessible Amazonian world that seemed unable to contribute significantly to the national economies.

The only important presence was the religious missions that maintained outposts of education in the Moxos and Chiquitos regions of eastern Bolivia, in Maynas along Peru's mighty Ucayali river, up the Napo and Pastaza rivers into Ecuador, and on the Putumayo and Caqueta of Colombia. Some of these missions had been founded two centuries earlier, and little had changed over the years. Roads linking the jungle to the highlands or the Pacific coast were little more than mule trails, and few and far between. The Spanish-American governments, whether colonial or independent, found no economic reason to invest in their Amazonian backlands, far removed from the existing population centers by the towering Andean cordillera.

* * *

This situation of abandonment changed dramatically during the second half of the nineteenth century when Amazonia experienced the natural rubber boom, which is the best studied and most colorful period in Amazonia's history.[21] As seen in Chapter 2, the boom brought a flood of immigrants into the region on a scale of the California goldrush. Rubber also brought Amazonia into the arena of international commercial rivalries, which produced new tensions among the countries of the region.

Even before the emergence of natural rubber as a major industrial commodity, foreign pressure to open the Amazon system to international traffic had begun to build. The Brazilian empire had maintained the Portuguese colonial policy of restricting access by outsiders to trade and travel in Amazonia, denying foreign-flag ships the right to transport cargo on the inland river.

International pressure for an "open door" policy for Amazonia did not take the form of a crude colonial adventure like the carving up of Africa or the imperialist wars over concessions in China. In Europe and the United States, the interest in the Amazon region was commercial, not territorial. Amazonia was seen as a source for sugar, cotton, and tropical commodities and as a promising area for the application of new technologies of transportation, like the railroad and steamship, requiring ports, telegraph cables, and commercial services. All this raised the level of international interest in Amazonia at a time when none of the countries in the region had the financial resources or the political will to develop Amazonia.

The establishment of an "open door" policy in Amazonia became enough of an issue in the United States in the decade before the Civil War to provoke diplomatic alarm in Brazil and great expectations in Bolivia, Peru and Colombia. This was at a time when the United States had just given Latin America an example of "Manifest Destiny" by invading and annexing Mexico's northern territories, from Texas to California. There were well-known expansionist designs on Santo Domingo and Cuba, and there were southern pro-slavery senators and New York capitalists interested in Amazonian ventures.

These U.S. interests were matched in some of the Spanish-speaking Amazonian countries by government leaders who saw the shipping technology and capital of the advanced commercial countries as a way of developing their isolated tropical lowlands. There were members of Brazil's imperial elite who favored a more open commercial policy, but there was also strong nationalist opposition from leaders who feared Brazil would lose control of Amazonia.

One enthusiastic publicist of Amazonian development was Matthew Fontaine Maury, a U.S. Navy captain. Maury was a strong supporter of Commodore Perry's methods of opening up the Pacific to U.S. trade by sending gunboats into Tokyo harbor in what the Navy Department professionals called "a philosophy of action" to support American commercial expansion abroad. Maury found time, while directing the U.S. Hydrographic Office in Washington, to write tracts on the wonders of the "Amazon valley" and the need to open it to the commercial "spirit of the age" in the interest of "Christian civilization." He actively lobbied the U.S. Congress for subsidies for an Amazon shipping enterprise.

In a series of articles signed "Inca," Maury described the Amazon basin as "the finest country in the world . . . [with] immense treasures that lie dormant and undeveloped."[22] He waged a vigorous campaign in favor of bilateral shipping agreements with Peru and Bolivia that would permit U.S.-flag vessels to ply their rivers. He reported that Bolivia and Peru would pay $20,000 to the foreign entrepreneur who placed the first steamboat on their Amazon waters.

As Maury wrote his articles, he was in contact with two young American naval officers, Herndon and Gibbon, who were making confidential survey trips down the Amazon rivers of Peru, Bolivia and Brazil to prepare reports to the U.S. Congress on their navigability and economic potential. The great obstacle to this scheme, Maury maintained, was the policy of Brazil that reserved all of its Amazon rivers to its own vessels.

> So fearful has she [Brazil] been that the steamboat on these waters would reveal to the world the exceeding great riches of this province, that we have seen reenacted under our own eyes a worse than Japanese policy;

> for it excludes from settlement and cultivation, from commerce and civilization, the finest country in the world.[23]

In a foreshadowing of developmental strategies that would later come to Amazonia, Maury gave his scenario for progress:

> . . . all must admit that the valley of the Amazon is not only a great country, but it is a glorious wilderness and waste, which, under the progress and improvement of the age, would soon be made to "blossom as the rose." We have, therefore, but to let loose upon it the engines of commerce—the steamer, the emigrant, the printing press, the axe, and the plough—and it will teem with life.[24]

The Brazilian ambassador in Washington at the time reported in alarmed detail on the campaign in the United States and Europe to open the Amazon. The matter was treated at the highest levels of government in Rio de Janeiro. In his diplomatic dispatches, Ambassador Sergio Texeira de Macedo observed that Brazil lacked "the capital, the credit or the spirit of enterprise" necessary to provide steamboat transportation on the Amazon rivers where only a swarm of small, oar-driven boats then moved cargo. With prophetic foresight he warned that inadequate transport for the then incipient Amazonian rubber production would lead consumers to "find the way to cultivate it in other climates," as subsequently occurred, to the ruin of Amazonia. The ambassador's recommendation was that the interior rivers be opened to foreign ships, as the Atlantic ports of Brazil already had been since independence.[25]

But the imperial ministers in Rio thought otherwise, and when the United States presented a formal request in 1854 for navigation rights, it was turned down with the argument that the Amazon areas of Peru and Bolivia were being supplied "from the Pacific" and there was little trade in Brazilian Amazonia because it was unpopulated.[26]

It took only a decade for Brazil to reverse this isolationist policy, and the Amazon was opened to foreign-flag shipping in 1867, just after the end of the American Civil War. The compelling reason for the change of policy was new economic realities created by the rubber trade. Industrial demand for natural rubber was generating an explosive boom in this Amazonian commodity.

Brazil's attempts to bottle up the Amazon had provoked friction with its neighbors. The rubber-gathering activity extended from the middle Amazon up the tributaries to areas of Bolivia and Peru, which were demanding transportation outlets through the Amazon rivers. Brazil could not provide the shipping, and its policy became untenable. With the "open door" to shipping, British and American cargo ships were soon providing regular service to Liverpool and New York, not only from Belem, but from Santarem, Manaus, and eventually Iquitos. Everyone profited, including Brazil.

The fears of foreign intervention in Amazonia were not entirely groundless, however, as was seen at the height of the rubber boom with the creation of a so-called Bolivian Syndicate, put together by U.S. and British capitalists and the Bolivian government, to run the rubber trade and customs house in Acre, a territory then divided among Bolivia, Peru, and Brazil.

An armed rebellion in 1902 by Brazilian rubber operators in Acre against the Bolivian Syndicate ousted the Bolivian troops, seized the territory, and annexed it to Brazil, much as Texas became part of the United States.

The leader of the rebellion was Plácido de Castro, a former army officer from the southern Brazilian state of Rio Grande do Sul, who had come to Acre as a surveyor. News of the Bolivian Syndicate reached him while he was mapping a rubber property in the jungle. The entry he wrote in his diary that June 23, 1902, shows the nationalist motives behind the revolt.

> The terms of the contract that have been signed between Bolivia and the Bolivian Syndicate are a complete sell out of the people of Acre. The cruel idea came to me that the Brazilian homeland could be dismembered; because, as I saw it, this contract was no more than a wedge introduced by the United States to force its way into our rivers, using any resistance as a pretext to employ force against us, with which our misfortune would soon be sealed. I put away my quadrant, abandoned the survey and headed for the banks of the Acre.[27]

Arms had already been smuggled into Acre by Brazilians, and an "Independent Acre" force led by de Castro occupied Xapuri, a key river town, driving out the Bolivian soldiers. Some sharp fighting followed along the rivers and in the rubber-trading centers. The Brazilian and Bolivian governments, both anxious to contain the conflict, hastily negotiated a settlement. In the treaty of Teresópolis of 1903, Bolivia recognized Brazil's rights to Acre—adding 180,000 square kilometers to Brazil's Amazon domain—in exchange for $3 million in compensation and a promise to build a railroad, which now links Corumbá (Brazil) on the Paraguay river to Santa Cruz, in eastern Bolivia.

The rubber boom also produced friction between Peru and Colombia along the Putumayo river, where the Peruvian trading company, Casa Arana, exploited rubber stands in territory claimed by Colombia—with such violence to Indian communities that a report denouncing the violations was presented to the British Parliament. In 1932 Colombia occupied the area militarily, after expelling the Arana traders, and established an Amazon stronghold at the garrison town of Leticia, facing Peru across the Marañon river and on the border with Brazil.

These experiences contain many useful lessons on factors that still influence international relations over Amazonia.

1. The rubber boom that brought Amazonia into the world industrial economy would not have been possible without the "open door" for international shipping and commerce. The policy of exclusion generated tensions between Brazil and its neighbors, as well as the outside world, that led Brazil to change its initial closed stance. Freedom of navigation on the international rivers of Amazonia is now an explicit commitment of the Amazon regional pact, and commercial relations underlie all other forms of cooperation.
2. Brazil and the other Amazonian countries obtained from rubber a large flow of capital and export income from their least developed tropical regions, but they lost the opportunity to make this permanent because they did not generate the necessary technology. The rubber boom could have been an occasion for an attempt to create rubber estates and industrial plants in Amazonian countries. In practice, the "dirty" extractive end of the rubber business, which was severely exploitative of labor and Indian communities, was left to the Amazonian jobbers. Foreign interests became directly involved only after the raw rubber was delivered to the Amazon ports for export, and the value added in final products went to the owners of industries abroad far from any of the Amazonian countries.
3. The farsighted Brazilian ambassador in Washington, Texeira de Macedo, had warned that "countries that stand still, without developing their industries, accumulating wealth and exercising effective dominion over their possessions, become victims, like Mexico, of stronger neighbors." In the case of the Amazon, the more powerful didn't grab territory; they broke the natural rubber monopoly by taking seeds from the Amazon to plantations in East Asia and Africa to create an alternative source. It was a case of better science and technology.
4. In a larger sense, Amazonia's misfortune was a failure of sovereignty and defense of national interests. The famous act of botanical "piracy" by Henry Wickham, an Englishman who smuggled rubber seeds out of Brazil in 1884, is not the real issue; many cultivars, such as coffee and jute, were transplanted to the Amazon from other regions. In a world of active commercial, scientific, and financial exchanges, the substance of economic sovereignty—and not just the appearance—rests on the ability of nations to manage their resources in ways that serve their interests. In Amazonia, government officials, rubber traders, planters, and scientists were unable to efficiently organize the necessary production of rubber to meet world market demand. There was not even an attempt by the countries of the region to promote a cooperative program in defense of their common interest in natural rubber. This failure of agricultural research and market development dealt Amazonia a setback from which it took decades to recover.

* * *

The legacy of the rubber boom left Amazonia with a taste for development expressed politically in militant regional demands for help from the more developed economic centers of their countries. Roberto Santos has marshalled figures showing that during the boom years, Amazonian rubber exports provided 25 percent of Brazil's foreign earnings, while only a small fraction of this was spent on imports to the region. "There is little if any doubt, therefore, that industrialization of the south of Brazil was due to the exchange savings generated by rubber," wrote Santos.[28] The argument extends to tax transfers, since Brazil's export levies on rubber far exceeded federal spending in the region during the boom.

The Amazonian claim was pressed when a new constitution was drafted in 1945, at the end of the Getúlio Vargas dictatorship, and an article in the text earmarked 3 percent of national revenues for investment in the Amazonian region. This was never carried out, partly because of opposition in Brazil's southern and northeastern regions to putting Amazonia ahead of other claimants for public funds with more promising economic prospects.

The decision by President Juscelino Kubitschek to move Brazil's capital inland to the central plateau brought great dividends for Amazonia. The opening of the Brasília–Belém highway in 1960, followed by the Brasília Cuiabá–Porto Velho road, stimulated colonization and land speculation.

The military leaders who seized power in Brazil in 1964 decided that "filling the void in Amazonia" was a "national security" goal. General Oswaldo Cordeiro de Farias, as minister of interior, designed an "Operation Amazonia" that set up an Amazonian Development Superintendency (SUDAM) to coordinate public projects and provide tax incentives for private investments in the region.

The Superior War College, the Brazilian military "think tank," produced a document in 1968 titled "A National Security Policy for Amazonia" that called for "a national policy of planned and methodical occupation of the region to impose, on the full extent of the territory, the characteristics of our civilization and integrate it forever into our national structure."

From this to the decision in 1970 to begin the eastwest Transamazon highway and the Cuibá–Santarém highway into the Tapajós valley was just a matter of mobilizing the resources. When the first regional radar mapping survey, Radam-Brazil, revealed the mining and energy potential of Brazilian Amazonia, billions of dollars in public investments flowed into the Carajás mining district, the Tucurui hydroelectric dam, and the aluminum plants at Belém and São Luis.

Similar efforts to develop the Amazon interior were made by the governments of other Amazonian countries. In Peru, a highway linking the Pacific coast to Pucallpa, the gateway to the eastern rivers, was opened in 1943, when the jungle town on the upper Ucayali had 2,500 inhabitants; 30 years later Pucallpa's population had grown to 57,000. Iquitos, the main Amazonian port in Peru, and Puerto Maldonado, in the lowlands east of

Cuzco, underwent similar expansions as oil companies carried out major exploration programs in the selva, as the Peruvians call Amazonian jungle.

Ecuador experienced a similar cycle, with oil discoveries producing strong migration into the lowlands by ranchers and small settlers. Bolivia's entry into its Amazonian region began with the construction of the Cochabamba–Santa Cruz highway, which was followed by colonization and timber clearings that have deforested much of northern Santa Cruz and are now advancing into the Beni.

Colombia also undertook colonization of the upper Caqueta area in the 1960s, and when oil was found on the Putumayo, a pipeline to the Pacific was built. Later, as oil production declined, the situation turned violent. The new rural populations created in the lowlands by colonists brought from the highlands became supporters of guerrilla movements out of frustration over their economic privations and social abandonment. Then the region became a hotbed of coca production for the international cocaine traffic, which has installed numerous laboratories and airstrips in the Amazon lowlands of Bolivia, Peru, and Colombia. The guerrillas and cocaine traffickers became a major security concern for Colombia and the other Amazonian countries.

In Venezuela, the COPEI (Christian Democratic) party came to power in 1969 with President Rafael Caldera, who declared a "crusade to the south" that had as its objective strengthening Venezuela's presence on its deserted Amazon borders with Brazil and Colombia. With more ecological foresight than its neighbors, Venezuela did not encourage colonization and instead invested in scientific work on the upper Rio Negro through an agreement with the University of Georgia. But a highway was built to the Brazilian border at Roraima which connects with Venezuela's industrialized Orinoco region and the Caribbean coast.

* * *

Against this background, the Treaty for Amazonian Cooperation emerged as an instrument for diplomatic consultation reflecting the reawakened interest of the Amazonian countries in the development of their region. The treaty provides a framework for economic, scientific, and cultural cooperation, and it contains the embryo of a regional free-trade area.

Yet this treaty, as crafted by Brazil's foreign ministry, was not an economic pact, and much less a mechanism to coördinate regional action on local issues, such as the environment and Indian rights. Its original purpose, as one Brazilian diplomat has put it, was to unfurl a "juridical and political umbrella" against undesired external interferences in the region.[29] This was in response to the chorus of criticisms from international environmental groups that had been formed over the damages being caused by internationally financed development projects in the region.

When the Treaty of Amazonian Cooperation was signed in Brasília in 1978, President Ernesto Geisel of Brazil called on the signatory countries to "give life to the empty heart of the continent" and asserted that the development and environmental conservation of Amazonia was the "exclusive responsibility" of the treaty members.[30] Ramiro Saraiva Guerreiro, Brazil's foreign minister, said in 1980 that the treaty created a "common front" against external attempts to impose "abusive and illegitimate constraints" on the rights of the Amazonian countries to use their tropical forest resources for the economic development of their countries.

Since the original purpose of the Amazon pact was mainly symbolic, it was given as little to do as possible, and almost no operational capacity. It had neither a permanent secretariate nor a budget. The presidency rotated every two years among the member countries, and periodic meetings of foreign ministers produced little substance. What work was done was left to a low-level diplomatic committee that "coordinated" minor programs in border trade and local health services.

Brazil's initial concept of the pact was a new version of the nineteenth-century attempt to ban foreign-flag ships on Amazon rivers. In much the same way as that policy was abandoned because commercial partners and the Amazon neighbors demanded an "open door," the Amazon pact was forced to change direction by new realities.

The rejection of outside involvement in Amazonia raised a dilemma for Brazil and the other Amazon countries. The poor economic and financial situations of all the member countries required them to obtain substantial external financing to achieve the goal of developing Amazonia, as well as other sectors of their economies. How, then, could they reject the "interferences" of the environmentalists without placing at risk the access to international financing, which was subject to ever-stricter "green conditionality"?

The decision of the World Bank in 1986 and later the Interamerican Development Bank to suspend disbursement on loans to Brazil until environmental standards were raised for Amazonian projects sharpened this issue, which had been raised by a worldwide campaign against forest burning in Amazonia. Facing a virtual boycott on foreign lending, President José Sarney was forced to reconsider Brazil's negative approach.

A presidential meeting of Amazon countries in Manaus in 1989 was the occasion for a substantial revision of the pact's purposes and operating mechanisms. At the initiative of President Virgilio Barco of Colombia, the Amazonian presidents responded positively to international concerns and approved the creation of permanent committees, dealing with natural resources, the environment, and Indian affairs. Even without a full-time pact secretariate, these standing committees, each chaired by an individual country, have shown that they can take the initiative in developing projects (such as creation of conservation units) that can be presented for international financing. Some of these units are binational because the ecosystems cross political boundaries, as in the Serra Divisor National Park in Acre and the bordering area of Peru.

The shift toward a more affirmative role provides the Amazon pact with the opportunity to demand international financial cooperation. The Barco doctrine, set forth at the Manaus meeting, argues that "the ecological debt that the industrialized countries have with the Third World and Humanity should be paid . . . by providing the indispensable resources to finance sustainable development models in Amazonia."[31] This line was taken up by President Collor of Brazil, who called a special meeting of the Amazonian chiefs of state in Manaus to take a common stand on environment and development of Amazonia before the UNCED "earth summit" meeting in Rio.

The potential for concerted action in the Amazon Pact region was reinforced by Collor's commitment to "development with protection of the Amazon environment." The Brazilian Amazon rainforest project with the support of the Group of 7 will certainly be extended through the Amazon pact to the other countries of the region, particularly in scientific research and technical-assistance programs. Increased cooperation, going beyond small border projects, will offer the pact members possibilities for more ambitious programs, including economic integration.

PART II

The setting and history of Amazonia have been sketched in the previous part. What follows in this part is an overview of human actions that are producing contemporary changes in the region. The subject is people in Amazonia—how they live and why they behave the way they do. Their ways are closely tied to the technologies they employ to make productive use of the resources of the Amazon environment, so these are described in some detail. It will be seen that the natural resources of Amazonia are of enormous importance, not only to the nations of the region, but on a scale that influences world markets. It will also be seen that the environmental protection of Amazonia depends directly on the economic viability and social stability of the communities that are emerging as the centers of political and cultural change in the region. An approach that fails to integrate environmental concerns with the political, economic, and social interests of the Amazonians will not succeed. To "save" the Amazon, the people have to be saved as well as the trees.

FOUR

An "Old" Frontier

If something is unsustainable,
it tends to stop.
Herbert Stein

One of the "classic" concepts of New World frontiers is that the pioneer who clears a place in the wilderness is an individual who has cut loose from an old, established economic and social order in quest of a new life. The decision to make the move to the frontier is generally associated with strong characters endowed with courage, determination, and a spirit of adventure.

So it is with many on the frontiers of Amazonia. The new settlers are people who went there of their own free choice. They were not shipped as slaves or as prisoners sent to penal colonies, like some of the first settlers of the Americas and Australia. They did not migrate to escape pogroms or wars or famines, like Russian Jews, Armenians, or Irish peasants. The majority of the new people in Amazonia moved there in search of opportunities, a chance to own some land, the freedom to work for themselves—without a boss—or in the hope of striking it rich by luck or cunning.

Some of the Amazonian frontier people can be described as a rural proletariate, deprived of access to land and employment opportunities in their places of origin or "expelled" by modern mechanization of agriculture or rural political violence. Rapid population increases in Brazil's arid Northeast region and in the Andean highlands have pushed some settlers toward Amazonia, although not on the scale of migration into large cities.

It is not huddled masses or the refuse of distant shores, however, that make up the majority of the frontier people I have met during my travels in Amazonia. Many are persons of unusual enterprise, often with considerable experience in farming, animal husbandry, mechanical skills, and commercial practice. Life is hard on the frontier, sometimes to the point of despair; but these people are determined not to give up and turn back. They are forging the future of Amazonia.

Yet, these settlers have created an ecological conflict in which they are often the unwitting victims of their own actions. The "liberation" from social and legal constraints on the new frontier stimulates in some individuals a predatory drive for property and exploitation of resources that can be "antisocial" and environmentally destructive. Even more damaging are errors that are closely linked to experiences brought by the settlers from their places of origin. The methods they have employed to exploit natural resources, particularly in conversion of forest land for agriculture and cattle ranching, were based on practices developed in other regions that are often out of step with Amazonian economic and ecological realities. In wildcat mining it has been much the same.

For this reason, much of what appears to be new on the frontiers of Amazonia is in fact old and inadequate. The behavior of the settlers lacks the creativity required for a successful shift into what evolutionary biologists call a new adaptive zone. For instance, the most serious threat to Amazonia's environmental stability comes from settlers who have seen the forests only as an obstacle to be removed as fast as possible to make way for crops, pastures, and plantations. Although it may make sense elsewhere, this land use can be a serious mistake in Amazonia. The removal of the exuberant forest often exposes fragile soils where the nutrients are soon exhausted. Crops decline and the land reverts to shrubs and secondary forest.

This error arose in Brazil from the transfer to Amazonia of earlier experiences with forest clearing by settlers from São Paulo, Minas Gerais, Santa Catarina, and Paraná, states that once had as much as 90 percent of their land under forest and are now down to as little as 5 percent. In those states, cleared forest provided new land for coffee and sugar plantations, cattle and dairy ranches, food crops, and more recently orange groves and mechanized soybean, corn, and wheat farms. Good soils and rainfall, access to large markets, and modern technologies adapted to local conditions made agriculture successful in these regions.

Similarly, the Andean peasants who make up the majority of migrants to Amazonia in Bolivia, Peru, and Ecuador practiced an agriculture based on highland cultivars, such as potatoes and frost-resistant grains, that supported large Andean populations. But these cultivars don't grow well in the humid lowlands; other tropical cultivars had to be developed, such as rice and bananas. But these perishable products were no match in the marketplace for the leaf of a perennial bush that came into great demand—the coca plant.

Coca leaf for illegal cocaine production is now the biggest cash crop in sub-Andean Amazonia. Production is on small plots covering an estimated 200,000 hectares from the Chapare region of Bolivia to the Huallaga valley in Peru and on to plantations in Colombia's eastern lowlands. The coca plant adapts well to Amazon conditions, where up to four harvests of leaves can be taken off each year. It is not in itself damaging to the environment. But steep slopes cleared of forest for coca plantations are prone to erosion, and rivers are polluted by chemical wastes from jungle laboratories producing cocaine base. The biggest problem, however, is social, because where coca takes over, lawlessness soon prevails. Normal economic activities are corrupted and attempts to enforce laws, including environmental protection of forests, becomes highly dangerous work.

The Amazon settlers, large and small, can't be blamed for these results when the technicians who designed official land colonization programs were themselves unable to provide necessary guidance and control. In Amazonia, there were virtually no soil studies and appropriate technologies for productive land use were untested. Markets were far away and transport costs made basic farm products uncompetitive. Where official programs existed to support perennial tree crops, such as coffee, cocao, and rubber, the necessary credit mechanisms failed for lack of funding and extension services were insufficient. Inadequate rural roads, lack of storage facilities, and woefully weak health services undermined the best efforts of producers.[1]

Forest management was never considered. There was no tradition among the migrants of living in and from the natural flora and fauna of the forest. They were not familiar with silvicultural practices, such as rubber gathering, which do not alter the forest. Instead, the tree stands provided some quick cash from logging if timber was extracted. This kind of lumbering was an on-off thing, like mining, with no thought to forest maintenance and renewal.

Government regulations on land titling encouraged further waste of the forests. As the migrants poured in over government-built roads, the demand for land overwhelmed the officially directed colonization projects. Serious land disputes broke out between holders of often flimsy titles and hordes of squatters. In this struggle, cleared land was the sign of possession, and large land holders burned down thousands of hectares of forest to strengthen their legal claims. Brazilian statutes regulating titling of public land in Amazonia virtually required deforestation as evidence of "land improvement" by an occupant. There was no equivalent recognition before 1989 of extractivism or forest management without cutting the trees as the basis for granting permanent legal right to forested land.

The fundamental weakness in the recent settlement of Amazonia has been, therefore, the absence of a comprehensive concept of land use that defined suitable sites for agriculture, ranching, forest management, traditional extractivism, indigenous lands, and preservation areas for the protection of ecosystems. Such an approach calls for zoning, which takes a considerable investment of time and money by national and state governments in research, fieldwork, consultation with local communities, and legal actions to enforce zoning regulations. This important subject is expanded upon in Chapters 8 and 9.

Zoning alone is not enough, however. A cultural change involving many adaptive innovations will be necessary before a genuine "new frontier" emerges. This will have to take into account the existing indigenous communities and the practices of traditional extractors of natural forest products, who have accumulated a lot of ecologically sound experience in the forest. The rustic lifestyles of the forest people are inadequate, however, for the needs of the much larger population of new settlers, who have brought to the frontier a more advanced economic behavior as consumers and as cultivators and transformers of natural resources. Moreover, the forest gatherers themselves are no longer satisfied with their deprived living conditions; they are seeking new technologies that can add value to their forest pursuits.

Therefore, the settlers and the forest people are potential allies. They both want, above all, secure possession of the lands where they live and control over rural development programs in their areas. This is a struggle for political power in which settlers and forest people need to work together to influence public policies. They are the local political constituencies that can organize and educate communities for the sustainable forms of development that are appropriate for Amazonia.

The new directions must grow out of the trial-and-error experience of the people who are living the Amazonian realities. This learning process has begun. New alternatives are being tested in land use and forest management. As I look through my travel notes, many faces come to mind of people who are forerunners of this society. Their experiences illustrate problems and suggest possibilities. So I will give some examples, drawn from personal contacts, that are representative of the human experiences that are encountered on the Amazonian frontiers.

* * *

Evangelista Antonio Cantão, the 11th of 18 children of a landless Brazilian peasant, had no hope of ever owning land in the crowded state of Espírito Santo where he grew up. So he headed for Rondônia, where he had heard that the government was giving out land. Without waiting for an official assignment of land, Cantão staked out about 100 hectares of virgin forest 40 kilometers down a jungle track from the sawmill town of Rolim de Moura. When he had cleared 15 hectares, he obtained a provisional title from INCRA.

The land was cleared by fire in the slash-and-burn style of agriculture that Cantão had learned in Espírito Santo, where the lush Atlantic tropical forest that once covered the land is nearly all gone. Cantão's homestead in Rondonia produced rice, mandioc, beans, and corn sown between the burned-out tree stumps for the first two years of bare subsistence. When yields declined, Cantão sold some timber to a lumberman who cleared a pathway in the forest to drag out the logs to be trucked to sawmills in Rolim de Moura. This cash income helped Cantão clear more land where he planted 5 hectares of coffee—a cash crop—in addition to his food crops. He also put in pasture for ten head of cattle.

In 1985, four years after Cantão arrived, two nephews came from Espírito Santo to help him farm while they looked for land for themselves. Cantão alone could not have done the annual forest clearings and coffee planting. His children were still too young and neighbors along his road, called line 51 from a surveyor's map, did not have the habit of sharing work. During all this experience, Cantão never received any credit, technical assistance, or medical attention.

When I met Cantão in a clearing by the rutted dirt road, he and his nephews were out with their dogs hunting for wild pig, an important supply of protein. They were also on the lookout for an *onca*, a large spotted jaguar, whose roar had been heard by the children walking through the forest to and from their one-room schoolhouse, three miles away. The stocky peasant had a day's growth of beard on his haggard face and was recovering from a bout with malaria, but he managed a broken-toothed smile when asked about the change in his life: "I didn't want to be a peon anymore, always working for somebody else. I am never going to get rich here, but this is mine and it will belong to my children."

There were at least 120,000 settlers like Cantão in Rondônia by 1986, according to INCRA. Fewer than half of these had obtained land through official colonization programs. Many were squatters without titles. Among both the official settlers and squatters, many abandoned their land after clearing a piece of the forest, selling out to new settlers or to ranchers. This impermanence is sometimes described as a settlement failure, but that is an inadequate description of a process that involves capitalization. The sellers pocket some cash for the free land they have cleared. Some move into the new towns springing up and become petty merchants, buy a taxi, open a stall in a market, or get a public job. Others use their capital to buy cattle or a piece of farm equipment, like a chain saw, and keep going farther into the forest to occupy new land.

The World Bank discovered, after lending Brazil $280 million to pave a highway through western Mato Grosso and Rondônia and $192 million for agricultural settlements in these states, that only 20 percent of the colonists in official projects were still occupying their lots eight years later. A World Bank technical review concluded that the original owners had chosen "to realize a capital gain on [their] 100 hectares and then move on to a new frontier in the hope of repeating the process. Buyers have tended to be local entrepreneurs or neighbors who have often accumulated several lots to form a small ranch."[2]

When I returned to line 51 three years later, Cantão was no longer on his lot. There was a tenant on the property who said Cantão had sold the land to a wealthy storekeeper in Rolim de Moura. The new owner had increased the cattle herd. Local beef consumption was rising in Rolim de

Moura, which grew, in less than a decade, from a crossroad hamlet with a dozen sawmills to a robust commercial town served by interstate buslines and a telecommunications satellite station. Cantão was said to have bought a heavy truck with which to haul logs to the sawmills from a new settlement front, where he staked a new claim.

This process has inherently damaging effects. The itinerant slash-and-burn settlers are a lethal force unleashed against the forest because they obtain a capital gain from cutting down the trees. Instead of taking root in one place, the settlers engaged in this land-clearing "industry" keep moving into virgin areas. As these forest sappers advance, cattlemen follow, buying up and concentrating frontier land in larger holdings. This has proved to be an obstacle to settling new migrants on suitable lands. With the better soils occupied, the newcomers spread into poorer quality lands which never should have been cleared. Soon they were looking for new land, which led to invasions of Indian reserves, or a search for nonfarm income from illegal logging and wildcat mining.

Another factor contributing to instability of tenure was the land speculator, who went to the frontier as a capitalist gambler, not as a producer. The financial stakes in this game were high. Land values rose fast as new settlers poured into Rondônia. During the decade from 1970 to 1980, Rondônia's population rose from 114,000 to 491,000, and in 1984, when the paved highway was completed, 25,000 migrant families entered the state in one year. But, as the World Bank reported in 1987, they were no longer able to find good land because "much of the better land . . . remains underused in the hands of speculators."[3]

Miguel Feitosa, a young São Paulo financier, made his move in Rondônia after a banking apprenticeship in the United States. When I met him in 1985 in Ji-Paraná, the biggest town in central Rondônia, he was conferring with the administrator of two ranches he had bought, totaling 22,000 hectares. One property was involved in a land dispute with small settlers. Conflicts over land were then frequent, and large proprietors often hired local gunmen to expel squatters from lands on which titles were far from clear.

Feitosa said he had bought the land for $20 a hectare before the paved highway was completed and hoped to sell it for $250, with a planned profit of $5 million. "That's what inflation does for land values," he said cheerfully. When inflation rates reach triple digits, as it did in Brazil in the 1980s, land investment is attractive for speculators who buy property they never intend to develop. Feitosa probably did not make as much as he had imagined because deforested land in Rondônia was selling for about $150 a hectare in 1990, but he has done well. He now runs a big private investment firm associated with a Pittsburgh bank in São Paulo.

Cantão and Feitosa came from two widely different social worlds, but they had at least two things in common: a strong acquisitive urge and the skimpiest knowledge of the Amazon environment. They both proceeded in time-honored ways of occupying land in Brazil, where frontier advances have been a constant for centuries as coffee, cacao, cattle, cotton, and food crops pushed agricultural populations into new areas. Recent expansions, such as the opening of southern Bahia for cacao and of Paraná for coffee, took place within living memory, and they were accompanied by enormous speculative gains in land values. The migrants who followed the new routes into Brazil's Amazon region were largely in this tradition.

Mining, as much as land, draws the ambitious and adventurous to Amazonia's frontiers. Simon Camelo, as an itinerant trader from Belém, started coming up the Madiera river to the Rondônia territory in the 1950s to buy Brazil nuts and rubber. Now he is a leading gold dealer in Porto Velho, supplying diving equipment and credit to thousands of miners working the rivers. He flies his own airplane and builds dredges with which the miners, known as *garimpeiros*, scour the riverbeds for gold and tin. Camelo is a nabob, with rings on his fingers and gold chains around his neck. He holds court in an air-conditioned office behind his supply store, where raw gold is the common currency. "The free miners, not the mining companies, are the future of Amazonia," he said.

But on the Madeira, the bonanza days are over. The 1,200 dredges that swarmed like sampans on the broad brown river when the riverbed produced up to 20 tons of gold a year had been reduced to fewer than 500 by 1990, when new finds turned scarce and the international price fell. Newspaper in Porto Velho advertised large dredges for two kilos of gold, worth about $25,000, or a new pickup truck. Many of the Madeira miners had left for other sites. In their wake is a river severely polluted by tons of mercury dumped by the dredges.

In other Amazon regions, access to mining locations is only feasible by air. Bush pilots, who are paid in ounces of gold, make fortunes flying *garimpeiros*, food, equipment, and gold in and out of dirt airstrips chopped from the jungle. There were at least 700 aircraft, mainly single-engine planes, and 1,000 pilots serving the miners in Brazilian Amazonia in 1990. About 75 planes are lost annually.

José Altino is a pioneer in this activity. He came to Amazonia from Minas Gerais, where he studied business administration, to work on an aerial mapping survey. He now operates a gold camp in the Tapajós river area and owns an airtaxi service in Manaus, where he has a palatial residence. He buys aircraft, paid for in kilos of gold, in deals made in hotel suites in Rio or São Paulo, and has an apartment in Brasília where he lobbies congressmen on behalf of "free mining" rights.

Altino and another swashbuckler, Elton Rohnelt, a ruddy-faced southern Brazilian who was a mercenary soldier in Africa before taking up logging and mining ventures in Amazonia, promoted the invasion by *garimpeiros* of the Yanomani Indian lands in Roraima. Gold and tin were extracted in great quantities by over 50,000 miners until 1990, when President Collor of Brazil ousted the miners from Yanomani lands in response to an international uproar over deaths of Indians from contagious diseases. An indigenous reserve covering 9 million hectares was officially decreed as the homeland for the 10,000 nomadic Yanomanis who live between Brazil and Venezuela.

The full extent of this wildcat activity has projections that go beyond Amazonia, affecting metal markets around the world, as will be seen in Chapter 7. In local terms, mining is a mixed blessing. There are at least 300,000 persons employed in placer mining in Amazonia. They pump cash into local commercial circuits, which supply the miners with food, fuel, equipment, and transport. Mining provides employment opportunities for migrants who have not found land on which to settle.

For the environment, however, wildcat miners are devastating. With dredges, high-pressure hydrojet pumps, and chainsaws, they destroy riverbeds and vegetation along waterways. They dump tons of mercury used in gold recovery into river systems, with danger of poisoning the food chain for fish consumed by urban populations. The lure of getting rich quick with a lucky find at a mine has also disrupted labor markets, drawing off farm workers from more stable agricultural activities.

The plight of Edson Moreira Alves illustrates this widespread problem. Edson is on his second frontier, but this time as a proprietor, not as a peon. He was born in rural Bahia and emigrated as a young man to Paraná in southern Brazil, where labor was in demand at coffee plantations then being opened in the forests there. For 12 years he worked as a tenant on other people's land tending coffee bushes he would never own. In 1980, INCRA was promoting land settlement in federally owned lands in Rondônia. Candidates for free land were recruited among landless rural workers in Paraná.

Edson was selected on a point system recognizing his farm experience and as a head of family with two children. When he arrived by bus at the Marechal Dutra colonization project near Ariquemes and was assigned a 100-hectare lot of dense, virgin forest, it was the first time he had laid eyes on Amazon conditions. Eight years later, Edson had 13 hectares planted in robusta coffee, shaded by interspaced mango, rubber, and brazil nut trees, and 18 hectares more had been cleared for rice and pasture. Sitting on a huge rotting log, a remnant of the old forest, Edson pushed back the ragged brim of a straw hat shading his swarthy face and gave a general appraisal of his situation. "My family and I went hungry in the first years and we have all had malaria attacks. But I prefer malaria to the curse of always working for the benefit of some big planter and not for myself. I had enough of living in someone else's land. At least I have something here to leave my children," he said.

When Edson described the current fruits of his efforts, however, the results were not encouraging. He reckoned that in 1988 his gross income had been about $10,000 from the sale of 500 sacks of dried coffee beans, some rice, and a couple of steers. He said his coffee harvest barely covered his annual farm costs because the price paid for coffee by merchants in Ariquemes, his only market, was 30 percent below the price in Paraná, while the cost of fertilizer and insecticides in Rondônia was twice as high. As a result, he said he could not make the fertilizer applications he knew were needed to raise his coffee output.

But his major problem was labor. To get helpers, Edson said he had given two families the right to clear and plant some land in his forest reserve, which is supposed to be maintained under INCRA regulations at 50 percent of his lot. Most of his forest reserve is gone in exchange for labor needed in his coffee grove for weeding, pruning, pest control, and harvesting. If it were not for these tenants, who have cleared 10 hectares of forest for their subsistence plots, Edson said he would face an insoluble problem. "Nobody around here wants to do farm work. Whatever labor is available is attracted to the mining camps not far from here. They give people free food and a share of any tin or gold that they find. It's like a lottery; only a few win, but that's where the young people go," he said.

Few settlers have had more diversified experience on the Rondônia frontier than Hugo Frei, a pioneer cacao producer in Ariquemes, 150 kilometers east of Porto Velho. Frei came from Santa Catarina, a coastal state where German immigrants, like Frei's grandparents, opened agricultural colonies in the forest and prospered. Frei had a construction company before heading for Rondônia in 1972. He owns 4,000 hectares now on two farms, including 400 hectares planted in cacao and 800 hectares in new rubber trees. Frei is a producer, not a speculator, and seated by the tiled swimming pool in his spacious townhouse, he looks prosperous. But Frei's diagnosis of Rondônia's agricultural future was grim.

"The situation is catastrophic," he said, describing his cacao operation. From a high of 4,000 bags of dry cacao beans in 1983, production plunged to 1,700 bags in 1986 due to irregular rainfall and a severe attack of "witches' broom" disease. "I got production back up to 3,000 bags, and made some money. But the international price has fallen sixty per cent and I am getting out of cacao. I have started planting citrus and am going to put 300 hectares into a new variety of high-yielding mandioc, which is consumed here as flour. Something has to change because it makes no sense to destroy the forest to plant things that don't produce," said Frei.

He said he was tearing up his rubber plantation because the seedlings distributed by the government's rubber agency of a supposedly disease-resistant hybrid variety had failed. "The rubber trees didn't grow. Forty thousands hectares are being pulled up in Amazonia because after five years the trees are like broomsticks," he said. Yet in other areas of Brazil, such as São Paulo and Mato Grosso do Sul, rubber plantations planted at the same time are doing well and do not suffer from the fungal leaf blight that has demolished monoculture rubber plantations in Amazonia since Henry Ford's failed experiment at Fordlandia on the Tapajós river in Pará.

The economic failures that lead to frequent turnovers in land ownership in Amazonia may be a necessary stage in the establishment of a more experienced and more stable rural population. An outstanding example is Walmir de Jesus, who came to Rondônia, like Cantão, from the Atlantic coastal state of Espírito Santo. He grew up there on a small cacao farm in the Rio Doce valley. His family had the means to send Walmir, the oldest son, to a technical high school in the state capital, Vitória, where he got a job in a chocolate factory.

Walmir became a union leader and lost his job when he helped organize a strike in 1980. At 22 years of age, he faced an unpromising future. "My name was on a blacklist and I couldn't get a job, so I decided to change my life. I went to Rondônia, saw that the land was good for cacao, and made the move," he said.

With a loan from his father, Walmir bought 100 hectares from a settler in a colonization district 30 kilometers off the main highway near Ouro Preto, a big town in the center of Rondônia. The settler had planted 3.6 hectares of coffee with a bank loan, and he was about to lose his land because he couldn't keep up his loan payments.

Since buying the property, Walmir has been joined by five brothers who now own lots in the same area. In 1989, he sold 250 sacks of green coffee beans, which provided cash income for himself, his wife, and two children while he waited for 10 hectares of cacao he planted to come into production. Walmir has installed a sawmill, where he produces boards, building materials, and furniture; keeps bees; has milking cows, pigs, and chickens; and seems self-sufficient in the well-carpentered house he built with the help of his brothers. They also brought the know-how and labor that went into the cacao plantation, where the trees, fragrant with white flowers and ripening yellow fruit, are carefuly pruned to eliminate the "witches' broom" fungus disease that can destroy a poorly managed cacao farm.

Yet, when production began, the international market for cacao was glutted. Buyers in Rondônia were offering one-third the price being paid in 1990 by export firms to producers in Bahia, the main growing state. This was a severe setback for Walmir and all cacao growers in Amazonia. "We have no power in the marketplace out here," said Walmir.

Walmir is a convinced nature conservationist who has begun to see that the forest may be his main asset. He said he watched the Rio Doce valley, where he grew up, be stripped of its abundant forests by lumbermen and industrial charcoal makers, leading to problems of soil erosion and flooding, and he intends to keep most of his forest reserve intact. "It will be worth more in fifty years than anything I could do with the land now," he said. Walmir has been careful to maintain stands of native forest between lots cleared for cacao to reduce the danger of propagation of disease and to retain clear streams for year-round watering.

With his previous experience as a union organizer, Walmir is active in the Ouro Preto rural workers' union. This local has 800 members among the region's small farmers and is affiliated with the national central workers union (CUT), the labor arm of the Catholic-left Workers party (PT), which has made peasant orgnization a major activity. Walmir believes the best way the rural unions can defend small farmers is by organizing production and marketing, and he has begun a community bee-keeping cooperative backed by small projects grants from Canadian and Dutch aid funds.

Walmir's grassroots initiative underscores the need for cooperatives and other forms of small-settlers' associations on the Amazon frontier. There is a dearth of public agencies that can develop and transfer agricultural technology, organize credit mechanisms, and provide marketing services. Without research and extension, many settlers make unproductive use of land. Without credit, most settlers can't put in perennial crops, such as cacao or rubber trees, which take several years to produce any income, much less practice agroforestry, which takes even longer. Without a marketing system, the isolated settlers are at the mercy of buyers in unstable markets. The settler's only capital is cleared land and cattle, which often is the only source of income. Land sold off in pieces to other settlers or to ranchers is usually shorn of the remaining forest.

This unstable rural situation contributes not only to environmental degradation, but to urban poverty, which is as much a part of the frontier as forest depletion. Half the people in Amazonia live in a few large cities, such as Belém, Manaus, or Iquitos, and in hundreds of new towns that have erupted in colonization areas. In Porto Velho, a city of 300,000 people, there are large slums of rural migrants who live in squalid shanties built on flooded land. But there is enough wealth for a booming demand for housing. Antonio da Silva drove trucks into the Rondônia territory when mud holes in the jungle track swallowed up whole vehicles in the rainy season. That was before the paving of BR-364, the highway that connects Rondônia with the industrial south, was completed in 1984. Now da Silva owns an auto rental agency, a garage, and a construction firm. He built a development where the first 15 houses were bought and occupied before there was water or electricity. "A housing development here is a slum before it opens, but people have to live somewhere," said da Silva.

The housing market includes people like Odette Ferreira, an enterprising woman who came to Rondônia with a bit of money salvaged from the sale of a boutique in her hometown in Minas Gerais. She opened a shanty restaurant selling beer and ox-tail stew to miners and prostitutes in a brawling camp on the Madeira river. The day I met her the camp was occupied by heavily armed state police. A prostitute had been knifed to death on one of the floating brothels on the river. "This is no place to raise kids, but the money rolls in," commented Odette. She said she was going to move to Porto Velho, open a variety shop, and put her ten-year-old daughter in school. Her problem was finding a place to live in the city that she could afford.

The urban food markets have to be supplied, which creates demand for local beef, fish, fresh vegetables, and fruit. Driving east from Porto Velho, one passes through a dismal swamp, where Rondônia's first hydroelectric dam has penned up the Jamari river in a reservoir full of dead trees; then, like an oasis, there is a cluster of flourishing truck farms at kilometer 70. One of these is owned by Fernando Ponce, a Chilean agronomist who came to Rondônia after working near Manaus on the installation of a rubber plantation that failed. On three hectares of rich, red soil, Ponce is getting rich growing lettuce, bell peppers, onions, cucumber, parsley, cabbage, and papaya with an intensive system pioneered in Brazil by descendents of Japanese immigrants.

"All it takes is the right technology, a lot of work and good salesmanship," said Ponce. His garden rows are enriched with rice husks, cacao compost, and some nitrogen fertilizer. Weeds and pests are controlled with chemicals. With a pump-driven sprinkler system for irrigation, a power tiller, three workers, and a truck to carry produce to Porto Velho, Ponce nets more than most ranchers make with herds of 300 head of cattle.

Even when Amazonian agriculturalists overcome bugs, fungi, soil fertility, and storage problems, they still face severe economic handicaps for almost anything they produce for sale outside the region. Whether measured from Belém or Porto Velho, the distances to major consumption centers such as São Paulo or Rio de Janeiro are over 3,000 kilometers by truck. Only products that are in high demand can pay such freight. Except for black pepper and natural rubber, Amazonia is a marginal producer in all agricultural commodities. Coffee bushes produce vigorously in Amazonia, and can be made to yield well for up to ten years, and cacao is of high quality; but when international prices fall sharply, producers in Rondônia, Mato Grosso, or Pará can't compete in price with coffee or cacao in south-central Brazil. They are simply pushed out of the market by freight costs.

In contrast, the Amazonian frontier is completely dependent for almost everything it consumes on distant suppliers. BR-364 brings trailer trucks with fuels, cement, chainsaws, outboard motors, tractors, rice hullers, aluminum skiffs, refrigerators, stoves, plastics, bluejeans, boots, and every other manufacture that is demanded by farmers and miners. Truckers who come to get Amazonian lumber, beef, fish, and minerals often bring food staples, such as beans and mandioc flour, as well as refrigerated eggs, chicken, and even tomatoes because local production and storage facilities are insufficient to cover local demand. Information is also imported. The government invested heavily in telecommunications, so towns in the remotest corners of Brazilian Amazonia see the same television programs that are on the air in Rio de Janeiro or Brasília, and they have direct-dial telephone service to Brazil and the rest of the world.

Because of all these linkages, the frontier does not stand alone in splendid isolation. Amazonia is very dependent on the larger national societies it is a part of and is unusually sensitive to reduced flows of public funds. When national governments are undergoing economic, political, and administrative crisis—as has been the case in Brazil and most of the Andean countries—Amazonia is strongly affected. The expansion of the Amazon frontier got moving in a period of relative prosperity in the national economies and then slowed down as growth declined and inflation and fiscal anemia undermined new public investments.

Amazonia's raw materials are battered by inflation and economic recession in the national markets. A sharp fall in housing construction in Brazil after an economic boomlet in 1986 closed down hundreds of sawmills in Amazonia. A credit squeeze in 1990 that ended subsidized government financing for agriculture in Amazonia took 500,000 hectares of recently cleared land in Mato Grosso out of soybeans and into cattle. Lack of federal and state funds cripple maintenance on major highways, where hundreds of heavy trucks frequently are blocked when torrential Amazon rains wash away bridges and asphalt. Electric shortages constantly black out towns and industries because public power companies, short of capital, cut back Amazon investments.

In short, as the decade of the 1990s began, the occupation of Amazonia had lost much of its early elan and momentum. There was perplexity over the conflicts created at home and abroad by disorderly development, and greater concern over the ecological consequence. Economic problems made official planners turn a critical eye on assumptions that had directed billions of dollars in public monies to Amazonia. Cost–benefit analysis demonstrated poor economic returns on investments in colonization and livestock projects, which led to suspension of tax and credit incentives for clearing the dense tropical forests. External resources that financed big infrastructure projects, such as highways and dams, dried up, and local governments could barely cover payrolls from meager fiscal resources. Even the price of coca leaf collapsed under the combined pressures of excess production and antinarcotics campaigns financed by the United States.

It was a time for rethinking Amazonian development policies. Rules of economic rationality apply, as anywhere else, and the region is paying a heavy price for mismanagement of the two natural resource areas in which it has a clear comparative advantage—forestry and mining. The waste of wood, natural forest products, and unique genetic resources through extensive land clearing is incalculable. The loss of potentially recoverable gold and tin, in the way these minerals are mined in Amazonia, is in the billions of dollars. Agriculture and livestock are appropriate activities in suitable parts of Amazonia, but this cannot be the main direction of development. Experience has shown that tropical commodities, such as coffee and cacao, carry high risks in international markets, and bulk grains are uncompetitive without a reduction of freight costs through heavy investments in agro-industrial processing in the region and a railroad system that replaces trucks.

In consequence, the forests and the mineral resources must be the foundations for a new strategy of economic development of Amazonia based on productive activities that are profitable, socially stabilizing, and compatible with the life-sustaining ecosystems that are Amazonia's most valuable natural resources.

* * *

Sustainable development of Amazonia's renewable, nonmineral resources comes down to choices between different forms of forest management and use of land for agriculture and livestock ranching. The forestry sector, which is the subject of the next chapter, has still to be developed. Agriculture and ranching, on the other hand, have been the most active areas in the initial cycle of frontier occupation, with great waste of both forest and land resources. For a sustainable future, Amazon farmers and ranchers must change their extensive, "cheap land" mentality and center their efforts on obtaining higher yields, on less land, reducing the pressures for further extensive clearing of forest. This is a new cycle in which the choice of appropriate lands and products can be more selective, drawing on the experience of what has worked and what has failed. Ecological safeguards and sound agro-economics are not at odds.

This can be illustrated by the southern Amazon uplands that extend in an arc from Bolivia through Rondônia and northern Mato Grosso to the Araguaia–Tocantins river valley of northern Goiás, Tocantins, eastern Pará, and southwestern Maranhão. This has been the most penetrated area of Amazonia, with highways on both flanks—the Brasília–Belém on the east and the Cuiabá–Portovelho (BR-364) on the west—and the half-finished Cuiabá–Santarém (BR-163) through the center.

These uplands, lying between 14 and 8 degrees of latitude south of the equator (see map in front matter), are a transition zone of open forests, woody brush, palm stands, and savannas known in Brazil as *cerrados.* They present a mozaic of ecosystems that are different in vegetation and in potential land uses from the closed humid evergreen forests that occupy the interfluvial lowlands closer to the equator and the sub-Andean piedmont. Variations in the soils and drainage affect the vegetation, but the main cause of difference is the amount and spacing of annual rainfall. Where rainfall is year-round and exceeds 2,000 millimeters, the vegetation that predominates is unbroken canopied forest—the Amazon *hylea,* in the Hellenist terminology of von Humboldt.

The *hylea* is picture-postcard Amazonia, with towering trees and galleries of lesser species intertwined with vines and ferns in a dense closed rainforest that occupies 2 million square kilometers, or about 36 percent of the Amazon watershed. The *hylea* is much larger now than it was at the end of the last ice age, 15,000 years ago, when the Amazon basin was drier and colder. It is also the Amazon ecosystem least altered by the modern occupation, which has taken place mainly in the transition zone. The *hylea* is richer in species than the *cerrado*, so there are enormous areas of pristine native forest that contain Amazonia's greatest biodiversity reserves.

The relationship between *cerrado* and *hylea* is fundamental for sustainable occupation of Amazonia. Each area has a different potential, but they are interdependent. The *cerrado* transition zone has been occupied and cleared by ranchers and cultivators whose production of crops and pastures depends on abundant, timely rainfall. The *hylea* not only contains unique genetic resources and the world's greatest tropical timber reserve, but it is the ecosystem that guarantees sufficient rainfall for the uplands. Scientists believe that serious alteration of the *hylea* could reduce rainfall, crippling agriculture in the transition zone.

Until recently, *cerrado* soils were considered useless for agriculture and the poorest kind of cattle range; but after a long research effort, it was discovered that soybeans, corn, cotton, and rice varieties adapted to the humid tropics could provide high yields on *cerrado* soils treated with limestone to reduce acidity and fertilized with phosphates. This converted the huge undeveloped upland arc, with its good rainfall, level topography, tillable soils, and year-round sunlight, into an agricultural frontier that expanded until it penetrated the adjoining Amazon rainforest.

As mentioned earlier, Olacyr de Moreas, a São Paulo builder who has invested his wealth in agro-induistry, pioneered the technology for growing crops in the transition area at the *fazenda* Itamarati Norte at Diamantino, 300 kilometers west of Cuiabá. The research conducted at this ranch turned 60,000 hectares of what had been considered useless land into a national center of soybean production. The experience at Diamantino shows that cattle pasture lands give way to higher-value crops when the tree stumps are cleared and the land cultivation is mechanized. This depends on lowering the costs of transportation—by highway or railroad—to make bulk transport of grains and necessary inputs, such as lime and fertilizer, economic. In this respect, the eastern Amazon transition area is much the same as the American Middle West in the 19th century.

Clearing of *cerrado* and transition forest land for mechanized farming spread rapidly in Brazil and eastern Bolivia. The new proprietors, who came from southern Brazil with machinery, capital, and agricultural knowhow, cleared about 15 million hectares from southern Pará to Rondônia. Much of this land went into pasture, but at least 3 million hectares was planted in crops. In 1988, Mato Grosso alone produced 3 million tons of soybeans, providing 15 percent of Brazil's soybean exports, which are second only to the United States. *Cerrado* pastures, improved by nitrogen-fixing soybeans and durable tropical grasses, such as *brachiaria*, carry millions of beef cattle. There is also significant production of rice and corn, planted in rotation with soybeans.

For the first 600 kilometers of BR-163 north from Cuiabá, vistas of planted fields stretch to the horizon, like in Nebraska. The gently rolling plateau land has been cleared of its forest cover by tractors that plough along the contours of the land. Beans and rice are planted between wind rows of burned logs and slash that reduce erosion. Aluminum warehouses for soybeans, rice, and corn and farm-equipment dealers are spaced about every 75 kilometers around new towns called Nova Mutum, Lucas de Rio Verde, Sorriso, Sinop, Claudia, and Carlinda.

Beyond the present line of agricultural occupation, further advances from northern Mato Grosso into central Pará pose a threat to a major bloc of intact hylea between the Xingu and Tapajós rivers. This forest sanctuary covering 500 million hectares is still largely intact. It contains major Indian reserves, including an area of 9 million hectares set aside by the Collor government as reserves for various clans of Kayapo Indians on both banks of the middle Xingu. This area is already facing inroads from the east, along the BR-80 road that cuts through the Xingú National Indian Park, and from the northeast by settlers along an active stretch of the Transamazon highway between Altamira, on the Xingu, and Itaituba, on the Tapajós.

Some commentators dismiss the Transamazônica as a failure of "official colonization" because fewer than 2,000 families were settled, and many abandoned their lots. The original dirt "highway" has crumbled back into a jungle rut in little-used sections. It is not the great East-West integration highway it was once supposed to be; but it is far from dead. Spontaneous settlement has placed a population along the Altamira-Itaituba section of 200,000 people, who are downing the dense forests and mining the rivers. Satellite images show ranches being opened far off the highway. Altamira, a lively commercial center, is supplied by a modern trucking service that brings trailers and containers by barges from Belem to a roll-off port on the Xingú. Cattle, cacao, lumber, rice, and other products go out to Belém the same way. If the 7,000 megawatt Bela Vista hydroelectric dam planned for the lower Xingú is built some day, Altamira is going to be the powerhouse for western Pará and Santarém. (See Chapter 10.)

As originally planned, the Cuiabá–Santarém highway BR-163 was supposed to continue through the wilds of the Serra do Cachimbo to the upper reaches of the Tapajós river at Itaituba and from there to Santarém, a large port on the Amazon. The pavement ended at kilometer 640, however, when money ran out, and the next 900 kilometers are a wretched dirt track, impassable in the rainy season. Mato Grosso ranchers, soybean farmers, and lumbermen keep demanding an export route to Santarém, which is half the distance to São Paulo. The governors of Pará and Mato Grosso issue joint declarations whenever they meet, promising to complete BR-163. They can't do it because they don't have the money; but the project is the dream of land speculators who have already bought up most of the properties fronting on the unfinished road.

Completion of BR-163 will depend ultimately on Brazil's national plans for Amazonia. Further extension of major highways, like BR-163 to Santarém or BR-364 from Rondônia through Acre to the Peruvian border, puts sensitive tropical forest ecosystems at risk. While there is still time, such highways should be preceded by agro-ecological zoning that regulates the use of Amazonian forest land according to technical criteria. There is enough experience now to make sensible decisions about zoning that can protect large areas and channel settler energies, public investments, and private enterprise toward land uses that are compatible with ecological stability.

The main problem for zoning is the private ownership of large undeveloped properties and private colonization projects in Mato Grosso, Pará, Rondônia, and Acre. Some of these properties are in *hylea*, others are in the transition zones, and others are in both. Precise lines will have to be drawn on small-scale maps that define the appropriate land use based on soil, topography, vegetation, rainfall, and other ecological criteria.

Amazon zoning will need legislation that imposes restrictions on private use of forested property. The Brazilian constitution of 1988, in its strong chapter on environment, declared the Amazonian forests a "national patrimony" and established the requirement of environmental impact reviews prior to any forest alteration.[4] These are good principles, but they will not be a working policy until Amazonia is zoned. This requires not only technical definitions, but high-level political decisions affecting powerful interests. President Collor ordered the completion of "macrozoning" studies in time for the UNCED ECO-92 meeting.

The private ownership of millions of hectares of untracked Amazon forest began in Brazil in the 1950s when the state governments of Mato Grosso and Pará, followed later by Maranhão and Amazonas, put up for sale huge tracts of state-owned land. The local political chieftains saw massive land distribution as the way to raise revenues and attract settlers to their backward regions. This was not a Homestead Act granting public lands to family farmers. It was an opportunity offered to big landowners from southern Brazil, who had capital, to buy land at a few dollars per square kilometer.

The result was an enormous concentration of public property in the hands of a few private owners for token payments—an early example of wasteful use of Amazon resources. A survey in 1966 of land ownership in a 24 million hectare area of Amazonia identified 3,668 properties of between 1,000 and 10,000 hectares representing 41 percent of the surveyed area and 223 properties of between 10,000 and 100,000 hectares representing another 21.7 percent—in all 15.7 million hectares for less than 4,000 proprietors.[5] (There were another 16 megaproprietors with 3.5 million hectares, but these were primarily old-time Amazonians who had claims to natural rubber forests, where rights to 1 million hectares were common.)

Experienced land speculators from the south knew that road construction would make their Amazon lands soar in value. The buyers included politically influential figures, like Adhemar de Barros, several times governor of São Paulo, who was one of many politicians who got land early in Mato Grosso and Pará. Their ambitions began to come true when President Juscelino Kubitschek completed the pioneering Brasília–Belém highway in 1960 and the first road from Mato Grosso to Rondônia, then a federal territory. This was long before military planners dreamed up "geostrategic" highway projects, like the Transamazônica, that later brought a new surge of land acquisitions.

Even with pioneer roads into the region, everything remained to be done if the Amazon lands were to be settled. The lift-off for Amazonian development was provided by the military governments that took power in Brazil in 1964 under President Humberto Castelo Branco and his successors. Amazonia's big landowners lobbyed successfully for a private investment scheme based on corporate income tax deductions that subsidized their Amazonian projects in the name of regional development. Instead of paying taxes, corporations put monies into cattle ranches, cacao and rubber plantations, sawmills, breweries and bottling plants, industries, trucking lines, hotels, and riverboats approved by the federal Amazon development agency (SUDAM).

These incentives were later extended to private land colonization companies that were created to induce tens of thousands of settlers from the south to buy uncleared land in the wilds of Mato Grosso and Pará and establish rural communities. The promoters of these ventures saw themselves not as land speculators, but as founding fathers of new frontier towns. They were mainly sons of Italian immigrants who had become wealthy developing land in south-central Brazil. Having prospered as colonists, some of these entrepreneurs, like reborn Roman publicans, turned their money and abilities toward organizing new colonization ventures. They often are living legends, although not always examples of shining success.

The stellar figure of private colonization in Amazonia was Ariosto da Riva, the founder of Alta Floresta. In a decade, starting in 1976, da Riva transformed a wilderness in northern Mato Grosso into a thriving agricultural and mining community with a population of over 150,000. The municipal seat is a well-planned city with 35,000 people. There are 13,000 rural properties, from big ranches to small holdings. With six satellite towns and several gold-mining centers, Alta Floresta is the third largest tax collector in the state.

Da Riva was a charismatic figure, over six feet tall, still going strong with full head of white hair and a straight back at close to 80 years of age. He spent his life on the frontiers and he had a simple formula for opening settlements: "Put a farmer with know-how and some capital on good soil, with a clear title and some technical assistance, and provide him with roads, schools and a church, and he will create wealth."

He grew up in Marília, São Paulo, when immigrants went there to work in the new coffee plantations. His father, an Italian musician, led the town band but made no money. As a young man, da Riva went to Minas Gerais to dig for diamonds and semiprecious stones. He became a gem dealer and acquired capital. Then he set about his real vocation as a land developer.

Da Riva developed the town of Navaraí in southern Mato Grosso, now a prosperous agricultural center, in association with Jeremiah Lunardelli, an Italian immigrant who developed so much land during the coffee boom in northern Paraná that he came to be known as "the king of coffee." Hermínio Ometto, whose family made a fortune in sugar in Sao Paulo. hired da Riva to organize a colonization project within his 600,000-hectare Suiá Missu ranch on the border of Pará and Mato Grosso, but they had a falling out and da Riva decided to do a development on his own.

Following the westward course of the Telles Pires river, da Riva found what he considered "very good land for small farmers" on the west bank of the river. It was a completely unsettled forest, but a physician in Rio de Janeiro already held title to 400,000 hectares and da Riva bought this claim. In 1974, the state of Mato Grosso put up 2 million hectares of public lands at auction, and da Riva bought another 400,000 hectares, adjoining his earlier purchase, at $4 a hectare. This block of 800,000 hectares became the basis for the Alta Floresta colonization project, with an outlay for land of less than $3 million.

Da Riva's development company, INDECO, capitalized at $9 million, opened an access road to Alta Floresta without any government subsidy. Antonio Severo Gomes, a bush pilot and surveyor from Goiás, who had scouted the route for both the Brasília–Belém and the Cuiabá–Porto Velho highways, led an experienced team of road builders that bulldozed 80 kilometers of raw roadway from kilometer 642 on the Cuiabá–Santarém highway to the Telles Pires, where they installed a barge hooked to a steel cable that hauled D-8 tractors, and later trucks and buses, across the river. The roadway reached the town site of Alta Floresta in late 1975, and the next year colonists began to arrive.

In line with da Riva's philosophy, the settlers were recruited in southern Brazil among farmers who had some capital and wanted more land. A settler who sold 20 hectares at Parana land prices could buy 100 hectares in Alta Floresta, for which INDECO charged about $80 per hectare in cash. Larger ranch units of up to 5,000 hectares were also available. All the properties were connected to a main road by feeder roads. INDECO built schools and community centers, maintained the roads, and provided coffee plants from a demonstration farm and other technical services to the settlers. The federal cacao agency, CEPLAC, set up an office in Alta Floresta, distributing selected plants and technical advice. A big São Paulo farm cooperative, COTIA, put in a purchasing office and a warehouse.

After more than 9,000 farm lots were sold and Alta Floresta was established as a thriving commercial town, da Riva's dream was a reality—but it had cracks. The inflationary economic crisis that overtook Brazil after 1986 slowed the flow of well-capitalized settlers, who preferred to invest in overnight bank certificates at high interest than run risks in frontier land ventures. INDECO reduced the size of its lots from 100 to 30 hectares to attract poorer settlers; but after clearing land, the settlers found that the basic perennial crops being promoted by INDECO, coffee and cacao, were not recovering production costs because of low yields and a fall in international prices. COTIA's coffee purchases fell from 50,000 sacks in 1987 to 15,000 in 1989. "Our presence here depends on the success of agriculture, and the outlook is not bright," said Geraldo Teruo Gabara, COTIA's manager.

A survey of settlers found that a great majority, after bad experiences in agriculture, thought raising cattle was the only way to make money since labor costs were minimal and, as one said, "the cows grow while you sleep."[6] Many lots have been converted to pasture or sold to ranchers. The growing concentration of land in larger properties is not what da Riva intended, but once INDECO sells a lot the owner does as he pleases.

Neither was it foreseen by da Riva that Alta Floresta's main economic activity would be gold mining, with thousands of placer miners destroying the banks of the Apiacás, Paranaita, Peixoto de Azevedo, and other rivers that feed the Telles Pires to get at gold in rich alluvial deposits. Alta Floresta's central mall is dominated by shops of gold buyers and supply stores displaying pumps, powerpacks, plastic hoses, chainsaws, shovels, and "everything for the miner." It is gold, not agriculture, that is making money for storekeepers, and many settlers, or their sons, are working in the rivers and not in the fields. Wealthy goldminers and brokers are investing in the lots that settlers are selling in large numbers. One of these miners, nicknamed Ouro Zezinho, or Little Joe Gold, owns a gaudy gold buying shop, one of the biggest supermarkets in town, and a 600 hectares cattle ranch.

It came as an unwelcome surprise to da Riva when Alta Floresta was identified on international television in 1988 as a flagrant example of torching the Amazon rainforest. An INPE report claiming that 80,000 square kilometers of Amazonian primary forest had been burned in 1987 had caused a national and international uproar. *The New York Times* declared in an editorial that Brazil's "heedless development" of Amazonia would "wreak havoc with the global environment."[7]

Alta Floresta was targeted as a critical area, and INPE researchers, accompanied by colleagues from NASA, made a field trip to check on the ground the evidence shown by satellite images of extensive forest clearing. They reported that 160 square kilometers had been burned by settlers along a new colonization road built by INDECO, suggesting that Alta Floresta was one of the worst examples of Amazon forest destruction.

Da Riva became acutely aware that Alta Floresta had an "image" problem. At a meeting in São Paulo of the Amazonian Entrepreneurs Association (AEA), a lobby representing the big landowners from the south who

The great migrations into Amazonia from all parts of Brazil are producing a remarkable ethnic and cultural mix. Settlers from southern Brazil of Italian, German, and Polish stock are neighbors with migrants from central and northeastern regions, where black and indian strains are strong. These children are from Nova Brazilaudia, Rondônia. (1988) *(Ricardo Azoury, TYBA Agencia Fotographica)*

New cultivars introduced to Amazonia include black pepper, brought by Japanese immigrants. Jorge Itó, a pepper farmer in eastern Pará, rests on his tractor at the Tomé-acu cooperative. (1990) *(Ricardo Azoury, TYBA Agencia Fotographica)*

The Tucurui reservoir was created by the first major Amazon hydroelectric project in 1985. Tucurui's "alligator men" salvage trees for charcoal and lumber from underwater by using a waterproof chainsaw.

LEFT: Timber felling, largely as a by-product of land clearing, supplies over 3,000 sawmills operating in the Brazilian Amazon. Sawn boards are trucked to southern Brazil or exported. This mill is near Paragominas in eastern Pará. (1990) *(Ricardo Azoury, TYBA Agencia Fotographica)*
BELOW: The conversion of Amazon forests into charcoal is stimulated by pig iron smelters in eastern Pará. The crude process reduces wood, often waste from sawmills, in mud ovens. This charcoal went from Rio Maria to a smelter at Marabá. (1991) *(Rogerio Reis, TYBA Agencia Fotographica)*

LEFT: A rubbertapper at the Cachoeira *Seringal,* where Francisco "Chico" Mendes began his struggle to create extractive reserves. Cachoeira is now part of the "Chico" Mendes reserve covering 980,000 hectares. (1990) *(Ricardo Azoury, TYBA Agencia Fotographica)*

BELOW: Indian militancy in defense of ancestral lands takes many forms of activism. Here, Kube-i, a chief of the Kayapo in eastern Pará, films his tribal people during a demonstration at Altamira against a proposed dam on the Xingú river. (1989) *(Ricardo Azoury, TYBA Agencia Fotographica)*

ABOVE: The world's largest iron mine at Carajás exports 35 million tons a year over a 900-kilometer railroad through Amazon forest. This mining center is surrounded by devastated forests occupied by cattle ranchers and small settlers. (1987) *(Luca Martins, TYBA Agencia Fotographica)* BELOW: The famous gold pit of Serra Pelada, where 80,000 wildcat miners once worked. More than 100 tons of gold were extracted by hand before flooding and slides closed the mine. (1985) *(Claus C. Meyer, TYBA Agencia Fotographica)*

ABOVE: A cattle drive blocks the highway in Acre, near Xapuri, where tensions are strong over land occupation by ranchers. Amazonia has about 15 million head on its pastures. (1990) *(Ricardo Azoury, TYBA Agencia Fotographica)*
BELOW: Satellite images of intense deforestation in the Carajás corridor of eastern Pará where highways and railroads brought rapid occupation. Settlers have cleared in ten years more than 50 percent of the primary forest in a 40,000-km² area between the Araguaia river and the Carajás mines. (Segments delineated in white are deforested.) Only the Carajás mine protection area and adjoining indian reserves remain largely intact. *(NASA Landsat; INPECURD)*

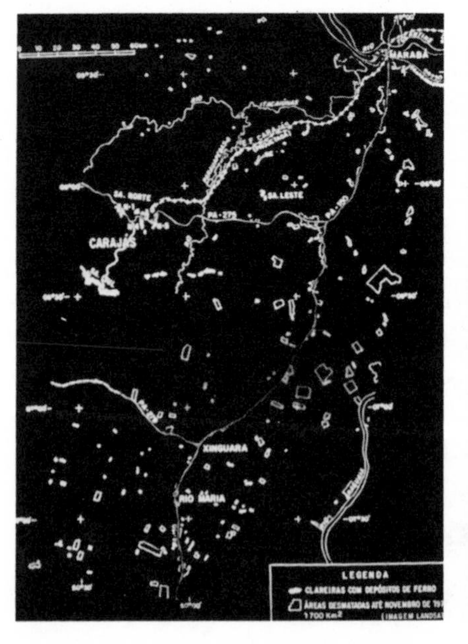

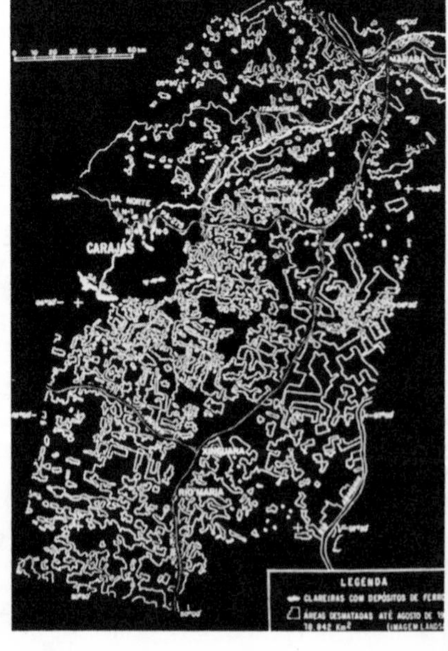

have ranches and land companies in Amazonia, da Riva said the criticisms of ecologists and the public outcry over massive burning could no longer be ignored. He said he was worried by a public opinion survey saying that big Amazonian investors, many of whom received tax incentives for their projects, were perceived as "social parasites who destroy the ecological equilibrium of the region with little or no benefit for society."[8]

The story of Alta Floresta's economic and environmental problems raises many of the major questions about sustainable development of Amazonia. Has INDECO's promotion of forest clearing for coffee and cacao plantations been an appropriate land use? Or are the problems mainly commercial because of depressed international markets and domestic inflation? The municipality of Alta Floresta, which is the size of Belgium, straddles the borderline between transition *cerrado* and closed *hylea* forest. Should Alta Floresta be zoned by law in a way that reserves the remaining dense forest, either to be managed for forest products or as an ecological reserve on the banks of the Telles Pires? Should INDECO provide settlers with a new technology based on enrichment of the forests on their lots with extractable species, like Brazil nuts or rubber? Or should the current preference for a combination of perennial crops and cattle be continued, with improved pasture management and breeding?

Clearly, the future of Alta Floresta, and of INDECO as an enterprise, depends on a combination of economic results on land that has already been cleared and on the uses made of the remaining forest. Of INDECO's original 800,000 hectares, da Riva said that only 150,000 had been cleared by settlers, so there is still a lot of forest left. Some moves are under way toward restoration and protection. Da Riva's demonstration farm began distribution to settlers of Brazil nut trees and fast-growing native hardwoods that can be planted to shade cacao. The value of the flora and fauna of the native forests has been recognized by da Riva's daughter, Victória, and her husband, Edson Carvalho, who operate a modern hotel in Alta Floresta that has housed international ornithologists and botanists studying the Telles Pires region. Carvalho said that an ecological park on the Cristalino river, where they have a lodge, may be a better way to use the forest than agriculture.

The example set by INDECO will be of major importance for the entire region. Alta Floresta has been the gateway for numerous other large ranches and colonization projects that go much farther west, across the Juruena river to the Aripuanã, a tributary of the Madeira. This is an area of 5 million square kilometers that contains transition forest and *hylea* in quantities that have to be measured in quadrants of longitude and latitude. Private properties occupy millions of hectares of undeveloped land. One of these is Colniza, which is a Lunardelli family project on 400,000 hectares of forested land that is 350 kilometers west of Alta Floresta. An airstrip and a town nucleus were built, a grid of dirt roads was opened through the towering forest, and family lots were put up for sale. The first settlers set up a pharmacy, a brickyard, and a sawmill, but only a trickle of land buyers followed. The economic crisis that slowed Alta Floresta brought Colniza to a halt, along with many other projects in northern Mato Grosso.

The only big sale at Colniza was to a wealthy lumberman from Santa Catarina, Orestes de Bertol Fattia, who took 17,000 hectares. "This land will be developed for the trees, not cleared for cattle," said Fattia, who had experience with a 2,000-hectare cattle ranch east of the Telles Pires. "That was a lesson. I have learned that the trees that we burned would be worth a lot more now than the cattle," said Fattia.

Fattia's option in favor of timber extraction is a lumberman's preference that has not been widely shared in the region—at least until recently. The option preferred by most settlers in Amazonia, after coffee and cacao fail, has been cattle. There is no package of forestry or sylvaculture technologies available to settlers that offers an economic alternative to clearing, burning, and planting. Lumber has not been worth enough to make trees seem valuable. Cows are seen as one of the few things that make money.

There were probably close to 15 million head of cattle in Amazonia in 1990. The last Brazilian farm census of 1985 reported 14,410,000 head in Amazonia, an increase of 34 percent in five years from 11,700,000 in 1980.[9] In addition to the growing Brazilian herd, there were another 2 to 3 million head in eastern Bolivia, Peru, Ecuador, and Colombian Amazonia. Cattle ranching is, therefore, one of the few activities in Amazonia that has shown steady growth over the past two decades.

Despite the large land holdings in Amazonia, the sociology of Amazon cattle ranching is more "democratic" than it is made to appear by those who see it only as a game for big ranchers or absentee land speculators. The Brazilian farm census of 1985 showed that the largest growth in cattle numbers in Amazonia between 1980 and 1985 was on small and medium farms of 20 to 500 hectares, which were carrying 65 percent of the cattle in Rondônia, 45 percent in Acre, and 34 percent in Pará, a state identified with large ranches. Even when cattle are on poor range, they survive and provide small settlers with a secure source of cash, some milk, and an on-the-hoof savings account that pays "interest" now and then in the form of a new calf. Small settler farms in Rondônia of 100 hectares, with 30 hectares of pasture carrying 25 cows, were sold near Ouro Preto in 1990 for the equivalent of $25,000. Similar lots with abandoned pasture, no animals, and 50 hectares of virgin forest were being offered near Pimenta Bueno for less than $10,000, with few takers.

The manifest preference of Amazon settlers for breeding and fattening beef cattle collides with one of the most militant tenets of Amazonian environmentalists, which is that cattle ranching is a malignant threat to tropical forests and to the people who live in them. Robert Goodland, a senior ecologist at the World Bank, rates cattle ranching as "the worst possible land use" in Amazonia.[10] The environmental case against conversion of forest to pasture is backed up by expert opinion that cattle ranches in Amazonia are not only ecologically destructive but "inherently uneconomic," and therefore an unsustainable land use.[11] Moreover, say the global ecologists, each new forest clearing means burning of biomass, which adds to the emission of climate-warming "greenhouse" gases.

Agronomic studies of large Brazilian ranching projects, particularly around Paragominas, Pará, which is in *hylea* forest, have shown that pastures failed rapidly because of low soil fertility and uncontrollable invasions by weeds and insects. This has been said to maintain an endless cycle of clearing of new forest for new, short-lived pastures, with destruction of wild species, genetic resources, and eventually an entire ecosystem. This folly, goes the argument, is economically irrational because Amazon ranchers lose money without government tax breaks and credit subsidies that were created to promote ranching.[12]

A Brazilian planning ministry study in 1986 of 90 Amazonian agroranching projects, out of a total of 556 that were financed by tax incentives and cheap government credit, found that only 3 showed a profit. Herds were less than half of what had been planned, and sales were 16 percent of project targets.[13] Some large corporate investors, including Brazilian subsidiaries of international companies like Swift and Volkswagen, tried Amazon ranching but sold out after burning down huge forest areas for pasture. The multinationals withdrew partly because the ranches were not producing expected profits and partly because of "image," or political reasons. Big Amazon ranches came under attack by environmentalists for burning forests and from human rights groups for land conflicts with Indians, squatters, and forest gatherers. None of this was good for the corporate image.[14]

Fiscal incentives for cattle projects are yet another example of waste of resources in Amazonia, in this case tax revenues that could have been used for better purposes. For instance, Hermínio Ometto, one of the early land barons, received $3 in SUDAM incentives for each $1 he invested of his own money in the Suiá Missu ranch—and he still went bankrupt. The incentive system was environmentally irresponsible since the choice of ranch locations was left up to the project promoters, without any requirement for a prior evaluation of the environmental impacts. Moreover, the system was corrupt; the top management of the Bank of Amazonia, which administered the incentive funds, was removed under charges of embezzlement, and Henry Khayat, the superintendant of SUDAM under former President José Sarney, was forced to resign after federal investigators uncovered kickbacks for approval of projects. In 1988, under pressure of international outcry and domestic scandal, Sarney suspended fiscal incentives for new ranch projects in dense forest areas. President Collor, on taking office in 1990, banned any new incentives and ordered SUDAM to recover federal monies in projects where there had been fraud. With the end of the incentives, SUDAM no longer hands out glossy promotional brochures to prospective investors predicting that ranching will "turn the Amazon valley into one of the world's main cattle-breeding centers."[15]

But the recognized failures of the fiscal incentive system in cattle projects does not permit a final judgment on the sustainability of breeding and fattening cattle in Amazonia. For each project that received fiscal favors, there are thousands of private cattle operations that received no incentives and have made headway. The 1986 study on cattle ranch incentives said that the causes of failure were "lack of knowledge of the appropriate breed, the most suitable forage, levels of pasture fertilization, management of herds and pastures, and scarcity of local labor."[16] This suggests that lack of know-how was the limiting factor, which is not the case with many ranchers who have learned from earlier failures and are continuously adopting new ways of producing under Amazon conditions.

The observations that follow are not presented as a justification for further conversion to pasture of dense tropical forest in the Amazonian *hylea*. There is no need for further expansion into dense forest areas because there is ample room in open savannahs, as in Bolivia's Beni department, and in the *cerrado* and open forest transition zones of Brazil for cattle ranching. The evidence I have seen on many ranches suggests that cattle breeding and fattening in certain definable areas of the Amazon basin is an appropriate and sustainable activity that can be a stabilizing influence on land use when pastures and herds are managed efficiently.

The more zoning limits access to new land, the more efficient will have to be the use of existing pasture, which is the key to reducing pressure on the dense forest ecosystems. There are recent studies that suggest that degraded pastures can be restored in areas that were considered beyond recovery.[17] Examples of how an appropriate technology and management can change the cattle-carrying capacity are discussed in Chapter 9, in relation to eastern Acre where the political aspects of this problem will be examined. There are also some other cases that need to be studied more closely.

Fazenda Mogno, which means mahogany in Brazil, is a property covering 300,000 hectares on the Apiacás river, 80 kilometers west of Alta Floresta. It is owned by the Ferruzzi Group of Milan, which has an empire of chemical, food-processing, and trading companies in Europe and ranches in the United States, Canada, Australia, Argentina, and Brazil. The creator of this global landholding strategy was Serafino Ferruzi, the former managing director of the family company. During a visit to Brazil in 1975, Ferruzi told Ariosto da Riva he had $50 million to invest, and asked him where to find the best land in Amazonia. When Ferruzi saw Alta Floresta, he bought the biggest spread he could put together from several owners. Da Riva says he got it all for less than $10 million.

At that time Ferruzi owned a cement factory in São Paulo, which was later sold, and the firm now owns CICA, one of Brazil's largest food distributors. Mogno is a sideline for Ferruzi, but its management is professional and the property is an important example of careful, long-term agricultural investment in Amazonia. Only 20,000 hectares, or 6.6 percent of the total property, was cleared in the first 12 years of operations, and no land was cleared after 1987. There are 1,000 hectares in robusta coffee 1,000 hectares in cacao, and 17,000 in pasture. "Serafino Ferruzzi had an instinct for land and he chose this site because it was fertile soil. He paid more than I would have, but it has proven to be a good choice," said Giuseppe Parini, who was Mogno's manager until his retirement in 1989.

Parini's family came from the Po river valley of Italy to settle land in Paraná in 1948. He worked on land colonization projects in Paraná and southern Mato Grosso, where he owns a ranch, before moving to Mogno in 1979 when the virgin forest was first being opened.

Mogno's diversified production strategy, based on coffee, cacao, and cattle, is supposed to deal with the ups and downs of internal and foreign markets. Mogno's coffee sector produced 66,000 bags from 1.5 million bushes in 1988, which is a good yield of about 4,000 kilos of dried beans per hectare. Cacao did even better, with yields of 1,200 kilos of beans per hectare, which was 50 percent above the average in Bahia, where Brazil's cacao industry is based. But prices for coffee and cacao have been low, so the cattle sector made the difference between profit and loss. In 1988, Mogno made a profit, Parini said, but the annual report for 1989 showed no profit to distribute.

Renato Pavesi, a veterinarian from the University of Bologna, with 15 years experience in Brazil and 5 at Mogno, is in charge of the breeding and fattening for a herd based on nearly 12,000 nelore cows. The calving rates at Mogno are spectacular. In 1988, 11,069 calves were born from 11,813 cows—a rate of 93.7 percent. In 1986, the yield was 92 percent and in 1987 it was 93.4 percent.

Pavesi attributed the high calving rates to "an unusually favorable combination of climate, water, and pasture" at Mogno and close management control of the herd and pastures. There are pastures ten years old where the hump-backed cows gallop like gray ghosts through green waves of guinea grass, known as *coloniao* in Brazil. The specialized literature on Amazon ranch failures reports that weed invasions degrade *coloniao* pasture, which has to be abandoned after a few years. At Mogno, this is considered a pasture management problem. Pastures are fenced into lots of 50 hectares each, and the cattle are rotated according to a strict plan that avoids overgrazing, which exposes the grass to weed invasions. Groundwater that keeps pastures green and growing during the three-month dry seasson is protected by corridors of virgin forest a kilometer wide that run between grassy lots.

The payoff on this was the sale in 1988 of 11,573 three-year-old steers and infertile cows to packing plants at Cuiabá, with a gross return of over

$3 million. In 1989, more animals were kept on pasture and Mogno's sales fell to about 9,000. The plan is to hold the total herd at about 32,000 head, which is the limit without clearing more forest. "We didn't clear any new land after 1987 because of all the adverse publicity over forest burning," Pavesi said.

Mogno's corporate development plan originally called for expansion of the present ranch to 75,000 hectares of cleared land, with a forest reserve of 75,000 hectares, a colonization project for small settlers on 40,000 hectares, and sale of the remaining 100,000 hectares to buyers of large units. These plans were all on hold, awaiting policy decisions by the Brazilian government on future Amazonian development. The same is true of every other large holding in northwest Mato Grosso, where owners have been awaiting long-promised roads and hydroelectric power plants for years. Now they wait for a new definition on what regional development policy will be.

"When I came from Italy to open up the north of Paraná, the world thought that pioneers opening new lands to feed humanity were doing something good, even heroic. It is hard for me to accept, after being a founder of Maringá, of Rondonopolis and now of Alta Floresta, that this is all wrong," said Parini. "I don't think I am wrong, but if the humid Amazon should remain untouched in the interest of humanity, as some argue, then the world should pay for that which is in the global interest," he added, giving his gray moustache an angry twist.

Most Brazilian cattlemen would agree with Parini. They are investing in Amazonia now with their own money, without subsidies, and they have accumulated a lot of know-how about Amazon ranching. They are working to adapt to the Amazon conditions what they know best, which is tropical cattle. Brazil has the world's second largest cattle herd (after India), with 150 million head based primarily on Nelore and Gyr stock from India that Brazilian breeders adapted to the American tropics as beef cattle. These hardy, loose-skinned *zebu* cows, which are tick-resistant and aquatic, are designed by nature for the humid tropics. When they are placed on well-managed pastures they fatten to commercial weight in three years, and under good sanitary control they breed well. Increasingly, these hardy animals are being crossbred with Hereford and Sania Gertrudis stock, with strong weight gains.

Mogno, because of its size, is not a typical Amazon ranch, and small family operations don't have the land or capital to maintain large herds. But better pasture management practices, including introduction of legumes that improve feed, and genetic improvement of even a small beef-and-dairy herd are possible, and nearly everywhere that cattle ranching is practiced in Brazilian Amazonia, one finds old pastures being reseeded with improved varieties, cows being vaccinated and provided with vital mineral salts, and hybrid bloodlines from expensive bulls improving herds.

Many properties that were opened and rapidly degraded in the initial pioneer cycle have changed hands and are now being run by technical managers with better methods. Ranches, large and small, are full of cattle in eastern Pará from Marabá, in the *hylea* west of the Araguaia river to the Carajás hills, and from there south to Redencão and east to Acailândia, which is in the transition zone between Maranhão and Pará. At the other end of the southern arc, in Acre, where cattle are brought from Bolivia to be fattened on the year-round pastures, it is the same. Everywhere, along with the usual gripes about low beef prices, lack of roads, and the high cost of credit, one gets the same answer from the enterprising ranchers: "We wouldn't be investing our money if we didn't expect to make money."

The importance given to cattle ranching in these comments on the behavior of Amazon settlers implies no bias in favor of this activity in comparison with other, more traditional Amazon alternatives, such as natural rubber extraction, gathering of Brazil nuts and palm hearts, or river fishing. These are the domain of the people of the forest who also hunt wild game, harvest turtle eggs, shoot alligators for their hide, trap monkeys and parrots for international pet shop traders, and in many other ways live from the forest in a predatory way. Just as the harmful activities of forest people that destroy the species on which they prey must be combatted by wildlife police, the destructive clearing of dense forest for cattle pasture must be prohibited by zoning laws designed to protect the environment. But there is room for both extractivism and cattle ranching in a sustainable Amazonian economy. This is more a political than an ecological controversy.

Cattle ranching may be considered an exotic intrusion in the Amazonian environment, but it is a long-established form of rural social organization in the Amazon countries that brings an element of order to a chaotic frontier. Critics of Brazilian cattlemen have created a political aura of violence around these landowners that presents their national organization, called Union Democratica Ruralista (UDR), as a tropical Ku Klux Klan, terrorizing the hinterlands with gunslingers. There is violence in disputes over land on the frontier, as has been noted earlier, but the presentation of a whole class of rural proprietors as criminals grossly overstates the situation.

Economically, cattle ranching is a stable form of production when the land, water, and native forest are not abused. Cattle make an important contribution to the household income of many settlers and to the economic life of many communities. With respect to social development—and this may be the most important aspect—the involvement of Amazonian settlers in the cattle "culture" brings to the frontier an established, socially created institution that is symbolized by the annual cattle fairs that have become the major event of the year in many Amazon towns. Even rodeos and country music are social steps by which the frontier stops being a wild west, without law or community, and becomes civilized. That is a step toward forming communities that will be protective of their environment.

FIVE

A Social Forest

When the Portuguese discovered the Atlantic coast of South America, they came upon a redwood tree called *pau Brasil* that rendered such a fine red dye for tanners and weavers in Europe that they named their new colony Brazil in honor of the tree. They then cut down the forests with such gusto that *pau Brasil* is now an endangered species, rarely seen outside botanical gardens. The same depradation befell *pau rosa*, the rosewood tree, which was destroyed by extractivists for a delicately scented oil that was much in demand in the European perfume trade.

Tropical forests encountered elsewhere by the Portuguese, before they reached the New World, had fared no better. When they came upon the island they called Madeira (which means wood) in 1418, they set fire to the dense forests, and Madeira is said to have burned for seven years. Once the land was cleared, colonists brought sugarcane from Sicily for sugar plantations and grapevines from Crete that gave rise to the strong, sweet Madeira wine for which the island is famous in England. In exchange for the native forest, Madeira acquired not only a sustainable export economy, but a new ecology. The island now receives 300,000 tourists a year who are drawn to what is described as a "vast botanical garden where indigenous plants and flowers thrive alongside species introduced from distant continents."[1]

Colonial sugar and tobacco plantations in Brazil were the killing grounds for the coastal forests that once stretched from southern Bahia to the grasslands of Rio Grande do Sul. Trees that were abundant, like the dark *jacarandá* used for heavy hacienda furniture, and the blond *araucaria* umbrella pine of Parana, have been severely depleted. The lumbering plundered the better wood, an d the rest of the forest was demolished to make way for the coffee, cotton, and cattle that led the advance of the agricultural frontier. The forests of São Paulo, Minas Gerais, and Rio de Janeiro, which covered 80 percent of these states at the start of this century, were consumed and in their place are the mechanized farms, highways, power dams, steel mills, auto industries, and chemical plants that ring modern Brazil's major urban centers.

Does Amazonia, the world's largest tropical forest, stand a better chance of survival? The history of western man's encounters with tropical forests suggests that prospects are poor—unless the people of Amazonia begin to see greater value in the forests and their products than in land-clearing alternatives. Some traditional gatherers already do use forests as a renewable resource that can sustain various forms of livelihood. Native Indians know much about how to live off the forests, and extractors of natural forest products such as rubber have a stake in protecting the tree world. But under the new conditions that have exposed Amazonia to the pressures of increasing population and external market demands for resources, modern forestry, aiming at sustainable production of wood, offers the best hope of forest conservation by enhancing the value of growing trees rather than alternative uses of the land.

There are market forces working in the direction of forestry. Demand for Amazon wood has increased sharply, mainly in domestic markets. Brazilian Amazon producers earned $1.1 billion from the sale of logs, charcoal, and fuel wood in 1987, when at least 29 million cubic meters of industrial logs were produced, compared with 10 million cubic meters in 1970.[2] There was a corresponding expansion of wood processing in Amazonia. More than 3,000 sawmills installed in Amazonia produced 14 million cubic meters of wood products in 1986, compared with 5.3 million cubic meters in 1980.[3] The shift to Amazonia came as timber production from native forests in southern Brazil fell 70 percent in ten years. Brazil's internal market of nearly 150 million people has come to depend increasingly on Amazonia for sawn wood for construction and furniture and for plywood and fiberboard.

Foreign markets are as yet of secondary importance for Amazonian lumbermen. Only 10 percent of the industrial wood produced in Brazilian Amazonia in 1986 was exported, bringing in $245 million.[4] This makes Brazil a large producer and consumer of tropical wood, but a minor exporter compared with Malaysia and Indonesia. These two countries harvested 51 million cubic meters in logs in 1980 and exported 38 million cubic meters, or 75 percent of their production, worth over $3 billion.[5]

Until recently, therefore, Amazonia's forests were not undergoing any severe pressure from foreign demand. They provided less than 2 percent of annual sales in the $8 billion-a-year international tropical wood market. Neither wood nor cattle is a major export product for which Amazonia's environment is being sacrificed to pay foreign debts, as is frequently, but mistakenly, asserted. The major exporters of tropical woods are, by far, the southeast Asian countries, led by Malaysia, Indonesia, Thailand, and the Phillipines, which have 85 percent of the international market for tropical hardwood logs, sawn wood, panels, and veneers. West Africa is a distant second.

But as the more accessible forests of Asia and Africa are cut and replaced by rubber and palm oil plantations, the eyes of the major consumers of southeast Asian wood—Japan, Singapore, South Korea, and Taiwan—turn toward Amazonia's rainforests. An increase in Amazonian exports seems to be only a matter of how soon and how much. For Duncan Poore, a forestry consultant who has surveyed production prospects for the International Tropical Timber Organization (ITTO), the handwriting is on the wall.

> . . . the immediate market reaction to shortages or very steep price increases in timber from traditional sources is likely to be a movement away from the countries where supply has declined to those which have largely untapped forest resources; Grainger (1987), for example, has predicted a strong shift from South East Asia and Africa to South America. The temptation for these newly favoured producing countries will be to follow the downhill path of resource mining which has been persued by many of their predecessors, largely because they have a resource which they consider to be infinite.[6]

Amazonia's timber resources are not infinite, but they are certainly the largest that will be on hand in the humid tropics as the world enters the twenty-first century. Some estimates place the standing timber of Amazonia as high as 80 billion cubic meters with a commercial value of more than $1.5 trillion.[7] Even on the basis of a conservative estimate of 60 billion cubic meters, if only 10 percent of this timber were converted into commercial wood products at the very low price of $50 per cubic meters, with a 50 percent rate of conversion of logs to sawn wood, the gross value would be $150 billion, or more than the total foreign debt of the Amazonian countries. Measured in another way, the untapped standing forests of Amazonia represent 1,000 years of tropical timber exports at the 1980 annual rate of 61.2 million cubic meters.[8]

These are dazzling commercial figures, and many modelers of Amazonia's future have declared forestry to be the region's natural vocation. "Of all the natural resources of Amazonia, the forest resources are undoubtedly those which offer the best possibilities for immediate utilization," said a report in 1988 from Brazil's Forestry Development Institute (IBDF), which has since become part of the Brazilian Environment Institute (IBAMA). The report forecast that domestic and export demand for Brazilian industrial wood would exceed 120 million cubic meters in the year 2000, with most of the increase in supply coming from Amazonia.[9] At that level of production, Brazil would become an industrial giant in wood processing in competition not only with southeast Asia, but with northern hemisphere forestry exporters such as the United States, Canada, and Sweden.

The prospect of Amazonia's becoming a significant source of wood for the international timber trade is viewed with horror by the numerous rainforest alliances, defense councils, and action networks around the world for whom tropical trees are the primal stuff. Their fund-raising efforts are oriented toward donors who feel strong emotions when they get newsletters describing an ecological holocaust in Third World tropical forests where the native peoples, furry animals, millions of yet unknown species, the global climate, and Nature itself face extinction.[10] This is a strong message, but it is one-sided, no-compromise "Big Green" environmentalism that offers no political or economic answers to the problems of people in Amazonia who are trying to find the way toward sustainable development.

The South American humid tropics have unusually favorable natural conditions for rapid creation of woody biomass, and the timber resources of Amazonia are a primary economic asset for a strategy of sustainable development of the region. Until now, it is an asset that has been wasted or underutilized. With rational management, it can generate a higher return on capital than cattle or agriculture. Timber, in contrast with coffee or cacao, is a product in which Amazonia has a growing market. Extraction of natural forest products can be an important source of income for local gathering communities, but it can't compare with wood production in creating income, jobs, and local and state taxes that can finance badly needed public administration.

Belatedly, Brazil recognized the importance of stimulating forestry. Reversing policies that had encouraged land clearing for agriculture and ranching, the new agricultural law of 1991 defined, for the first time, sylvacultural management of native forests and reforestation as "productive utilization" of land.[11] This classification provides incentives, in the form of a lower land tax for proprietors who practice forestry and extraction of forest products such as rubber, and a legal safeguard against expropriation for agrarian reform. Previously, native forest was considered evidence that a property was unproductive and therefore subject to expropriation. As a result, many proprietors burned off forest and put in pasture just as a protection against possible expropriation.

Successful forest management can produce a forest culture that will change frontier behaviors that have been hostile to the environment. The forest culture makes people think about having trees and their products tomorrow; it is not short-term gain at whatever cost to the environment, but long-term husbandry of forest resources, that brings stability. This outlook has been brought to the frontiers of Amazonia by Nicholas Burch, an adventurous English forester who has started a forestation and forest management demonstration project in Rondônia. He has convinced a few ranchers to plant quality trees in their forests on a small scale as a test of forestry as an economic option. Burch is also working with an organization of small-settlers' cooperatives with the aim of making them timber producers. If the idea takes hold, it would bring a big change to land use habits.

Burch turned up in Brazil as a buyer of mahogany and Brazil nuts for export after managing forest estates in England and Scotland. He lives near

Pimenta Bueno, a town on BR-364, where he built a house made of many-hued tropical woods beside a river which he shares, from his veranda, with otters, wild ducks, turtles, jumping fish, and glorious sunsets. Observing the devastation of the forests around Pimenta Bueno, a ranching area, and the predatory logging of rich mahogany sites near the Guapore river, Burch decided to start a nursery and plant mahogany, cedar, and other native hardwoods in degraded plots, and to offer a model of "sustainable forestry" on a 6,000-hectare property he bought in an area of virgin forest. Burch says he has invested $150,000 of his own money, and he tells visiting tropical forest experts and members of foreign aid missions that he will need grants to keep the project going. He justifies this on environmental and cultural grounds.

> By developing more sophisticated management, harvesting and marketing techniques for tropical forests and their renewable resources I hope to create an affinity between Rondonian people, from whatever origins, with the native jungle which surrounds them and eliminate the common concept of the jungle as a barrier between man and his execution of conventional agricultural practices. . . . I have begun to interest other landowners in re-afforestation and general forest management as a practical alternative to jungle clearance. . . . I have developed a relationship with the Surui Indians which their chiefs have valued sufficiently to invite me to manage their forests and assist them in the extraction, industrialization and marketing of their forest products. . . . [This] will help them in their struggle to preserve their environment and develop their commercial and social relationship with the "white man" in a dignified and equalitarian manner.[12]

The Burch project is not quixotic; it is an experiment in Amazon forest management from which Burch says he hopes to "make a million and retire at fifty." But, Burch was bankrupt in 1991, in trouble with his commercial partners, and in poor health. The mahogany, cedar, Brazil nut, and rubber trees that he has planted in lines cleared through partially logged native forest are thriving on sunshine coming through the canopy gaps and control by machete of competing growth. But this is expensive, and it is far from clear that forest enrichment by planting seedlings is economically viable without long-term cheap forestry credit, which does not exist. Bankers know that many foresters question whether wood yields will provide an adequate economic return on the heavy investments required. These doubts have to be overcome by practical results because only the lure of long-term profit will bring ranchers or lumber companies to invest in management of large forests for wood, or attract small settlers to cultivate trees as a cash reserve, as they now use their cows.

Despite the uncertainties, long-term management of Amazonian forests for profit is an indispensable tool for sustainable development and environmental protection. Research and trial projects should receive large infusions of long-term concessional credits and tax incentives as a socially beneficial activity. Without forest management, many Amazonian ecosystems will remain exposed not only to land clearing, but to the widespread practice of selective logging of a few high-priced species, such as mahogany. This disorganized, often illegal "creaming" of the forest degrades the native biomass and produces invasion by loggers of Indian reserves and ecological preservation areas, like the Guapore reserve near Burch's property. Pathways cut by loggers into the forests often bring settlers who clear patches of land for subsistence crops. This process of "shifting" agriculture has destroyed as much forest as cattle ranches in areas of Bolivia, Ecuador, Rondônia, Pará, and Maranhão, where peasant populations have increased rapidly.

In addition to timber, native forests contain huge amounts of biomass fuel, building materials, fibers, fruits, nuts, natural rubber, gums, and organic compounds, as well as the habitats for wild game and fish that are important sources of food for local people. Many residents of Amazonia live from these natural resources, gathering forest products for consumption or sale, clearing small plots of land for crops and animals, and using wood sparingly for fuel and construction. These uses all have economic value, and the traditional rights of forest people to their ways of life must be respected. But they can't compare with timber production under forest management systems as the basis for a modern, competitive Amazon economy that can generate stable employment, higher income, and social development for millions of people. There is room in the Amazon forests for all these uses, traditional and modern, if there is enforced zoning on economic and ecological grounds that assigns appropriate spaces to each activity.

Before turning the region into a lumberyard, however, it must be recalled that the foundations of sustainable devlopment rest on the conservation of the ecosystems within which resources are used, as was said in Chapter 2. Amazonia's solar-driven water and energy cycle takes place in an environment in which the vegetation and the land act like a sponge. How many trees can be extracted, over how large an area and with what frequency, without altering the capacity of the "sponge" to retain the water that is the essential medium for photosynthesis and other chemical processes that drive the exuberant growth of the forest? How many spatial gaps can the forest take before the ecosystem climate changes on a large scale? Scientists such as Eneas Salati, director of the Brazilian Amazon research institute (INPA) in Manaus, and Thomas C. Lovejoy, international director of the Smithsonian Institution in Washington, have organized research to answer these questions, but no one is yet sure.

Until researchers acquire better scientific understanding of these and many other ecological questions, the alteration of the remaining primary forests in Amazonia must be approached with caution and on an experimental basis. In the case of timber, this calls for tight controls over the areas of extraction, and policies that favor forestry management practices that regenerate and renew the forest resources. There are several large experiments under way on "sustainable forestry" in Amazonia, and major projects for recovery of degraded lands through reforestation are looking for international finacing. Some of these are described later in this chapter; but before looking at useful new directions for the future, it is necessary to examine the central problem for Amazonian forestry, which is deforestation.

* * *

The deforestation of Amazonia is a direct consequence of the heavy investments that all the Amazonian countries made in highways that provided truck transport to what had been completely isolated regions. Brazil's federal highway system in Amazonia increased during the decade of 1965–75 from 1,786 kilometers to 9,000 kilometers.[13] In Ecuador, Colombia, Peru, and Bolivia, roads reached over the Andes and connected the populated highlands with the tropical eastern plains. The promoters of these land routes saw Amazonia as a "resource frontier" where oilfields, virgin lands, and mines invited human settlement. In the initial scramble for land, the occupiers gave little value to timber and other forest products, although it did not take long for the new highways to attract sawmill operators who became the principal employers in many frontier towns.

At that time, it was difficult for the general public to perceive the extent of forest that was being destroyed on the distant frontiers. Remote sensing by satellites was in its infancy and images of the region were not available. Immense fires and towering columns of smoke were visible to anyone flying over the area in the dry season, but it was assumed that the primeval Amazon forest was too vast to be endangered by settlers clearing a few tracts in the wilderness.

International concern over the loss of tropical rainforests accelerated rapidly during the 1970s, with emphasis on southeast Asia. As a result, a global survey was organized by by Jean–Paul Lanly, director of the forestry division of the Food and Agricultural Organization (FAO), who concluded that by 1980 about 150,000 square kilometers of tropical forests had been removed in all Latin America and the Caribbean. When the study was released in 1982, the estimate for Latin America was criticized as too low, which seemed to be confirmed by a follow-up report from Lanly saying that the rate of deforestation of the dense, broadleaf tropical forests of Latin America, of which 90 percent are in Amazonia, had averaged 40,000 square kilometers a year between 1981 and 1985.[14]

In Brazil, the national space research agency (INPE), headquartered at São José dos Campos, São Paulo, had installed a satellite reception station in Cuiabá that began to receive images in 1976 from the U.S. Landsat satellite orbiting over the Amazon region. From these images, in 1979 an INPE team led by Antônio Tebaldi Tardin produced the first comprehensive deforestation survey of Brazil's "Legal Amazon" (4,950,000 square kilometers) which showed that 54,502 square kilometers, or 1.1 percent, had been removed as of 1978.[15]

Tardin's figure seemed low, however, to environmental activists who believed the rate of destruction was much higher. For instance, Warwick Kerr, a Brazilian botanist who was head of *INPA* in 1979, told a Brazilian congressional committee investigating the military regime's development policies in Amazonia that "nearly all tropical lowland forest will be cut within one generation."[16] The same tone of catastrophism permeated a special edition, "Amazonia, 1984," of *Ciência Hoje*, the publication of the Brazilian Society for the Advancement of Science. "Amazonia is coming to an end," said an editorial, which attacked every officially promoted development project in Amazonia, from hydroelectric dams to highways and cattle ranches, and denounced the destruction of Indian cultures by "the national society."[17]

Initially it was politics, more than ecology, that provoked the public outcry in Brazil against Amazonian deforestation and its social effects. The new settlers brought land conflicts and rural violence in their wake. Frontier killings became an ugly symbol of an occupation promoted by military regimes. Roman Catholic missionaries and rural union leaders working among Indians and peasants organized these groups in defense of their rights, and the violence began to attract international attention from human rights organizations. The rubbertapper movement led by the late Francisco "Chico" Mendes in defense of the forests of Acre against large cattle ranchers grew out of this political resistance. It was only after leaders of the scientific community, who were active in the Brazilian opposition to the military, provided the ecological arguments against deforestation that environmentalism, as such, became part of the campaign against "authoritarianism."

After Brazil returned to elected civilian government in 1985, under President José Sarney, the level of frustration in the environmental movement rose because of official inaction at high levels in the face of massive illegal deforestation. Satellite images pinpointed the location of enormous fires set in virgin forests in gross violations of environmental laws and forestry regulations; but no measures were adopted to punish violators.

The issue exploded publicly in 1988 when Alberto Setzer (an INPE researcher) and a satellite team at the Brazilian Forestry Development Institute (IBDF) presented a shocking report. They said that 325,000 fires had been detected in Brazilian Amazonia by the infrared band of the NOAA-9 satellite. By Setzer's interpretation, each fire was presumed to have deforested a fixed area, which appeared as a pixel, or point, on the surface of the satellite image. Working late into the night, Setzer and his colleagues scanned 46 images over an 80-day period, counting the clusters of heat-activated pixels that glowed like fireflies on their screens. From the total number of fires, they concluded that in 1987 deforestation had demolished 80,000 square kilometers of native forest, an annual rate that doubled any previous estimate. The emissions of CO_2 and particles from the man-made fires were on the scale of a large volcano, said the report.[18]

Then, Dennis Mahar, a consultant for the World Bank who had been involved in the bank's Rondônia development project, issued a report stating that fiscal incentives and other government measures had produced deforestation of 12 percent of Brazil's Amazonia by 1987. The figure was a guess that lacked supporting evidence, but it gained wide circulation in environmental circles because of the World Bank's imprimatur.[19] The Mahar estimate seemed to discredit the official estimate of the Sarney government in 1988 that said only 5.1 percent of Amazonia had been deforested, adding fuel to the debate.

This controversy produced a major review of data on deforestation and a critical analysis of the accuracy of the different methods used to measure forest alterations. Research teams at INPE and NASA carried out a painstaking comparison of chronological sets of the 224 Landsat images that cover the entire region and reached very similar conclusions. The Landsat Thematic Mapper images, at a level of resolution of 1:250,000, provide much more specific measurement of forest alteration than Setzer's NOAA-9 fire counts, which detect a heat point but are not accurate on the area being burned, or the type of vegetation, since a pasture burn-off turns up the same as a primary forest fire.

The results of the review were summarized in a joint report by INPE's Tardin and Philip Fearnside, an American ecologist at INPA who in 1982 had projected the "end-of-the–Amazon" forests scenario by the year 2000. The new report put deforestation as of August 1989 at 299,079 square kilometers, or 6.1 percent of "Legal Amazonia."[20] In March 1991, the secretariate of science and technology added 4,711 square kilometers to complete the figures for 1989 and reported additional deforestation in 1990 of 13,818 square kilometers, which raised the total of "new" deforestation (since 1975) to 317,608 square kilometers.

The INPE methodology excluded 97,643 square kilometers of "old" deforestation in the Bragantina region of eastern Pará and in western Maranhão, where there was colonization in the nineteenth century. Much of this area has been reoccupied by secondary forest, but if it is included in the total altered area in Brazil's share of Amazonia, the total comes to less than 9 percent. Rondônia (13.2 percent) and Para (11.1 percent) have been the most heavily deforested states, and the least have been Amazonas (1.4 percent) and Amapá (0.7 percent). Acre has lost 4.8 percent.

As for the rate of current deforestation, it is in steady decline. "New" deforestation shows the dynamics of the "decade of destruction" between 1978 and 1988. During this ten-year period, the annual rate averaged 21,000 square kilometers, according to INPE, with a peak in 1987 of about 40,000 square kilometers, or half of Setzer's estimate for that year. Since then, the rate has been falling. INPE's figures for 1988 are 34,658 square kilometers deforested. In 1989, the total fell to 19,135 square kilometers, and in 1990 dropped again to 13,818 square kilometers, a decline of 27 percent from the previous year. Setzer's count of fires has also fallen sharply from the high of 325,000 in 1987, the year of the big burn, to 225,000 the next year and less than 120,000 in 1989.

Thanks to INPE's early start on remote sensing, the Brazilian sector is now much better studied than the Amazon areas of the Andean countries, where the level of deforestation is uncertain. To remedy this, an expanded satellite survey covering all of Amazonia has been set up at NASA's Goddard Space Center in Maryland under Compton "Jim" Tucker, an expert on interpretation of satellite images on Amazonia. FAO was also conducting a new global study of tropical deforestation.

Until a full regional survey is completed, the best approximation of the total area of deforestation in all Amazonia is about 520,000 square kilometers, based on INPE's figures, including old and new deforestation, and 100,000 square kilometers for the Andean areas.[21] That amounts to 8 percent of the Amazon basin as of early 1991. Without minimizing the amount of forest that is gone, the fact is that the heart of the Amazonian *hylea* is virtually intact. The depth and diversity of the remaining forest provides good prospects for a strategy of containment of further deforestation by managing the forest resources for their full economic and ecological value. Measuring deforestation of the past is important because it helps explain the dynamics of occupation. Far more important for the future is close monitoring of further alterations. Post mortem quantitative measurement of how much has been deforested needs to be supplemented by early-warning detection of forest felling, so it can be prevented. Satellite images can also provide qualitative analysis of the vegetation in ecologically sensitive areas that need to be permanently protected. These are missions that the "eyes in the sky" can fulfill if remote sensing programs for Amazonia are properly organized, adequately funded, and regionally coordinated. These technologies are only tools, however, and control of deforestation demands zoning and the political will to bring legal and administrative order to the economic use of territory and resources.

* * *

Forestry management has never been practiced in Amazonia, in the sense of sylvaculture that seeks sustained yields of wood from native forests. Traditionally, logging of trees was just one more form of extractivism that depleted the resource being taken from the forest without replacement. More recently, with highways opening the way into untouched upland hylea, the abundance of timber from land clearing made forest management seem irrelevant. In both cases, great forest potential has been wasted for lack of rational management.

In the old days, timber extraction in Amazonia was a rivermen's affair. Rivers were the only means of transportation and logs had to be harvested near water, in the seasonally flooded forests of the floodplain, known as *várzea*. Dominated by the annual rise and fall of the rivers, the *várzeas* are relatively fertile because of the annual deposits of fresh organic and mineral nutrients on the soils, and although the floodplains occupy only 2 percent of the Amazon basin, they hold 1.5 million people in Brazil alone.[22]

These riverbank people, called *ribeirinhos*, live in poverty despite the good soils they occupy. As a supplement to their fishing and farming, they go logging during the months of low water when forest stands can be reached by canoe. Trees are felled and left to dry until the annual flood makes it possible to float the logs out to the main river. There they are handed over to buyers who usually have advanced food supplies or bits of money. When enough logs are assembled, they are lashed together in rafts and floated downstream, the way logs were floated down the Ohio and Mississippi when young Abe Lincoln was a railsplitter. The logs are the main source of wood for sawmills and plywood industries along the Amazon river at Iquitos, Manaus, Itacoatiara, Santarém, and Belém.

The *várzea* extractive system is primitive and involves no reforestation. In many cases no one owns a *várzea* forest, and they are logged by the first comer. Most of the hardwood trees are considered worthless because only a few *várzea* species are in commercial demand, particularly the light, fine-grained *virola*, which is second only to mahogany in export value, and *andiroba*, a fine hardwood that also produces an oil seed.

For many years, easily reached *várzea* stands were sufficient to maintain the mills and plywood industries that include subsidiaries of some big foreign producers, such as Georgia Pacific of the United States and Atlantic Veneer of West Germany. Only a handful of wood industries invested in forests, such as Wagner S.A., a big plywood producer that has planted 1.8 million native trees since 1978 in land it owns on the Purus river, or Ovidio Gasparetto, a lumberman from Paraná who logs a forest on the Tapajós that supplies his modern Amazonex S.A. wood factory in Belém. But these are exceptions, and timber sources from the floodplains are being exhausted without replacement. "As it is practiced today, and with the ever increasing demand levels from the wood industry, the forest extraction methods in the *várzeas* are a destructive process of highly selective logging that depletes the major commercial species in large parts of the accessible Amazonian old-growth floodplain forests," says Paul Vantomme, a French forester who spent five years in Manaus as a wood industry adviser to INPA.[23]

The decline of the *várzeas* has been more than offset, however, by logging in the upland *hylea* which became accessible with the opening of new highways in Amazonia. With the exception of selective logging of a few valuable species, such as mahogany, the extraction of logs in the newly occupied areas of Pará, Mato Grosso, and Rondônia has been little more than a by-product of land clearing for agriculture, ranching, and power dam reservoirs. Sawmill operators followed the highways like scavengers, pulling logs out of the condemned forest one jump ahead of the slash-and-burn gangs. With a small investment in used sawmill machinery and rudimentary technology, they produced truckloads of rough sawn boards for buyers in southern Brazil, and made lots of money with little value added. As one analyst observed, "The timber industry has only contributed a very small part to [Amazonia's] gross product and has certainly not approached its potential as a force to promote long-term value for forests."[24]

Recent declines in land clearing, however, have reduced the windfall supply of logs, with a visible impact on the supply of commercial logs to the big sawmill towns, such as Paragominas in Pará, Sinop in Mato Grosso, and Ji–Parana in Rondônia. Amazonian lumbermen are having to think, for the first time, about investing in sources of wood for their mills. This is the first step toward putting into place a technically managed forestry industry that can provide a stable market for timber and an incentive to private landowners to invest in managing their forests, instead of burning them down.

For the long-term value of forests to be promoted, a major reduction in the deforestation of Amazonia is an indispensable condition. Forest management involves investments that are economically unattractive when timber is virtually a free good. While large-scale land clearing could be counted on by sawmill operators, the cost of getting logs was like sending a truck to pick up timber on the beach after a shipwreck. Now, with fewer logs and more demand for wood products, the economics of Amazon logging is changing.

The new situation poses a challenge for lumber towns like Sinop, where 600 whining sawmills turn gigantic logs into stacks of sawn boards and dunes of sawdust along BR-163. When Enio Pipino, a founder of agricultural towns, launched the colonization of Sinop in northern Mato Grosso in 1976, he thought it would be another coffee center, like Maringá in Paraná, where he recruited the first settlers for his new venture. The colonists went north on a busline Pipino set up between Maringá and Sinop. They occupied 600 family lots, cutting down the trees to plant rice and then coffee. Big ranchers soon followed and began deforesting blocks of land bought from the state.

Timber was abundant at this stage, and many sawmills sprang up as Sinop grew into a commercial center, with a main street full of stores, hotels, a courthouse, three bank branches, and numerous well-built residences on the side streets. Sawn wood was trucked back to Paraná, where lumber was becoming expensive. But Sinop considered lumber a sideline until the settlers discovered that the land produced poor coffee and tore up thousands of hectares of plantations.

Pipino then came up with a new scheme, an alcohol distillery using mandioc as the fermenting stock. Pipino got SUDAM approval and a promotional loan for a $25 million plant, and the Ministry of Industry's "alternative fuel" program provided a subsidized price for any alcohol the distillery could produce. President João Batista Figueiredo visited Sinop in 1980, shortly after the distillery was inaugurated, and extolled it as a model for Amazon industrial development.

Six years later, Pipino was bankrupt and the alcohol plant was a white elephant. To induce farmers to plant mandioc and supply his plant, he had guaranteed loans to the farmers from the Bank of Brazil. When these root crops also failed, Pipino could not pay and the distillery joined a long list of Amazon projects financed by SUDAM tax incentives that have gone sour because of bad management and untested technology.

Despite the distillery setback, Sinop is a prosperous township this is surviving on its sawmills and as a commercial center. Sinop, and its 50,000 people, no longer depends on Pipino, but they do depend on stable sources of timber for the sawmills, which employ over 8,000 workers and produce more than 1 million cubic meters of wood annually. With lumbering, Sinop makes money and pays town and state taxes, instead of throwing federal money away in subsidized coffee that has no market and alcohol ventures that fail.

Luis Carlos Favero, president of the Sinop Wood Industry Association, says that forest management is a must for the larger wood firms that have invested heavily in industrial machinery and in company housing for their large workforces. "We need to have control over forests that will produce over many years, or we will soon be finished," said Favero, who produces laminated wood. He said his plant processed logs from his own 71,000-hectare forest property, where he is harvesting 850 hectares annually with a moderate yield of 30 cubic meters of roundwood per hectare on the first cut. At this rate, he said his forest could provide wood for 80 years, after which new growth would be on the same land through reforestation.

Brazilian forestry law, which has been loosely enforced in the past, requires wood industries that consume over 10,000 cubic meters in logs annually to submit an "integrated forestry-industry plan" (PIFI) for approval by IBAMA. They are supposed to be able to show that 80 percent of the lumber they use is coming from "renewable sources." If they own forest, an inventory of species must be accompanied by an extraction and reforestation plan. Every mature tree that is cut is supposed to be replaced by six seedlings of the same species, or the equivalent in a royalty payment to a national reforestation fund.

The PIFI system is very unwieldy and hard to administer. "There are 600 plans awaiting approval in Mato Grosso because IBAMA doesn't have the staff to review them," said Favero in early 1990. To break the logjam, he proposed that IBAMA approve plans prepared by licensed forestry engineers automatically, and decentralize the reforestation program by placing administration of the funds, including nurseries and technical assistance, under a consortium of IBAMA, private wood industries, local universties, the state forestry institute, and cooperatives of small farmers interested in planting trees. Without this broad cooperation, Favero said "gypsy" sawmills, avoiding the costs of forest management, would continue to buy and process logs outside the law. "This can ruin those of us who are trying to conserve the forests," said Favero.[25]

The situation was much the same in Paragominas, the wood industry center of eastern Pará. This large town, founded in 1959, was settled by pioneer cattle ranchers who came ahead of the bulldozers that opened the Brasília–Belém highway. Settlers have burned off at least 8,000 square kilometers of virgin forest of the 27,000 square kilometers in the township, mainly to provide pastures for cattle. Degradation and later abandonment of some pastures made Paragominas an object lesson, for many academics, of the economic and environmental folly of stimulating cattle ranching, through tax incentives, in dense tropical forests. Few new ranches have been opened since 1980, but the ranchers have not disappeared; the last agricultural census reported 365,000 head of cattle in Paragominas in 1985, and the municipal cattle breeders' association put the total at 500,000 in 1990. Either figure makes Paragominas by far the largest ranching township in Pará, hardly the image of a cattle ghost town where the pastures all fail.

Ranching aside, the economic high ground in Paragominas now belongs to dynamic young lumbermen, mainly from the southern state of Espírito Santo, who took their sawmills and know-how to Amazonia as timber ran out in the South. There are more than 500 sawmills, plywood factories, wood-product industries, and charcoal producers in Paragominas that provide 14,000 jobs—by far the biggest payroll in town. The wood sector pays 80 percent of the town's local tax revenues, and much of the commerce and construction income comes from the lumbermen. Many of them have built solid homes with tall beams and carved balconies, like the two-story chalet where Sidnei Rosa lives with his wife and two children.

"When I first came to Paragominas in 1981, people would ask, how long will the timber last? Now that I have seen how strongly the forest comes back after a first cutting, I say, we have forest for a hundred years . . . or maybe forever if we treat it right," said Rosa, the president of the wood industry association. Rosa is a muscular, blond-haired entrepreneur in his early thirties who came to Paragominas from São Gabriel da Palha, a lumbermill town in Espírito Santo that was populated a century ago by Italian, German and Polish colonists. The native forests where Rosa grew up are all gone. Why should it be different in Pará?

Christopher Uhl, an ecologist from Pennsylvania State University who has spent years studying forest regeneration in degraded Paragominas pastures, sees an early end to the wood industry. "All the sawmills intend to pull out in five to ten years when the closest forests will be exhausted, as seems inevitable," said Uhl in an article on logging in Paragominas.[26] Uhl found that ranchers he interviewed were selling rights to their forest, or had become loggers themselves, earning $175 per hectare for timber sold to sawmills.

If this were the case, the responsibility for completion of the ecological ruin of Paragominas, initially blamed on the ranchers, would shift to the lumbermen. Rosa and other lumbermen who have invested heavily in modernization of their industries and formed an export consortium say it isn't so. This group has no doubt that the wealth of Paragominas lies in the rational management of the remaining forests that occupy two watersheds, the Gurupi and the Capim, that are within a 100-kilometer radius of the town.

"There are cattle ranchers who will continue to sell off timber so they can increase pastures, but we have discovered that the forests are more valuable than cattle," said José Carlos Gabriel, who owns a modern sawmill and three forest ranches covering a total of 105,000 hectares. Like Rosa, Gabriel is from São Gabriel da Palha, and he has a law degree. "No one has a bigger interest in the continuation of the forests than lumbermen who know the score. I have been to Europe, the United States and the Caribbean, and I have seen the potential demand for the fine woods we can produce," said Gabriel.

For example, a species called *tauari* (*couratari oblongifolia*) was not even harvested at Paragominas two years ago. Then it was discovered that this creamy white wood could substitute for Asian ramin if it were properly dried and finished. The export consortium installed an expensive drier, and *tauari*, used for interior panels, began to be exported to Italy at a price of $375 per cubic meter. A score of other species that were discarded before are now being sold as wood for a range of products from furniture to plywood fillers thanks to better wood-processing techniques. This is the road to increasing economic return from Amazonian forests.

As in Sinop, the lumbermen in Paragominas were interested in participating in the application of the reforestation fund. Rosa proposed that the regional Bank of Amazonia act as the financial agent for the fund and make loans for forest management projects to industries that paid their royalties. He said he had a project ready for regenerating a 4,500-hectare forest that had been logged. "The costly part comes in helping the best species grow back after the cut. That takes a lot of labor, and you don't see results for many years. So forest renovation credits have to be at low interest and long term," said Rosa.

The industrial side of the lumber business has an air of permanence, as does the township of Paragominas, which has a population of 100,000, three television stations, four banks, Rotarians and Lions, a paved highway to Belém or Brasília, and high-voltage transformers connected to the state power grid. The wood industry association has 147 members, about 10 of which are as big as Rosa Madereira S.A., which is a $10 million investment with 800 workers on the payroll and a logging fleet of 10 D-6 tractors, 2 skidders, 7 forklifts, a road grader, and 23 trucks. At Rosa's office, a fax machine ratcheted out waybills for cargoes of construction lumber for Caribbean points and furniture wood for Italy being loaded at the port of Belém. In the yards there were stacks of quality tongue-and-groove boards and plywood sheets ready for trucks going to Brasília and points south. On Rosa's desk was a wooden plaque bearing the company motto: "Generating Wealth without Destroying Nature."

Paragominas will never be an ecological showcase, but with rational management of its substantial forests it could maintain a sustainable economy based on the wood-processing industries. If these industries continue to be supplied by the renewal of the local forests, they will not move on to devour intact forests elsewhere. That is also part of environmental protection.

* * *

If the containment of Amazon deforestation has led some modern-minded lumbermen to conclude that forest management is necessary for their future, this pro-forest constituency would be strengthened by a clear national forestry policy that provided a regulatory framework, technical support, and adequate credit for this activity. A basic element would be ecological–economic zoning—establishing rights and prohibitions on what forest areas should be used for production—and laws providing secure property rights to private forests that are being managed.

In addition to legal and institutional guarantees and economic incentives, the attainment of sustainable yields in Amazon production forests will depend on the solution of ecological problems that are peculiar to these native forests. When tropical forestry experts are asked why Amazonia has been a minor exporter of tropical timber compared with southeast Asia, the usual answer is that the 3,000 species of trees in Amazonia are a dreadful problem. Loggers are generally looking for high concentrations of one or a few commercial species, as are found in the dipterocarp forests of southeast Asia and the coniferous forests of Central America. In Amazonia, they sometimes find 400 species per hectare, but only one or two grown stems of the species they want. "Species variety has been our natural defense because if it were easier to extract wood from Amazon forests they would all be gone," said Clara Pandolfo, director of natural resources of SUDAM.[27]

The variety of trees, with different densities, drying rates, mechanical properties, and visual qualities, has also been a commercial disadvantage. International buyers are bewildered by more than 80 different commercial woods on offer, from *acapu* to *ucuubarana*, and they concentrate on only a few they know. As a result, tree harvesting in Amazonia is costly because loggers take only a small part of the biomass and waste much of the forest's full potential. Without a management plan, there is usually no effort to protect growing stems or to leave mature seed trees to promote regeneration of desired species.

The design of forestry management technologies for tropical forests, a controversial issue for many years, is the principal focus of the International Tropical Timber Organization (ITTO). This organization began in 1983 as a modest commodity agreement negotiated under the auspices of the United Nations Conference on Trade and Development (UNCTAD) that went into effect in 1985 with the participation of 42 governments of tropical timber producers and consumers. ITTO operates from a small secretariate headquartered in Yokohama, Japan, where B.C.Y. Freezailah, a Malaysian, is executive director. The commercial interests of the tropical timber trade remain the central business, but ITTO has acquired a new dimension since 1987 as a forestry development center. It arranges financing for projects that seek practical answers to the problems of forest management in the humid tropics. ITTO has three major forest projects under way in Amazonia that are potentially important, as will be discussed later.

Research and pilot projects in forestry management are also being developed in Amazonia by official bilateral lenders, particularly the aid agencies of Canada, Britain, Germany, Switzerland, the United States, Japan, France, and the Netherlands. Major botanical gardens, such as in New York and St. Louis in the United States and Kew Gardens in Britain, and research centers like Woods Hole in Masssachusetts, branches of Germany's Max Plank Institute, and the Worldwide Fund for Nature (WWF) in Geneva support some forest research. Strong inputs have come from universities, particularly the University of Florida, with its graduate program on Amazonian studies concentrating on Acre, and more recently the University of Pittsburgh, with a program in the *varzeas,* and Princeton University, with a mahogany research project in Bolivia. Nongovernmental organizations that provide genuine expertise, such as World Wildlife Fund–USA and Conservation International, or private development agencies, such as the Ford Foundation, Oxfam, Cultural Survival, and Ashoka, combine on-the-ground personnel with funding. The NGOs are particularly important for forestry projects that have a strong social element, such as agro-forestry for small farmers and extractivists.

All of these laudable efforts can become as dispersed as the species of the tropical forests, however, without a strong framework of national forestry policies that make clear the role of forestry in relation to other land use alternatives, such as agriculture and energy, and promote the large investments that are necessary for sustainable forestry. In Amazonia, the forest sector has not had the political clout or the short-run economic appeal held out by oil exploration, highway construction, hydroelectric dams, cattle ranching, commercial farming, or even wildcat gold digging. As a result, forest resources have been wastefully expended with severe environmental damages because of mismanagement. Brazil, again, serves as the worst example.

For more than a decade, Brazil has had available a blueprint for Amazon forestry development that was never implemented for lack of political decision. The original proposal was presented in 1978 by an FAO forestry mission that worked in the Amazon for two years. It called for logging 122,000 hectares in the Tapajós national forest of Pará, which covers 600,000 hectares, over a 22-year period. The cutting plan was for an annual supply of 150,000 cubic meters of roundwood from about 30 species for a modern sawn wood and plywood mill in Santarém. With an investment of $22.7 million in the mill and annual costs of $6 million for forest management, including a 30-year cycle for regeneration of cut areas, the FAO mission concluded that the operation would earn at least $20 million annually, provide some 600 jobs in the mill and forest work, and supply growing domestic demand.[28]

The Tapajós production forest was presented as a model for the region of "sustained yield (forest) management " that would demonstrate "the most rational means of occupation of the Amazon, preserving the national patrimony and forests, by using it rather than wasting it by wanton clearing and burning."[29] The mission, led by M. K. Muthoo, an extroverted Indian, included many of the big names in tropical forestry, like J. A. Tosi of the Tropical Research Center, Costa Rica; A. J. Leslie of Australia; and Marc Dourojeanni, now environmental director of the Inter-American Development Bank. They made the significant finding that "there are no technical, economic or environmental obstacles to the industrial utilization of the Tapajós national forest and its management as a permanently productive and essentially natural system."[30]

Despite this endorsement and initial support from the Brazilian government, then led by General Ernesto Geisel, the Tapajós project floundered in the political cross-currents of the period. The Tapajós national forest was the first of 12 "production forests," with a total area of 392,530 square kilometers that had been proposed by Clara Pandolfo, SUDAM's director of natural resources, as part of an ecological-economic zoning of Amazonia. Geisel put the proposal before the Brazilian congress, and this unleashed an opposition campaign led by Orlando Valverde, an ultranationalist geographer who was leader of a movement called Campaign for Defense and Development of Amazonia (CNDDA). Valverde seized on a FAO consultant report, suggesting that foreign investors be offered long-term logging concessions, to whip up public opinion against handing over Amazonia's forests to "multinationals."[31] One of the most agitated protagonists of this campaign was José Lutzenberger, a former pesticide salesman who became an ecological activist after his conversion to the doctrine of Gaia. (President Collar named Lutzenberger to be his secretary of environment in 1990.)

General João Figueiredo, who succeeded Geisel as president in 1980, found himself facing not only the nationalist fulminations of the Valverde campaign, but more subtle pressures from powerful landholders in Amazonia who were opposed to SUDAM's forestry and ecological-economic zoning proposal. The CNDDA campaign provided a convenient political expedient for Figueiredo; he discarded the forestry and zoning proposals as a gesture to the nationalists, while providing Amazonian capitalists with a new round of tax incentives for ranching and more hydroelectric dams, such as the Balbina dam which has flooded 2,000 square kilometers of forest near Manaus. Without a forestry plan, the assault on the Amazon forests rose to new heights.

But after a decade of neglect and destruction, the Amazon forestry management proposals rose from the ashes of the forest and reappeared under the Collor government in an expanded version. Under the title National Program for Forest Conservation and Development, a comprehensive forestry sector plan emerged proposing investments totaling $3 billion in five years.[32] The plan was drafted by an interministerial working group, including many outside experts, under the direction of José Carlos Carvalho, a forestry engineer, who was the first secretary general of IBAMA and negotiater of the National Environmental Program loan signed with the World Bank in 1990.

The new forest sector proposal embraced many environmental concerns, including biodiversity and genetic resources. It proposed an expanded federal system of conservation units covering 163,642 square kilometers, including 34 national parks and 62 ecological reserves, and creation of 9 extractive reserves in Amazonia for at least 75,000 families of forest gatherers, protecting another 400,000 square kilometers of forest. The main objective, however, was an increase in wood production, for which the plan proposed two lines of action: One was massive reforestation of areas already cleared of forest; and the other was selection of native forest areas on public lands as sustainable production forests.

The plan report said that total demand from consumers of industrial wood, energy wood, and cellulose in Brazil had risen to 400 million cubic meters in 1990. Of this total, an astonishing 73 percent was consumed as charcoal and energy wood, with pulp and paper demand also rising quickly. To reduce the pressure of this demand on native forests, the plan proposed investments of $1.1 billion to reforest 200,000 square kilometers in 30 years, including plantations of eucalyptus, pine, and other fast-growing species in degraded pastures in Amazonia to produce charcoal and cellulose.

External demand for Amazonian timber would increase rapidly, the plan report predicted. "South-east Asia, the largest producer now, will undergo a decline over the next two decades . . . and customers will turn to South America, principally Brasil. We have to prepare ourselves to transform this tendency into an opportunity for development of a national forest industry in harmony with the indispensable protection of our environment," said the report. It outlined a strategy to obtain the necessary increases in wood production from native forests on a sustainable basis. For this it proposed investments of over $1 billion, including $520 million for ecological-economic zoning, $250 million for forest inventories leading to the identification and creation of 24 national production forests covering 222,975 square kilometers, implantation of the Tapajós project, and a major increase in forestry research and training of forestry engineers, technicians, and skilled workers for forest managment and industry.

This was, at least on paper, the biggest national tropical forestry plan in the world. But when a copy of the plan was circulated in Rome during the October 1990 meeting of FAO's committee on forestry, it was evident that what Brazil was proposing went beyond what the international forestry community could handle. Funds on the scale of $3 billion have to come from big lenders and private capital sources, and Brazil was locked in a confrontation with its creditors over refinancing its $110 billion foreign debt. As a form of pressure, the United States and other governments that control development lending had cut off new loans to Brazil.

The overriding international financial realities created the paradox that Brazil, proprietor of one-third of the world's tropical forest, was a distant spectator of the global emergency program to "save the tropical forests" called Tropical Forestry Action Plan (TFAP). TFAP was launched in Washington in 1985 under the co-sponsorship of the World Bank, the United Nations Environmental Program (UNEP), World Resource Institute (WRI), and FAO as an $8 billion, five-year effort to put most of the world's tropical forest on a footing of "sustainable development." Five years later, a review meeting in Rome of the 67 participating countries agreed that little had been accomplished, and least of all in Amazonia.

TFAP sent missions of forestry experts, organized by FAO, to Colombia, Peru, Bolivia, Ecuador, Venezuela, Surinam, and Guyana to clarify domestic forestry goals and identify projects for donor financing. A TFAP regional meeting for Latin America was held in Jamaica in 1989, and a mission met with the Amazon Pact as a group. But very few Amazon projects found TFAP donors. The reason was partly lack of good projects, as was recognized at a meeting of national TFAP coordinators in 1990 with Matt Heering, the Dutch forester in charge of TFAP.[33] But it was also pointed out that TFAP has no funds to finance project preparation or to support administration of national forestry programs that are often short of local funds, so there is a vicious circle. Little wonder that an independent panel of experts commissioned to review TFAP reported back to FAO in 1990:

> TFAP has bred a sense of frustration because it takes so long to produce any visible results; meanwhile more forest is destroyed. . . . If the tropical forests are to be saved, the international community needs to respond with more than gestures.[34]

But more than its ponderous beaurocracy and its financial limitations, what hobbled TFAP in Amazonia were political problems. There is very little that international cooperation can do, in tropical forestry or anything else, when countries reach an acute state of economic and political disorder. In Amazonia, Peru, under former President Alan Garcia (1985–90), provides an outstanding example.

Peru's forestry potential is second only to Brazil's. With 775,000 square kilometers of tropical forest, covering 60 percent of the national territory, the annual rate of deforestation between 1981 and 1985, as calculated by FAO, was a relatively low 3,000 square kilometers. Peru has a huge intact forest to be developed and the advantage of nearby ports on the Pacific. But this resource is neglected and wood exports were a negligible $2.5 million in 1988.

Peru gave TFAP an enthusiastic welcome. With the help of a Canadian forestry advisory team, it was among the first to complete the TFAP "country review" process. By 1988, a National Forestry Plan had been drawn up and circulated to potential donors, who met in February 1989 at the Lima Country Club to consider Peru's request for support for 72 specific projects.

The result was disastrous. The United States and the World Bank—two of the main potential donors—were notoriously absent. Under Garcia's "populist" leadership, Peru had unilaterally suspended foreign debt payments and was cut off from new loans by the International Monetary Fund and the World Bank. Peru's participation in the cocaine trade had embittered relations with the United States. Canada, Spain, West Germany, the Netherlands, Belgium, Switzerland, and the Inter-American Development Bank indicated willingness to support some of the 72 projects—if Peru could put up local currency financing to show its commitment to forestry. This condition could not be met. Garcia had plunged Peru's finances into such disarray that inflation had risen to 40 percent a month and public revenues didn't cover government wages.

A further complication was the unwillingness of the international aid community to send technicians to Peru after two French agronomists were killed by Maoist "Shining Path" guerrillas in the Apurimac valley. Barbara D'Aquile, Peru's leading environmental columnist and a promoter of TFAP, was killed by guerrillas on a trip to the highlands. Between guerrillas and heavily armed drug traffickers, the upper Amazon valleys where most of the coca was grown became a combat zone, with roads mined between large towns and guerrillas protecting smugglers loading aircraft with cocaine paste at jungle airports.

The Peruvian Amazon, as well as Bolivia's tropical Chapare and most of Colombia's eastern lowlands, was a place where tropical forestry projects, particularly involving international personnel, could only be carried out at high risk. Saving the tropical forest was not enough reason to take the chance.

* * *

Somewhat awkwardly for TFAP and its donors, in the few cases when national forestry plans have become a reality, they have met with hostile gestures from skeptical NGOs that are a vocal part of the international environmental community. A $135 million TFAP program in Cameroon to promote tropical timber exports, raise domestic fuel wood production, and combat desertification was labeled "a plan for environmental and social disaster" by Washington's Environmental Defense Fund.[35] The Cameroon program included a $30 million loan from the World Bank. This was one of a package of similar African forestry loans to the Ivory Coast, Ghana, Togo, and Zaire that were in the World Bank's lending target for 1988–92 of $2 billion for forestry, as a way of supporting TFAP.

All signs point to continuing opposition by ecological and anthropological activists to international lending for tropical forestry programs that involve wood production from native trees. There have been no development loans or large international investments for forestry in Amazonia as yet, but when the situation arises, it is bound to be controversial. The rainforest defenders will invoke arguments about the genetic, sociopolitical, economic, and global environments.

Are the species, the biodiversity, and the genetic resources of the forest exposed to extinction by the extraction of timber? Can humanity afford to lose the genetic "treasure" accumulated over eons for a bit of commercial wood or cellulose paste? Are forest people and their traditional ways of life exposed to destruction by intensive, mechanized exploitation of the forest and contact with "alien" cultures? Are local communities being consulted, participating in the benefits, and exercising political rights? Are the economic returns from tropical timber extraction likely to provide a good return on investment, including the cost of managing the forest for sustainable yields? Or is forest being cut for quick gain and foreign exchange income without considering long-term alternatives? Is timber extraction an environmental alteration that can affect water cycles and eventually produce global climate change? Will replacement of native forest with monoculture plantations kill an ecosystem?

These questions often contain a "zero option" premise against any alteration of the tropical rainforests, as if these Third World forests were so radically different that they cannot be managed for wood as are the temperate forests that are the world's main sources of timber. But they are valid questions that need to be answered objectively before forest managment projects can be presented for financing. Answers that are applicable to Amazonian realities are being sought, and a few examples will show how this is done on a case-by-case basis.

Case 1. Social Forestry at Palcazu

In September 1988, a column of Peru's "Shining Path" guerrillas operating south of Pucallpa, the gateway to Peru's Amazonia, entered the Palcazu valley. This is where the U.S. Agency for International Development (USAID) had invested $10 million in an important tropical forest management project with the local Amuesha indians. When reports reached the embassy in Lima that the guerrillas were threatening cattle ranchers around Oxapampa, the main town bordering the Palcazu forest region, all U.S. personnel were evacuated by air from Palcazu and project funding was cut back sharply. Only an $80,000 grant from World Wildlife Fund–USA helped the Yanesha Forestry Cooperative that manages Palcazu's forests survive the crisis. In 1990, the project was limping along on local sales of wood and posts.

The Palcazu project involves managing about 40,000 hectares of forest in a lush valley that covers 289,000 hectares. The valley adjoins the Yanachaga–Chemillen National Park, which was invaded by settlers and ranchers from Oxapampa who began pressing for a new road through Palcazu. The 6,000 Amuesha and Campa Indians in the Palcazu valley resisted the ranchers and their roads because they depend on intact native forest for game and fishing. They are the people in the region who have the biggest stake in defending the ecosystem.

In 1980, the Indians got help from friendly anthropologists at Cultural Survival who mobilized an international NGO campaign against an ill-considered Peruvian government road project into the area. The campaign forced USAID to transform the road project into an Indian community development project in which forestry and ecological components were combined. What emerged is an example of social forestry.

The Amuesha accepted a forestry management plan proposed by international consultants contracted by USAID because the project took into account their political, economic, and cultural interests. USAID agreed to finance a 100-kilometer road to Oxapampa and a sawmill and wood-processing plant at Palcazu, under control of the Indian community, only after the Peruvian government agreed to conditions that included formal titling for the Amuesha community lands; land use zoning of the Palcazu valley for natural forest production, not agriculture; and definition of ecological preservation areas. That is an example of site-specific ecological-economic zoning.

The social effects of this plan were described by Manuel Martin Lazaro, a slim, dark-eyed Campa Indian who is one of the Palcazu cooperative leaders:

> Our communities plant foodstuffs in lots cut in the forest and we also have cattle pasture. I saw that the forest around us was being finished and that we had to adopt a system to protect our environment and the rivers which depend on the forest. If everything was cut, it would be the end of the fish. It was also clear that we needed to create jobs for our younger people. When lumbermen come into the forest, they clear everything, damage the soil with their tractors, and afterwards there are no trees left, no jobs, and the soil is hard for planting. With the system we have adopted there is hunting, fishing and fruits from the forest, and we have more jobs for the communities.[36]

Robert Simeone, a forester who gathers maple syrup in northern Wisconsin when he is not an international consultant, worked for three years at Palcazu. In his view, Palcazu is a critical test of "sustainable yield management" of tropical humid forests, with relevance to all of Amazonia.[37]

The method of sustainable forest management adopted at Palcazu is the now widely studied "Strip Shelterbelt Model" that was designed by Gary Hartshorn, a botanist, now director of project development at World Wildlife Fund–USA, who was working under a USAID contract in 1982. In forestry terms, the system mimics the tree-fall, canopy-gap process that produces regeneration in a natural forest. Narrow strips of forest are clear-cut, removing all the useful wood, between standing strips of undisturbed trees from which seeds produce natural rengeneration in the cut strip.[38]

The Amuesha quickly grasped the idea of gap-dynamics because it followed the natural sequences of the forest. Indian community leaders chose the strip areas, and 8,500 hectares were set aside for strip cutting. The first strips were cut in 1985 and 1986, and were ready for evaluation by 1988. Hartshorn's model was judged by USAID to be a silvicultural success. Simeone wrote:

> Two years after harvest, richness of tree species on the first regenerating strip is double that of the tree seedlings found in the understory of the original forest prior to harvest. . . . The natural regeneration of trees on the first two strips not only confirms the attractiveness of the strip shelter-belt system as a viable tool for the sustained yield management of natural tropical forest stands, but it also appears to be an important means for maintaining biological diversity.

But Peru's chaotic reality intruded on Palcazu's sylvan world. The withdrawal of U.S. advisers and of Peruvian project personnel, due to budget cuts, produced a lag in the construction of extraction roads, which delayed the planned harvest schedules. Sawmill operations continued on a small scale, with lumber being sold for local building projects and in neighboring valleys. The Yanesha Cooperative was selling a bit of high-quality wood to a California furniture association, but it was clear that the future of Palcazu's social forestry project, and the desired marriage between between economy and environment, will depend on increased production and better salesmanship in the wood markets of Peru and the world.[39]

The political future of Palcazu is also in the balance. During one of his visits to Lima in search of wood buyers and financing, Lazaro gave this assessment:

> The Shining Path guerrillas don't bother us, we are not enemies. But if the project fails to produce income for the cooperative members, the communities will lose interest. Then, the lumbermen from San Ramon and Oxapampa will come offering gifts to get at the forest, and the drug people will try to turn the valley into coca plantations. If that happens, it is the end of the forest, and, who knows, young people may then begin to listen to radical ideas.[40]

Case 2. ITTO and Multiple-Use Forestry

The International Tropical Timber Organization (ITTO), as described earlier, is the focal point for cooperation between consumers and producers of wood from the world's tropical rainforests. In Amazonia, ITTO is supporting research projects in forest management at three locations that are strategic for future expansion of tropical timber supplies from the region.

One is at the Chimanes forest in Bolivia's Beni department, where headwaters of the Amazonian Mamore and Beni rivers descend through densely forested Andean piedmont. The Chimanes forest, covering 1.16 million hectares, is Bolivia's largest concentration of mahogany. It was almost untouched until 1987 when a

400-kilometer highway from La Paz brought in commercial loggers who now ship mahogany boards and finished wood products from the Pacific port of Arica in Chile. Chimanes is protected partially by the Beni biological station, the regional Yacuma park, and a state-owned protection forest; the rest is a "permanent production" forest where five Bolivian lumber companies with concession rights to 300,000 hectares can annually extract about 27,000 cubic meters of mahogany, cedar, and oak-like roble (amburana cearensis). Local Indian communities have rights to fish, hunt, plant, and harvest forest products in about 200,000 hectares, where there is also illegal logging.

Chimanes's production was programmed under a long-term management plan, designed by Conservation International (CI) consultants and approved by ITTO. The project was supported by $1.2 million in grants from Japan, Switzerland, and France. Supervision was given to Luis Goytia, a forester hired by ITTO, whose small staff of workers mark mature trees for cutting, control logging practices, and keep an eye on sawmill output. There are a tree nursery and a modest reforestation effort. A research team from Princeton University is following the growth of replanted mahogany. At this early stage it is too soon to say whether the Chimanes program will achieve "sustained yield"—a stable balance between what is extracted and what is added by regrowth—particularly after an unexpected crisis in 1990 put a new partner into the scheme.

The Chimanes forest had been used as a hunting ground by native Indians long before the loggers arrived. With the granting of the concessions, these indigenous groups organized a political movement that led to a dramatic march by thousands of demonstrators to La Paz, the capital. President Jaime Paz Zamora, who was elected with strong support from "progressive" sectors, put the Office of Indians Affairs at the Ministry of Agriculture and Colonization in the hands of Wigberto Rivera, an ecologist and activist for Indian rights from Riberalta. Rivera insisted that the Indians be given one-third of the Chimanes forest, based on tribal claims, and the project was reorganized, reducing the shares of the private loggers. Mauricio Hauser of the Fatima Lumber Company, which has one of the biggest concession areas, said the reduction "is not going to help manage this forest in a sustainable way because the Indian area will be logged illegally by outsiders."

ITTO ordered a review of the project by an "interim" committee, including two Indian representatives and six other representatives of the NGOs, the government agencies involved, and the loggers. The review was to establish the degree of "community support" for the project and whether the political conflicts could be resolved. Until these questions were answered it ordered increasing patrolling, with a small budget of $82,000, and a reduction in the levels of timber harvesting by the concessionaries.[41] The project was clearly in trouble, mainly because CI had failed to address the potential Indian problem in the early design. CI fired its Bolivia representative and reduced its participation to support for the biological reserve.

Another ITTO project has adopted the Tapajós national forest in Brazil, described earlier, as a timber production research area for the middle Amazon region, with financing from Britain's Overseas Development Agency. But the most important ITTO project in Brazil is the Antimari state forest in Acre, where an old rubber forest is serving as a laboratory for testing different ways in which logging and extraction of natural products, such as rubber and Brazil nuts, can be combined with timber extraction in forests that have large local populations.

Case 3. Trees and Politics

The $3 million Antimari project, financed jointly by Brazil and Japan, is in one of the most politically sensitive parts of the Amazonian *hylea*. In Acre, where the future of the state depends on whether cattle ranchers or forest-based interests dominate use of land, the ITTO project is a factor in this struggle. As presented by the Brazilian government to ITTO, the project is an attempt to establish "multiple-use forest-based development in the Western Amazon . . . [by] management of forest resources for sustainable production."[42] This means combining the extractive interests of the rubbertappers, who have obtained reserve areas covering 15,000 square kilometers for their activities, with lumber industry interests that want logging throughout the state's 150,000 square kilometers of forest.

The project was prepared by the Acre Technology Foundation (FUNTAC), under the able leadership of Gilberto Siqueiros, an economist, who was a close adviser to former governor Flaviano Melo. Melo and his successor, Edmundo Pinto, elected in 1990, represent the forest-industry sectors in Acre, which have as a strategic objective the opening of an export highway from Acre to the Pacific ports of Peru. The only economic justification for such an expensive investment would be a major expansion of forest exports.

Acre, with its Texas-like origins, thinks of itself as the hinge between Brazil and neighboring Bolivia and Peru, with which it shares the western Amazonia. It is too far, by 3,500 kilometers, from the Atlantic and southern Brazil to look to those markets, and as of 1991, Acre still awaited a paved highway link to the rest of Brazil by the extension of BR-364 from Rondônia. In Acre, many political and economic interests have their sights set on the Pacific and the Far East.

When Governor Melo took office in 1985, he had worked as a young engineer for a big contractor on BR-364, and he had seen the onslaught of settlers in Rondônia's virgin forests. Melo, from an old Acre family, opposed a repetition of the land rush in Acre, which has fewer than 500,000 people. He also saw little advantage for his state from expansion of big cattle ranches which had brought severe social conflicts with rubbertappers. So he turned to the "alternative model" presented by Siqueiros of forestry development, providing jobs in lumber industries as well as a continuation of rubber-gathering activities for 50,000 forest dwellers. Melo went to Japan to sell the Antimari project to ITTO as a technical step before moving on to a major forestry investment.

The modern forestry industry that Melo and Siqueiros visualized as the economic foundation for Acre requires a $350 million investment in a paved highway to the Peruvian border as part of a 1,700-kilometer export corridor. An agro-ecological zoning order would set aside 4.4 percent of the state's forest for managed timber production, which would extract about 1.2 million cubic meters of wood in 16 years. Antimari is supposed to provide techniques for native forest recovery, but it is admitted that the initial lumbering would involve "some ecological costs." A private sector investment of $85 million would be needed for two wood product industry centers at each end of the state. The forestry industry would create 15,000 direct jobs and provide the tax revenues needed by Acre's public administration.[43]

"In the year 2000 the Pacific highway will be the economic independence of Acre and a corridor for western Brazil to the Pacific," said Melo in an interview before leaving the cream-colored, gingerbread governor's palace in Rio Branco for a federal Senate seat in Brasília.[44] It may be a long time before Acre gets the money to build such a highway, but a forestry-based economy and a highway live in the dreams of many people in Acre, including some of the rubbertapper leadership, now that they have their extractive reserves. "We are not against the road, as such, as long as it does not bring destruction of the forests. I would like to see a paved highway extended to the Chico Mendes Reserve," said Osmarino Amâncio Rodriguez, secretary general of the National Rubbertappers Council (CNS).[45] Jorge Viana, the candidate for governor who was backed by the CNS, is a young forestry engineer who was in charge of the Antimari project at FUNTAC before his unsuccessful run. He lost narrowly and still has a political future.

The Antamari project was designed to test ways of profitably managing timber extraction from the 300 tree species in the complex Acre forest. This is of interest in Bolivia, Peru, and Ecuador, where there are similar forest conditions involving both timber and forest extractivism. The ideal technology would provide stable supplies of lumber for wood industries from "permanent production forests" that would also supply rubber and Brazil nuts. After two years, results were being obtained slowly at the Antimari, where a small, isolated natural extraction community of 80 families live in a largely intact forest of 67,000 hectares west of Rio Branco. With the support of ITTO, the forest has been inventoried, and the local community has obtained social services and contacts with the outside world that open new economic horizons. A consultative committee including CNS and other NGOs was involved in drawing up a $3.3 million plan that combines production with community development.

The unusual importance of Antimari is that it is a pilot project that will determine forestry policy in the whole state of Acre and beyond. The basic management plan prepared by FUNTAC, with the participation of international consultants, foresees maintaining production from over 6,600 castanheiras and continued tapping of seringa along 62 rubber routes, in the traditional way. The novelty is that a selection of mature trees from the 23 commercial species identified in the forest will be cut over a 25-year felling cycle. With a logging volume of 20 cubic meters per hectare, which is light, the harvested amount would be 52,300 cubic meters annually, starting with 4,000 cubic meters in the first year while logging techniques are developed and local workers are trained.[46] The production would make Antimari a major supplier of timber for the 64 sawmills located in and around Rio Branco, thereby generating an economic base for the protection of the forest and to maintain the roads that would be required. This would bring a major improvement in Acre's highways, including the paving of BR-364 to Porto Velho, if wood products are to be shipped to out-of-state markets.

Palcazu, Chimanes, Tapajos and Antimari have not come up with all the answers to "sustained yield" forestry in Amazonia, but they are milestones on a jorney of discovery that will continue.

* * *

A chapter on forestry in Amazonia cannot fail to make a closing reference to the famous Jari project, a $1 billion investment in a cellulose plant in the Amazon wilderness by Donald Ludwig, an American shipping magnate. In its time, Jarí was one of the greatest political and ecological projects in Amazon history. Before the dust settled, Ludwig pulled out, losing most of his investment, but Jarí marched on under new management.

The ecologists were right in their condemnation of a massive deforestation in a pristine area on the border of Pará and Amapá, where Ludwig bought a claim covering 60,000 square kilometers after seeing it only from an airplane window. The ever-vigilant Philip Fearnside at INPA in Manaus was among the first to raise the international outcry over ecology that matched an internal uproar over Ludwig's high-handed style as foreign investor in Brazil.[47] It was politics, not ecology, that forced Ludwig to give up his 120,000-hectare plantation of pine, eucalyptus, and gmelina trees; a private railroad network; a pulp factory and power station barged from Japan to the Amazon site; and the company town built at Monte Dourado. Ludwig's penchant for secrecy made the venture seem to critical eyes like a private empire, and therefore a "threat to Amazon sovereignty." A nationalist campaign led the military government to withhold promised financing for a vital hydroelectric plant on the Jarí river and consolidation of the Ludwig's land titles. Ludwig, feeling double-crossed, pulled out.

Jarí was handed over for $400 million in long-term notes to a Brazilian private syndicate that has kept the project afloat. The consortium of new owners includes some of the biggest private firms in Brazil, including the company that had pioneered in managanese mining in Amapa, where it also had large pine plantations. Under the new management, it was found that Ludwig, who changed managers every few months, had imposed arbitrary decisions on the plantation foresters without any technical studies. His wasteful style extended to the spare parts for equipment that were flown to Jarí from abroad at prohibitive costs, instead of being repaired in shops at the site. Reforms were introduced, and Jarí gradually began to break even. In 1991, the owners reported that it had made a profit for the first time.

Jarí has been compared with the Henry Ford rubber plantation ventures at Fordlândia and Belterra in the Tapajós region that were heavily damaged by South American leaf blight. The failure of the Ford operation, incisively analyzed by Warren Dean, was mainly ecological, although there were also management problems (not only in the Amazon, but in Detroit where Ford lost the U.S. market leadership to General Motors during the same period). What Jari and Ford have in common was the megalomania of their sponsors, not their ecology.

The Jarí project's main error of destroying pristine forest to be replaced by monoculture plantations will not be repeated. Foresters who have studied Jarí's rich experience with different species have learned valuable lessons that are now being applied in reforestation at Jarí. These lessons are also being applied in replanting pastures, such as the CVRD plantations at Acailândia. The main conclusion is that Jarí can be self-sufficient in raw material with the introduction of species of eucalyptus that are adapted to the seasonal rainfall shortages and soil conditions. The other conclusion is that a 120-megawatt hydroelectric dam on the Jarí river, for which the company obtained approval from IBAMA after presenting an environmental impact statement, will eliminate native forest cutting for energy wood and significantly reduce production costs.

Production at Jarí of 280,000 tons a year of cellulose pulp and the biggest kaolin mine in South America earn $300–400 million in annual export sales. There are over 25,000 people at Laranjal, a town across the river from Jarí's plant, whose main income is from jobs in the plantations and the commerce that Jarí's activities provide. Jarí's ecological sins of the past remain, but the company is now operating in a responsible way. Ludwig's experiment was reckless, but the lessons are invaluable.

SIX

Sun, Water, and Oil

The discovery of abundant energy resources in Amazonia has been a powerful magnet for integration of the region into the national economies of Bolivia, Brazil, Ecuador, and Peru. The tropical forests have been pierced by pipelines and high-voltage towers that carry oil and electricity to users far removed from Amazonia and its environments. This energy flow has created an interdependence that never existed before. In some cases, Amazon energy has become the most important national export; in others, it is the basis for industrial and agricultural strategies outside the region. As such, Amazon energy has been enfolded into the economies and external relations of the countries of the region, with the exception of Venezuela and Colombia, which have less relative interest in Amazonia precisely because their substantial energy resources are outside the region.

Yet, despite its importance, the way Amazonian energy is produced and distributed has contributed little to sustainable economic development at the local level. Like other Amazonian activities, energy production has been a form of resource extractivism with little thought to what happens if it runs out. The energy sector has benefited distant users without bringing stable growth or sufficient compensation at the source. Energy is one of Amazonia's proven marketable assets. A fair share of its value, including environmental costs, should be recovered as a royalty or tax on energy sales and plowed back into the region's long-term productive capacity. Otherwise, yet another Amazon resource will be squandered.

Amazonia's energy resources are on a scale that was not imagined when economic modernization plans began to be studied in the 1950s.Considering only conventional technologies, the energy sources are a combination of nonrenewable hydrocarbon fuels (such as oil and gas) and renewable power, generated either from the flow of water through turbines or from solar energy captured in biomass (such as firewood and charcoal). The amounts of each type of energy that have been quantified are approximately as follows.

Hydrocarbons. The proven oil and gas reserves of the western Amazon basin, between the Putumayo river on Ecuador's border with Colombia and the Santa Cruz gasfields in Bolivia, are equivalent to 3.5 billion barrels of oil, including the huge Camisea gas field in Peru. At $20 a barrel, that is worth over $70 billion. The volumes are small by Middle East standards, but new discoveries are being made and great areas remain unexplored. The locations of the fields that have been found favor an Amazon basin energy exchange that has already begun between Bolivia and Brazil.

Hydroelectric. Over 44 percent of Brazil's national hydroelectric potential is in Amazon tributary rivers, where 35,000 megawatts could be installed at generating costs of up to $40 megawatts per hour (1986 prices), according to Eletrobras, the state power company.[1] Ninety percent of Brazil's electricity supply comes from hydroelectric sources, which provide more clean, renewable energy than the national daily consumption of 1 million barrels of petroleum. The national power development plan, in addition to striving for petroleum self-sufficiency, includes new Amazonian hydroelectric sites, where costs per kilowatt are cheaper than remaining sites in south-central and northeast Brazil. Despite the cost of long-distance transmission, economics points toward development of large sites in Amazonia, on the model of Tucurui, a high dam on the Tocantins river that generates 3,900 megawatts and can be expanded to 7,000 megawatts.

Biomass. Amazonia's 500 million hectares of forests contain, conservatively, 50 billion cubic meters of wood, which is half carbon. (Coal is over 80 percent carbon.) The fuel potential captured by the trees from solar radiation is therefore enormous—somewhere in the order of 10 billion barrels of oil[2]—but it is far from free. Gathering wood to a facility where it can be converted efficiently into useful power has proved to be much more costly than hydroelectric generation. On a large scale, it would also be environmentally ruinous. Wood-fired power plants capable of generating the same 35,000 megawatts as commercially competitive hydroelectric sites would over 25 years consume an amount of wood that would require clearing 700,000 square kilometers, or 14 percent of the Amazon native forest.[3] By comparison, the forest area lost to reservoirs for the equivalent energy output would be at most 50,000 square kilometers, or 1 percent of the Amazon forest.[4]

This scenario shows that large-scale utilization of native forests for the energy contained in the biomass is uneconomic as well as ecologically suicidal. But wood remains a highly important source of energy in Amazonia for local users, particularly in the form of charcoal, and research may turn up sophisticated ways to tap solar energy from biomass conversion in the Amazon rainforest. For now, there are strong arguments for stimulating reforestation of degraded Amazon areas that can produce energy wood without additional environmental cost. Plantations may be the most effective way to reduce pressures on native forest from charcoal consumers, particularly if many small wood producers become involved in sustainable forestry.

The modernization of Amazonia has moved in tandem with the development of its energy resources. Even after the nineteenth-century rubber boom brought Amazonia into contact with the modern world, the region was still in the dark ages as far as energy production. When the first five-year development plan for Brazil's Amazonia (1956–60) was drawn up, the total electric generating capacity in the region for a population of nearly 4 million in 1955 was 180 megawatts, half of which was consumed in just two cities, Belem and Manaus.[5] What this dearth of energy meant for the region was isolation, no communications, no television, no refrigeration, no industry, and, therefore, no economic development.

It was against this deprivation that the regional firebrands of modernization rebeled, demanding a national crusade for the "recuperation of Amazonia." Artur Cezar Ferreira Reis, a historian-politician who was superintendent of the first five-year plan, said the regional objective was "the development of Amazonia in a manner parallel and complementary to the Brazilian economy."[6] The main instruments of this integration were highways and energy. The necessary capital was to come from the national treasury.

The first five-year plan provided for a modest 50 percent increase in installed capacity, almost entirely from widely scattered, unconnected small thermal plants; but it also launched the first hydroelectric dam in the region, a 25-megawatt plant in Amapá that made possible a private manganese mine. In 1970, this mine provided 39 percent of all the foreign exports of the Brazilian Amazon region, which at that time totaled a paltry $77 million. Since then, installed hydroelectric capacity has zoomed to 4,800 megawatts, and industrial exports have climbed to $1.5 billion a year, mainly from energy-intensive metallurgy, such as aluminum plants. Moreover, surplus electricity in Amazonia now fills a critical need for power in Brazil's arid northeast region, where 35 million people live.

In Ecuador, a tropical export economy based on bananas, coffee, cacao, shrimps, and fishmeal was transformed after oil drillers made major discoveries in Amazonia in 1970. Production reached 320,000 barrels a day in 1990, and exports of about 165,000 barrels a day earned $1 billion which was as much as all other exports together. With oil, Ecuador is in the happy company of Venezuela, Mexico, and Colombia as a western hemisphere oil exporter; without oil, Ecuador is the prototype banana republic. The World Bank warns that the oil may last for only another generation, but by then Ecuador hopes to develop other energy and foreign trade alternatives. Biomass fuel may by then be an economic possibility.

In Bolivia, the exhaustion of mineral production and fertile soils in the Andean highlands, which were the economic mainstays of that poor country, has made the discovery of oil and natural gas in the eastern lowlands the nation's new economic hope. Gas exports are Bolivia's largest exchange earner by far, and it has enough oil to be self-sufficient for many years. A binational project with Brazil for a 600-kilometer gas pipeline would fuel a 500-megawatt electric power plant at Corumbá, on the border, as the first step toward a transcontinental gasline to São Paulo's industrial areas. This $2 billion project is critical for Bolivia's economic future, and that future is largely Amazonian. Energy is the key.

In Peru, Royal Dutch Shell found a huge Amazonian natural gas field in 1987 at Camisea, on the upper Ucayali river, that could have saved Peru from becoming an oil importer after years of being an exporter. Instead, negotiations with the Garcia government broke down in 1988, Shell pulled out, and the gas remained untapped while oil production plunged below national needs. From a high of 195,000 barrels a day in 1982, when Amazon oilfields were expanding, output fell below 120,000 barrels a day in 1990.[7] As reserves sank from 835 million barrels in 1982 to 330 million barrels in 1990, it was clear the decline was due to lack of drilling, not to lack of oil and gas in Peru's Amazonia. President Alberto Fujimori, elected in 1990, promised to bring back foreign oil developers to avoid a further slide into poverty.[8] As in most areas, Peru went backward in energy instead of forward during the 1980s, and Amazonia was hard hit by the decline.

From these developments, it can be seen that Amazonian energy can be discussed only in the context of national energy strategies. The oil and hydroelectric resources are too big to be treated only as local assets. Their development requires capital and technology on an international scale, and when they are brought into production they have impacts that go far beyond Amazonia. The level of decision on how these energy assets should be exploited has not been local at all, although the effects on local people and environments is powerful.

For instance, the 3,900-megawatt power station at Tucurui on the Tocantins river, the first major Amazon hydroelectric project, cost over $4 billion, which was largely financed by French loans, and it will take another $1 billion to reach full capacity of 7,200 megawatts. This power is fed through high-voltage transmission lines that extend 1,600 kilometers from eastern Pará to the electric grid of the Northeast region, where one-quarter of the Brazilian population lives. Hundreds of thousands of jobs depend on this power.

This exchange also influences peasant migrations into Amazon forests. When the Transamazon highway was announced in 1970, it was supposed to be so landless peasants from the drought-stricken Northeast could homestead in the Amazon. "People without land, for a land without people" was the slogan for the strategy of reducing rural poverty by outward migration from the Northeast. The colonization contributed to ecological damages in Amazonian forests that were denounced by environmentalists who argued that land should have been distributed to peasants outside the Amazon region instead of sending them to burn up the Amazon frontier.

A strategy based on agrarian reform, instead of colonization, runs into the problem, however, that much of the land that could be divided into family plots in the Brazilian Northeast is useless without water. Pumping water for irrigation systems requires lots of electricity, and demand for electric power has far outstripped supply in the Northeast. Water supplies for dryland agriculture cannot fail without disastrous crop losses, so local electrical systems have to be backed up by reliable base load production from a grid. Surplus Amazonian hydroelectric power is, therefore, part of the answer for keeping Northeast peasants on the land. The old slogan now runs in reverse: "Energy from a place without people, for people without energy."

There are environmental costs in damming up Amazon rivers, as there are in drilling oilwells and extending pipelines through tropical forests. These potential alterations vary from case to case, depending on the characteristics of the river basin and the size of the reservoir that is created, in the case of dams, and on the location of the hydrocarbon structure and the manner in which oil or gas is transported, in the case of petroleum. There are enough examples now of how these damages can be controlled through sound environmental management to be able to calculate the costs of protective measures as a factor in the overall investment. Usually, the most serious limitation turns out to be people who live on or near affected land or waterways.

For instance, when the 1,500-megawatt Itaparica dam was closed in 1987 on the São Francisco river in the Northeast, the reservoir of 840 square kilometers inundated the lands of 15,000 small farm families, with great political turmoil and high compensation costs. After this experience, two low-cost, high-priority sites included in the Eletrobras 2010 plan for Amazonia at Santa Isabel on the Araguaia river and on the Ji–Parana river in Rondônia were scratched when it was determined that the inundated areas, totaling 4,700 square kilometers, were already occupied by thousands of small settlers. The amount of power programed for the two sites was 1,160 megawatts. In contrast, the best potential site in Amazonia is at Belo Monte on the lower Xingú river, which would generate 6,300 megawatts in the first stage and 11,000 megawatts at full capacity, with a reservoir covering a rocky area of only 1,225 square kilometers where 344 people were living in 1988.

Decisions on Amazonian energy development involve a complex balance of socioeconomic factors in which the direct environmental impacts of projects usually are not the most difficult problem. What is hard to manage is the attraction that oilfield operations and dam construction projects exercise on migrants, who come to work and remain in the region. As Nelson Ribeiro observed when he was secretary of industry of Pará, these megaprojects produce a different form of human settlement than the preindustrial migrations when newcomers to Amazonia scattered in the forests in search of rubber trees. "They learned to live in the river-forest environment, like the Indians, without destroying nature, because their existence as extractivists depended on the waters, the subsoil and the forests," said Ribeiro.[9] In contrast, the new migrants of the industrial period not only clear land and exploit natural resources, such as timber and minerals, in an aggressive relationship to the forests and rivers, but also create demand for roads and urbanized communities with basic services, such as electricity, for home consumption, communication, and industry.

From the standpoint of sustainable development, it bears repeating that stable modern communities are what Amazonia most needs. The energy producers in the region play a special social role in community development because they have legal authority over wide areas, money, planning skills, environmental concerns, and political interests in the locations where they operate. Community development in areas around big dams and oilfields does not mean isolated company towns. It means watershed management, with concern for land use, control of deforestation, and protection of fishing in river basins where there may be many old towns and indian lands. It also means working with local communities, preferably organized in municipal development associations, to supply local energy needs in a way that helps users develop urban water systems, small industries, cold-storage plants, hospital services, and schools. It should also involve responsibilities for Amazon environmental protection financed by special taxes on energy generated in the region.

The megaprojects are catalysts for such dramatic changes in Amazonian conditions that the executing and operating enterprises have to become directly involved in managing regional development. Amazonian regions lack the capital, technology, and legal ownership of energy resources to undertake autonomous programs. The revenues from energy projects assigned to state and local authorities have fallen far short of the needs of new communities, and the use of the resources has been poorly planned. The least the Amazonian regions can do is get a fair share of the value of the energy generated in their territories and more technical assistance from the big energy producers in administering the growth of their communities. Some examples of recent experiences show the way.

* * *

There are no great lakes in Amazonia, such as the massive freshwater bodies of Lake Victoria and Lake Tanganyika that feed the rivers of East and West Africa. The South American heartland for the world's largest river system has no natural reservoirs, unless the whole spongy network of water, land, and vegetation is seen as a hydraulic holding system. The rivers plunging out of the Andes come down so steeply through narrow, unstable valleys that advantageous locations for dam construction are rare, such as the Mantaro dam in Peru.

Lower in the Amazon basin, the flat alluvial topography of what was once a great lakebed makes for meandering rivers with ill-defined central channals that drop less than a meter every 100 kilometers. In their natural state, therefore, most of the Amazon rivers do not offer good sites for generating electricity despite the great volume of water flowing in them. But there are important exceptions to these topographical and geological limitations, and at such locations engineers can build some of the largest power stations in the world. Such is the case of Tucurui on the Tocantins, the first high dam in Amazonia.

Tucurui was a light-year leap from Amazonia's permanent brownout to another world of high-voltage power. This Amazonian hydroelectric station is second only in Brazil to Itaipu, the world's largest, with 12,800 megawatts of generating capacity on the Paraná-Paraguay river system. Tucurui was the initial step in tapping the Amazon system under an energy strategy that was based on experience in Brazil's rivers of the South and Northeast. As has been noted, Tucurui was part of a national energy plan that integrated Amazonia into the national economy, both as a power source and as a producer of aluminum and other mineral products requiring large amounts of cheap energy.

This hydroelectric strategy, based on high dams and big central generating plants, has been questioned on both economic and environmental grounds. The capital costs of installing new electric generating capacity have risen sharply, and many analysts claim that investments in end-use energy efficiency and smaller, decentralized generating systems are a better way to go than increasing base load. Opponents of the Eletrobras 2010 strategy claim that energy from natural gas and nuclear power are desirable alternatives to damming great tropical rivers where ecological alterations are inevitable. Nonetheless, none of the alternative proposals excludes the need for further development of hydropower from Amazonia if Brazil's future power requirements are to be met, particularly in the Northeast region.[10]

In analyzing the alternatives, it must be kept in mind that Brazil, with its great rivers, is eminently a hydroelectric country, and Amazonia is the largest source of river water power left in the world. To avoid petroleum dependency, Brazil increased the electricity component of its total energy consumption from 20 percent in 1970 to 40 percent in 1990. This extraordinary shift was achieved by raising installed hydroelectric generating capacity in 1990 to 40,000 megawatts, which provides 90 percent of the country's electric power.

The hydroelectric expansion was mainly in the South, where the rivers flowing west from the states of Minas Gerais and São Paulo into the Paraná–Paraguay system provided cheap, abundant electric power for Brazil's industrial development and urban growth. The only big river in the Northeast, the São Francisco, was harnassed in 1959 by the 3,900-megawatt Paulo Affonso dam, which is the key to that watershed's power and irrigation potential. With the completion of Itaipu on the Paraná and Xingo on the São Francisco, all the large hydroelectric sites are gone, and only small locations remain, mainly in populated farming areas. Coal-fired plants are being expanded and nuclear power was given a try at three plants; but only one was operating in 1990, and antinuclear public sentiment is strong.

Eletrobras has a recognized bias in favor of big dam projects, and the best major hydroelectric sites still to be developed are on the rivers of Amazonia. These have a rated generating capacity of 60,000 megawatts (excluding the Amazon itself), or half of Brazil's remaining hydroelectric potential. The Eletrobras plan forecasts the need to increase installed electric generating capacity by the year 2010 to 70,000 megawatts. Some of this increase may be avoided by energy economies, but some Amazonian dams cannot be ruled out if Brazil is to continue to generate an adequate supply of low-cost power. Obviously, this is as essential for Brazil's export strategy as a producer of metals, chemicals, paper, and industrial products as cheap power is for competitors in Canada, Norway, France, or Australia.

Tucurui offers a good starting point to examine the economic, social, and ecological implications of harnessing Amazon rivers. It is a test case because Tucurui was the first high dam on a major waterway in Amazonia.

As is the way with large hydroelectric dams, the engineering and marketing studies for Tucurui began a decade before the decision was made to begin construction in the 1970s. Brazil was under military governments at that time, and significant policy decisions were made by closed circles of technocrats without benefit of congressional hearings or critical appraisals by scientific or ecological groups. The II National Development Plan (1975–79) simply announced that Tucurui would be in operation by 1981, along with four other dams that were to supply electricity to Manaus, Porto Velho, the territory of Roraima, and a big bauxite mine on the Trombetas river. In fact, by 1990 only the Balbina dam for Manaus and the Samuel dam for Porto Velho had been built.

Eletronorte, a state power company created just for Amazonia, went ahead with Tucurui after the federal government came up with a $2 billion loan from France, whose electrical industries received assurances of obtaining the contracts for Tucurui's generators and turbines. Eletronorte commissioned a consultant report from Robert Goodland, an ecologist familiar with the Amazon, who pointed out precautions that should be taken, but he agreed that Tucurui filled a "clear need" to increase Brazil's electric supply.[11] Only after all of this had been worked out, with construction and supplier contracts signed, did the opposition to Tucurui from environmental scientists and opposition political forces break out in full cry. At that point, it was too late to have much influence on the design of the project, much less halt it. Too much was riding on it by then.

The basic economic rationale for Tucurui was that it would generate the power necessary to implant an aluminum industry based on Amazonian bauxite. A high-quality mine had been located on the Trombetas river, 600 kilometers up the Amazon from Belém. The bauxite would be barged to smelters near Belém and to the deep-water port of Itaqui at São Luis, Maranhão, and transmission towers through the jungle would provide the electric power from Tucurui at low rates. Secondary goals were power for a new iron mine being opened at Carajás and urban power for Belém, where consumers and local industries were starved for electricity. The construction took three years longer than originally planned, and the cost has been estimated at $4.5 billion, representing substantial overruns. But when Tucurui's power began to flow in 1984, Brazil possessed an aluminum industry on a world scale that is Amazonia's major export.

At Rio Trombetas, the Rio do Norte Mining Company opened the world's third largest bauxite mine, with an annual capacity of 8 million tons. This mine is a $500 million joint venture between Cia. Vale do Rio Doce (CVRD), Brazil's state mining company, and five other large aluminum companies, including Alcoa, Alcan, and Shell-Billiton. In addition to 4 million tons of bauxite exports, the mine supplies 2.5 million tons annually to domestic smelters that have made Brazil self-sufficient and an exporter of aluminum worth $1.2 billion a year. The Albras smelter at Barcarena, near Belém, produced 165,000 tons of aluminum in 1988, and CVRD with its Japanese partners began an expansion of the refinery to 345,000 tons. At São Luis, the $1.6 billion Alumar smelter, a joint venture of Alcoa and Shell-Billiton, produced 245,000 tons in 1988 and began expanding to 340,000 tons.

The Amazonian aluminum industry, with investments of $2.5 billion, created more than 6,000 direct jobs and is a major source of tax revenues for Pará and Maranhão. The aluminum companies lost money in 1985–86 when bauxite and aluminum prices fell 50 percent worldwide, but prices recovered in 1988, leading to the strong expansion of Amazon aluminum operations. With these results, it cannot be argued on economic grounds that Tucurui was a misguided investment, particularly if its supply of electric power to the Northeast is considered.

Environmentally, however, Tucurui raised a lot of criticism. The reservoir formed behind the dam, rising 72 meters above sea level, inundated 2,430 square kilometers of forested land, including a partly demarcated indigenous area for some 500 Parakana Indians. The rising waters also displaced 5,021 families who were mainly recent settlers. Goodland's exhaustive study in 1978 had warned of the possible corrosive effects on water-driven turbines from chemical decomposition of the inundated forest, the disruption of fisheries for old riverside populations, and loss of wildlife and valuable Brazil nut trees, and he had recommended a forest inventory to interest loggers in a timber salvage plan. He also stressed watershed management by Eletronorte, in close cooperation with local populations, to protect Tucurui's reservoir from silting.

Eletronorte executives, under pressure to finish the job as fast as possible, paid little attention to environmental warnings, however, until a political storm broke over Tucurui as the waters in the reservoir began to rise. Clearing of the forest from the reservoir area had been awarded to a company formed by executives of a military pension fund, who had no lumbering experience but who had obtained a $100 million loan to do the job from the same French banks that financed the dam. When the timber-removal operation failed, the fund declared bankruptcy. Eletronorte decided it couldn't delay the two-year period for filling the reservoir, and it closed the gates. The press then reported that the waters were covering not only trees, but unremoved drums of powerful chemical defoliants left behind by the loggers.

Rumors spread all over the lower Amazon of poisoned drinking water and contaminated fish. In Belém, environmental groups got a court injunction—later overturned—ordering a halt to Tucurui's alteration of river flows, claiming that saltwater would invade the Amazon estuary. The Brazilian Society for the Advancement of Science (SBPC) adopted resolutions at its annual congress in 1983 condemning the military government and Eletronorte for "endangering Amazonia's ecology."[12] The Roman Catholic Indigenous Missionary Council (CIMI) said Eletronorte had betrayed the Parakana, who were forced out of their reserve area altogether by invading colonists. Displaced settlers and indians staged violent protests over delays in receiving new land or indemnities, as they had been promised.

The uproar went silent, however, after Tucurui was inaugurated in 1985. Out of the white mists that swirl at dawn over the green jungle, the sun rises now from behind a massive concrete barrier that spans the Tocantins river. Gone are the Tucurui falls, where old-timers tell stories of trading *machetes* for diamonds with painted indians in canoes. In their place stand a two-mile-long dam and a powerhouse with twelve generators, each spinning out 330 megawatts of electricity, not tall tales from a past that is no more.

A visit to the site shows that none of the disasters predicted for the project have taken place. No one was poisoned. There has been no increase of salinity in the Amazon estuary. The acidity of Tocantins water has remained neutral, with no corrosive effects from tree rotting because the flow of water (11,000 cubic meters per second) is so strong that the lake content is replaced 11 times a year. Fishing was disrupted during the initial filling of the reservoir, and riverside communities complain that there are more mosquitoes and the water doesn't taste the same as before. But the fish catch in the reservoir has grown to 140 tons a month. "Since 1986 the catch has been more abundant than it ever was before on the river and the fishing lasts all year, when it used to be seasonal. We now sell not only in Belém but to buyers who come from São Paulo and Brasília with refrigerated trucks," said Octâvio Farias Gonçalves, president of the Tucurui fishermen's colony. There are 2,600 licensed fishermen in four colonies around the lake, all registered for health insurance and pensions, something unknown before. Farias said they have a stake in keeping a "stable level" of fish in the reservoir and would prevent overfishing by "adventurers."

Timber extraction in the reservoir area has come to life. The reservoir created hundreds of islands and provided access by water to forest areas that had been unreachable for loggers. Tucurui's "alligator men" cut trees even under water, like Juarez Cristiano Gomes, who invented a waterproof chainsaw. The loggers dive to the lake bottom and cut hundreds of submerged trunks of mahogany, ipe, castanheira (Brazil nut), and other valuable trees that can easily be floated and towed down the lake to embarkation points. Sawmill operators with heavy trucks wait at waterside, cash in hand. In one quick deal I saw a mahogany log, 12 meters long and 1.20 meters thick, sold for $25 a cubic meter, when the foreign price for this wood was $900.

Much of the less valuable wood is turned into charcoal under contracts with pig iron foundaries at Marabá. This Tucurui charcoal comes mainly from sawmill residues, which are the main source of energy wood for the foundaries. This has reduced pressure for cutting primary forest for charcoal. The multimillion-dollar timber-clearing operation that failed before the dam was closed is now being done spontaneously by hundreds of Amazon loggers, with their own capital and technology. It is a good example of Amazonian problem solving by private initiative.

The Tocantins river, the inundated forest, and the area around the lake—occupied by thousands of ranchers and settlers—have undergone irreversible changes. In a study of habitat disturbance at Tucurui for World Wildlife Fund–USA, Andrew Johns presented evidence that many monkeys (eight species were found at Tucurui) were trapped by rising waters and probably drowned. Feeding grounds for some of the 200 recorded bird species were lost. An unmeasured loss of floral species was symbolized by the tens of thousands of Brazil nut trees that were submerged, producing an uncompensated economic loss for the region's extractivists.

A last-minute "faunal salvage" operation, financed by Eletronorte to counteract bad publicity over drowning animals, saved 12,296 howler monkeys and other species that were caught and released in forests around the reservoir. This was good public relations, but Johns concluded that the $30 million spent on the operation could not be justified because "the vast majority of animals escaped unaided [increasing populations in lake fringe areas] . . . which suggests that releasing captured animals on the lake shore is worse than useless." Johns suggested that the monkey money would have been better spent creating and managing forest reserve areas protecting the animals, the flora, and the reservoir.[13] This Eletronorte has not done, leaving the reservoir unprotected against deforestation and siltation.

Yet, whatever value is assigned to the loss of Brazil nut production, or to the reduction of wild game for local hunters, or to genetic impoverishment through loss of endemic species (for which there is no inventory), the combined economic and ecological balance sheet does not, by any reasonable measure, make the Tucurui dam an environmental disaster. The generation of electric power and the activities that have sprung up around the dam site are sustainable, without further ecological alteration, if

Eletronorte and state forestry authorities adopt measures to protect the Tocantins watershed against deforestation. Dryland logging and agriculture taking place on the fringes of the Tucurui reservoir and on large islands in the lake are producing noticeable erosion. Similar deterioration is visible along many rivers in the region, particularly those where gold mining takes place.

As the regional power utility, Eletronorte should be vitally interested in preserving watersheds on which its generating capacity depends. Laws to protect waterways exist, but funding is minimal and enforcement has been weak. A "green" tax on electric power from Amazonia should finance this environmental protection under the responsibility of Eletronorte, in close coordination with federal, state, and, above all, municipal authorities in its areas of operations.

The direct environmental impacts of Tucurui are not as hard to manage, however, as the serious social problems created by the project. Eletronorte's engineers and technical staff live in a well-groomed company town on a bluff overlooking the river, where lakefront villas are surrounded by dairy farms, vegetable plots, and fruit orchards producing for the local markets. But this orderly town is an enclave of 6,000 people in the middle of a demographic explosion. The townships of Tucurui, Jacundá, and Itupiranga, which ring the lake, have a population of 250,000 where a 1977 census counted only 3,500 people. "When the dam was being built, there were 30,000 jobs and people came with a one-way ticket," said Armenio Barreirinhas, the first elected mayor of Tucurui township.

Barreirinhas, an enterprising Portuguese immigrant, became wealthy while the dam was being built as the owner of Tucurui's first filling station. But as mayor, he found the municipal treasury had very little money because Eletronorte's charter exempted it from payment of most local taxes. "If they paid what they should for the electricity generated here this would be the richest municipality in Brazil," said Barreirinhas, whose dreams of more revenue began to came true when the new Brazilian constitution of 1988 increased the municipal share of general taxes, and established a royalty payment for power generation. Barreirinhas said his main priority was a new urban water system. The existing system was built for 4,000 people, but the town population had reached 40,000 people in 1990.

Most of these people are migrants who came to work on the dam. With the end of heavy construction came unemployment, crime, and beggars in the town. Except for fishing and lumbering, there were few new work opportunities. One exception is an energy-intensive industry based on silicon. Camargo Correa, the prime contractor for the dam, had uncovered an ancient riverbed containing acres of egg-size pebbles of high-purity quartz. The pebbles are now the raw material for a factory east of the dam that produces 36,000 tons a year of metallic silicon in four high-voltage furnaces. Shiny metal sheds containing the furnaces are surrounded by a

15,000-hectare plantation of eucalyptus that supplies charcoal used in the reduction process. The product is a lustrous gray metal used in aluminum engine blocks and chemical industries. Major buyers are from the United States, Japan, and western Europe. The silicon plant was an unforeseen spinoff of Tucurui that provides a foretaste of the metallurgical industries that are expected in a region where minerals, biomass, water, highways, railroads, and electricity are all in one place, as is described in the next chapter.

Another set of small Eletronorte dams respond basically to local energy demands. This is the case of Balbina, a hydroelectric dam across the sluggish Uatuma river 175 kilometers north of Manaus. No Eletronorte project has been as roundly condemned on both economic and environmental grounds as Balbina. The plant cost $700 million and can generate only 250 megawatts from a shallow reservoir that inundates almost as much forest as Tucurui. It would have been cheaper to build a thermal plant with the same capacity with no loss of forest. Eletronorte officials admit that Balbina would not have been constructed under present environmental impact criteria for Amazonia, but it was built to satisfy the political demands of the state of Amazonas for electricity for the then emerging industrial district in the Manaus "free-trade zone," a tax paradise where light industry and electronic assembly plants provide 50,000 jobs.

Balbina went ahead over the militant objections of an alliance of environmentalists, church groups, and anthropologists whose cause was the Waimiri–Atroari Indians, a once fearsome tribe now reduced to about 2,000 people. Their tribal lands along the Alalaú river have been divided by an interstate highway, redrawn to make room for a mining company, and partially flooded by the Balbina reservoir. A manifesto signed by international and Brazilian environmentalists, including José Lutzenberger, later President Collor's environmental secretary, appealed to the World Bank to "take emergency action with the government of Brazil to stop the flooding."[14]

Despite the protests and "scientific" predictions that the reservoir would not fill and that polluted water would kill the fish, Eletronorte closed the floodgates and the reservoir filled up as scheduled. In February 1989, the Japanese-built turbines began to generate electricity for Manaus, a power-short city with over 1 million inhabitants. There was no unusual fish mortality on the Uatuma, where about 100 families live. Gilberto Mestrinho, the Amazonas politician most identified with the Balbina project, successfully ran for governor in 1990 campaigning against the "ecologists." Mestrinho declared that his loyalties were "with the working people of Amazonas, not with the alligators."

In 1991 Mestrinho went a step further in challenging the environmentalists. "Amazonia is not untouchable, it is not a museum," proclaimed Mestrinho[15] as he convened a conference of eight Amazonian governors in Manaus to draw up a regional strategy against zoning limitations on use of nonfederal lands. Most of the governors felt Mestrinho had gone too far; Governor Jader Bardalho of Pará said he was in favor of creating large conservation units and exchanging them for a reduction in Pará's $50 million foreign debt. But Mestrinho's campaign won public applause from high-ranking army commanders in Amazonia and Brasília who denounced foreign "interference."

Environmentalists and defenders of tribal people lost the battles of Tucurui and Balbina, but the war continued. The same week that the first turbines began to spin at Balbina, a conclave of international celebrities including the rock musician Sting, folk singers such as Canada's Gordon Lightfoot, environmental activists from the world over, and tribal leaders in warpaint and feathered headwear were gathering at Altamira, where the Transamazon highway crosses the Xingú river. For two days, with intense media coverage, Altamira was where the action was for the world tropical rainforest movement.[16]

The Xingú, a powerful river that begins in Mato Grosso and runs 1,800 kilometers to the Amazon, is in Kayapo Indian country. The Kayapo are a bigger, better organized, and politically more skillful group than the other tribal people Eletronorte encountered in the past. After Tucurui, Eletronorte's best site for large-scale power from Amazonia is at Belo Monte on the Xingú river, just north of Altamira. There, the Xingú falls 40 meters in a U-turn over rocky rapids that have historically protected the interior from upriver navigation. A dam rising 30 meters above the river's bedrock would create a reservoir covering 1,225 square kilometers, only half as much as Tucurui, but generating 50 percent more electricity. At an estimated cost of $5.8 billion, twenty 550-megawatt generators would produce 11,000 megawatts of cheap power by the year 2005 for transmission to power-hungry southern Brazil, explained Eletronorte's president Miguel Rodrigues Nunes in 1989.

This high dam, known as Belo Monto or Cararao, would extend the national power grid 400 kilometers deeper into the Amazonian forests and create a new pole of industrial activity at Altamira, midway between Marabá on the Tocantins river and Itaituba, a gold-mining center on the Tapajós river. This project would probably reactivate the Transamazonian highway, now in serious disrepair but still the main overland route between Altamira, Itaituba, and Santarem. The energy would probably be extended to Santarém, which would become an industrial center and transport terminal at the mouth of the Tapajós in the mid–Amazon. As has been pointed out earlier, this is a strategic area for ecological-economic zoning and forest protection, and the Xingú is where the future of eastern Amazonia will be decided. The decision may be governed more by financial limitations than by strictly environmental considerations.

Since Tucurui became a spillway for controversy, no new dams in Amazonia can go forward without running a gauntlet of scrutiny from economists, alternative-energy promoters, environmentalists, defenders of

Indians rights, and critics of "developmentalism" as a threat to the rainforests. The criticisms from these sectors weigh heavily on sources of international financing for expensive energy investments, such as the Xingú project. The financial element is what has given environmentalists the leverage to force energy developers to consider more than kilowatts in their project proposals. The higher the cost, including environmental protection, the harder it is to finance the project.

Brazil began negotiations with the World Bank in 1986 on a $500 million loan for its energy sector. The proposal initially involved financing for four Amazonia dams plus studies of others, including the Xingu project. Resistance to damming the Xingú had begun to develop among Kayapo Indian leaders who were in touch with NGO environmental groups around the world. This carried Brazil's plans for hydroelectric development in Amazonia to the halls of the U.S. Congress, where it met a hostile reception.

Senator Robert Kasten, a Wisconsin Republican, who was chairman of the powerful foreign operations subcommittee of the Senate Appropriations Committee, became the point man for stopping the loan. Kasten's committee controls U.S. funding of international development banks, and environmental NGOs have cultivated his staff. In 1986, Kasten, whose voting record is against foreign aid in any form, notified then Treasury Secretary James Baker III that the U.S. Congress was not prepared to allow "another half billion dollars to be wasted on environmentally unacceptable projects."[17] The U.S. executive director was ordered to vote against the loan. But the proposal resurfaced in 1989, with the addition of an environmental "master plan" for Eletrobras projects.

The World Bank has never directly financed hydroelectric projects in Amazonia, but it loaned Eletrobras $500 million in 1985 to finance generating and transmission projects outside of Amazonia while Eletronorte pushed ahead with Tucurui, which the World Bank had turned down, with French supplier credits and private bank financing. Kasten moved to plug that hole.

The new $500 million was supposed to be the first in a series that would later finance long-term electric power needs, as foreseen in Plan 2010, including the Xingú dam at Altamira. Kasten again attacked the whole package on environmental grounds, relying on information provided by environmental NGOs in Washington who had taken up the Kayapo cause on the Xingú dam. "This loan is like a Macy's department store of environmental disasters," said David Wirth, legal counsel for the Natural Resources Defense Council, a major U.S. environmental group. In January 1989, Kasten wrote a threatening letter to Barber Conable, Jr., president of the World Bank, claiming that the Eletrobras projects would displace thousands of tribal people for whom there were no resettlement plans. Kasten warned that there would be "a storm of controversy on Capitol Hill and in the world" if the bank went forward with the loan.[18]

These charges were backed up by Paulinho Paiakan, a young Kayapo leader who visited Washington and denounced the dam project as a threat to his people in meetings with Senate staffers, World Bank representatives, the State Department, and the Treasury. He then went on a swing of western European capitals with the same message and finally called a meeting of Brazilian tribal leaders at Altamira, which became the occasion for the international "happening" on the Xingú. The Indian leaders recalled that other hydroelectric reservoirs had flooded indigenous areas, such as the Parakana at Tucurui and the Waimiri–Atroari at the Balbina dam near Manaus. They denounced the Brazilian government for inaction on punishing invaders of Indian lands, such as 50,000 goldminers who overran the Yanomani reserves in Roraima. They demanded prompt demarcation of tribal lands that were recognized as traditional homelands in a new constitution adopted in 1988. As for Eletronorte's electricity plans, they turned down any hydroelectric development of the Xingú.

The Eletronorte reply was that the Altamira project would provide better attention to community interests than Tucurui had done, and officials presented plans for turning Altamira from a frontier town into a modern city, with benefits to the rural areas. As for Indians rights, Eletronorte said it had conducted a survey of the reservoir area that showed only 344 persons would have to be relocated, among which 286 were Indians who were living outside any indigenous areas claimed by the Kayapo.[19] It was further stated that Eletronorte did not intend to develop the Babaquara site on the Xingú, upstream from Altamira at the confluence with the Iriri, which would affect the northern end of a Kayapo reserve. There are 14 Kayapo communities in reserve areas on both sides of the mid–Xingú that cover over 3 million hectares from Gorotire to the Xingú national park in Mato Grosso where some Kayapo live. The nearest reserve to Altamira is 150 kilometers away.

Walking the red mud streets of Altamira, one sees almost no Indians. The few to be found are at a trading post for native articles on the outskirts of town. When one talks with people at the bustling markets, cacao warehouse, busline terminals, government agencies, and rural workers' unions, it is hard to find anyone in this frontier township of 40,000 people who does not support the Xingú dam.

"We are all waiting for the dam," said José Fernandez Canales, who sells fish at the municipal central market where one sees the typical Brazilian frontier mix of blond-haired migrants from southern Brasil and straw-hat peasants from the Northeast. Fernandez has a cold storage plant and supplies ice to eight boat owners, including a strapping young Kayapo who bring him fish caught up the Xingú by local fishermen. "From what we have heard about Tucurui, there will be no lack of fish, and there are going to be a lot of hungry construction workers when the project begins," said Fernandez. He came to Altamira in 1981 from Rio Grande do Sul, where he was a wheat farmer, and now owns a 500-hectare ranch with

150 head of cattle at a location near Altamira overlooking the majestic sweep of the river. "This place has a great future," he said.

Eletronorte has had an advance project team at Altamira for several years preparing a "municipal plan" to go along with the dam construction that foresees an Altamira population of 160,000 in the year 2010. "We will build a construction villa outside the town to house workers, but the town has to be reorganized. It will need more local food supplies, a new transportation system, electricity service and more housing, or there will be a collapse," said an Electronorte engineer. Who would pay for this was not clear.

The Kayapo Indians are not without influence in Altamira, however. They are backed by the Roman Catholic bishop in Altamira, Edwin Krautler, whose uncle, an Austrian missionary, was the first bishop of the Xingú, the biggest diocese in the world. Krautler followed in his uncle's footsteps and became head of the Brazilian Indigenous Missionary Council (CIMI), a powerful arm of the Roman Catholic National Council of Bishops.

The Kayapo also have close ties with Brazilian and foreign scientists, such as anthropologist Darrell Posey, who came from the University of Georgia to Kayapo country in 1977 to do research on insect classifiction. "The Kayapo can identify 56 species of bees, of which nine were unknown to western scientists," said Posey, who is from the bluegrass farmlands of Kentucky where the Cherokee and Shawnee Indians once roamed before the pioneers who followed Daniel Boone's trail into the territory drove them out. Posey became an authority on Kayapo ethnobotany at the Goeldi Museum in Belem, where he organized a magnificent Kayapo exposition that presented the environmental lore of these Indians on soil management, medical uses of forest plants, and their cosmology, inspired by the beehive, in which man and nature live in harmony.

There was no harmony between Posey and the Brazilian government, however, after he accompanied Paiakan and Kube-i, another Kayapo, on their trip to Washington and translated into English Paiakan's attack on the World Bank energy loan at a televised press conference on Capitol Hill. On their return to Brazil, Posey and the two Kayapo were summoned to testify before a judge for "interfering in Brazil's internal affairs." Charges were dropped later, but Posey lost his job at the Goeldi Museum and left for Europe on a research fellowship. The last laugh in this conflict may well be with Posey and Paiakan, however. Their story was turned into a screenplay by Mike Austin, who did the script for "Greystoke, The Legend of Tarzan," and the project has been described as an "environmental cult" movie on the Kayapo.[20]

The Kayapo of Pará are a tenacious tribe of about 3,000 who are the survivors from a population of about 30,000 a century ago, according to Posey. A traditional wing of the Kayapo led by Chief Raoni of the Txucarramae clan live in the Xingú National Indian Park, a forest reserve for Indian tribes. Raoni accompanied Sting on a worldwide tour to raise funds for creation of a new Kayapo reserve; they were received by Pope John Paul, Prince Charles, and President Mitterand. President Collor has visited Raoni at the Xingú park and signed decrees giving the Kayapo part of the expanded reserve.

Another, more militant Kayapo group lives in a dozen villages around Gorotire, east of the Xingú. The Gorotire community, where Paiakan is a young leader, wants full control over an indigenous area that includes all the upper Xingú watershed. Only part of this enormous forest claim covering 30,000 square kilometers has been demarcated, and there are conflicts over land rights with large ranches in southern Pará. The Kayapos are men of action, and they dominated by arms an invasion by gold prospectors at the Maria Bonita and Cumaruzinho placers on the Fresco, inside their reserve. With mediation of the dispute by the National Indian Foundation (FUNAI), a deal was struck whereby the miners stayed and the Gorotire Kayapo receive a 10 percent royalty on gold extracted, of which 5 percent supports FUNAI services and 5 percent goes directly to the tribal council. This revenue amounts to hundreds of thousands of dollars annually, according to Salamão Santos, the FUNAI district chief in Belém.

Suddenly rich, the Kayapo bought diesel generators to light their villages, where they watch national television in modern brick-and-mortar houses. For a time they rented an airplane, with a pilot, to fly back and forth to Belém. Some own pickup trucks and drive to Imperatriz to sell gold and buy supplies. Gorotire has a communal cattle ranch. Some Kayapo chiefs, such as Toto Pombo, are loggers and gold traders who are critical of Raoni for his traditional outlook. Paiakan drives a new car during his visits to Belem, where he once worked for FUNAI, and he has taken up filmmaking with a high-tech video camera. "When I began to work with the Kayapo in 1973, they were treated like children. A lot has changed. Now they quarrel over money. They have lost their innocence," said Santos.

The younger Kayapo and other acculturated Indians in Amazonia are obviously interested in more than the birds and the bees. In their relations with the encroaching "national societies," they want their rights to land and their cultural heritage protected; but in most cases they also want access to what they find attractive in modern society, from antibiotics to television. World concern over the Amazonian environment has given the tribal people support they didn't have before, and they have learned how to use it to advantage. It is important to understand, however, that the Indians see their reserves not only as an environmental defense, but as part of their larger social and political objective of obtaining legal rights to property and resources that will give them security and a fair share of the economic progress taking place around them. Their goal is not preservation of the primitive, but respect from the national society. Mining, logging, and dams are negotiable.

In the modern energy age, hydrocarbon fossil fuels such as coal, petroleum, and natural gas are a more important source of primary energy worldwide than is hydroelectric power. The search for oil in South America has disclosed a geological seam rich in hydrocarbons where the ancient tectonic plate that broke away from Africa collides with the newer Pacific plates that have pushed up the Andes mountains. From Tierra del Fuego to the Caribbean, the sub–Andean eastern piedmont and the adjacent plains have produced major oil and gas deposits. In Amazonia, the main discoveries have been in eastern Peru, Ecuador, and Bolivia, and more recently in western Brazil's Juruá river basin. The Andean countries are net exporters of energy, led by Venezuela and Colombia; Brazil, which has the biggest national market, is a net importer. Therefore, a potential market for energy exchange exists within the Amazonian region.

The best alternative to hydroelectric power within the region may be natural gas, which is clean and suitable for efficient decentralized generation of electricity. Bolivia has been exporting natural gas to Argentina for many years from Santa Cruz, and expanding production (reserves are 1.5 billion cubic meters) has made enough gas available to turn toward the larger Brazilian market through a pipeline to Corumbá in southern Mato Grosso. The Bolivian gas would fire a 500-megawatt thermal power plant and provide raw material for a fertilizer industry. Electricity and fertilizer are in demand on the expanding agricultural frontiers of Mato Grosso's cerrado lands.

The significant growth of population in Amazonia and new sources of oil and electricity in border areas increase the possibilities for energy exchanges. The presidents of the member countries of the Amazon Pact talk about regional integration whenever they meet. An Amazonian energy market would be one of the components of a regional trading area. But except for the Bolivia–Brazil deal, nothing has been done. Peru probably holds the key to potential exchange with its large undeveloped oil and gas fields in the Amazonian region.

Since exploration began in Peru's eastern "*selva*" in the 1930s, oil and gas have been found in the Ucayali river basin, in the Huallaga basin near Pucallpa, in the northern sector of the Santiago and Marañon rivers, and most recently in the southern sector along the divide between the upper Ucayali and the Madre de Dios river basin east of Cuzco. This new discovery by Shell at Camisea is estimated to hold 450 billion cubic meters of gas and 970 million barrels of condensates. It cost Shell $200 million and five years of exploration before the first successful well, and it would cost an estimated $1.8 billion more to develop the field and build pipelines to the Pacific coast. Camisea is still a sleeping beauty, because of the political problems already described. Until agreement has been reached with Shell or other investors, the most important project in the history of Peru's oil industry will hang fire.[21]

Peru's governmental affairs were so chaotic under President Garcia that oil production fell below export levels while the finance ministry withheld payments to the oil companies that were trying to increase output. Price controls on refined fuels decapitalized Petroperu, Peru's state oil company, which was unable to expand activities for lack of financing. A Peruvian government capable of implementing economic development would put a high priority on the proven oil and gas potential of the "*selva*" on which the country's energy supplies, and the eastern region's prosperity, depend.

When negotiations with Shell were suspended in 1988, Peru invited Latin American national oil companies to form a consortium to finance development of the field. Only Petrobras of Brazil, which has an international exploration program, showed interest. The Brazilian market would be a possibility for the Peruvian gas particularly if a pipeline were linked to a highway that has been discussed from the Brazilian state of Acre to the Pacific coast, passing close to the Camisea field. The right of way would have to cut through dense forests east of the Ucayali, but gas-fired electric power generation in western Amazonia would be an environmentally desirable alternative to hydroelectric dams on the Madeira river system, for which Eletronorte has a study.

In the recent past, petroleum has led to armed conflicts in Amazonia. Peru and Ecuador fought a war in 1941 over control of the lower Napo and Putumayo rivers, where oil had been detected. The conflict ended with Peru's occupying Guayaquil, Ecuador's main port on the Pacific, and depriving Ecuador of direct access to the Amazon. Subsequent discovery of large amounts of oil in the Amazon regions of both countries contributed to an arms race. Recurrent border "incidents" sparked outbreaks of nationalist fervor and movements to populate their Amazon territories. The perennial conflict (Ecuador has refused to recognize a border negotiated in 1942 with mediation by the United States, Brazil, Chile, and Argentina) is an obstacle to regional cooperation.

Since the Middle East war in 1973 demonstrated the political vulnerability of the oilfields of that region, oil companies have been hedging their bets by searching for new sources in the western hemisphere. Mexico, Venezuela, and Colombia are the established leaders, but a lot of interest has been shown in exploring the less-known Amazon regions of Peru and Ecuador. Occidental in Peru and Texaco in Ecuador have been successful and are established producers, and their example has brought in at least 20 other companies besides Shell. Maps of the eastern regions of Peru and Ecuador are offered to bidders in blocks of up to 250,000 hectares that cover the Amazon region like a counterpane quilt. In Peru, from Occidental's Block 1-AB to Shell's 42, where Camisea is located, to Mobil's 53, where Shining Path guerrillas attacked the camp in 1990, the "*selva*" is all up for exploration. In Ecuador, the petroleum map offers more than 20 blocks that go from the Putumayo river on the northern border with

Colombia to the Pastaza river to the south, right up to the disputed border with Peru.

Ecuador became the smallest member of OPEC (Organization of Petroleum Exporting Countries) just before oil prices began to rise in 1973. This brought boom times to this small Andean country as production rose to 325,000 barrels a day. But in 1987 Ecuador suffered a major setback when an earthquake wrecked the pipeline Texaco had built to bring oil out of the Amazon region to the Pacific coast for export. When the pipeline was restored after a six-month interruption, the world oil market was glutted and prices were down 50 percent from what they had been in 1987. But Ecuador was hooked on oil, and the desperation to increase output led to offering even more exploration contracts in the Amazon region to foreign companies. There was strong interest from state companies from Brazil, Spain, Chile, Italy, France, and Canada, as well as international giants like British Petroleum and Atlantic Richfield.

Probably the most successful of these exploration sites has been Block 16, where Continental Oil (Conoco) of Houston, a company owned by the chemical giant E. I. Du Pont of Delaware, is the developer. By 1990, eight wells had been drilled and seven struck oil, establishing a commercial field with an estimated recoverable 250 million barrels of heavy oil. That was good enough for Conoco, which had invested $90 million in Block 16, to decide to build a pipeline that would deliver oil to the Transandean pipeline. The catch was how to build a pipeline into Block 16 without opening a land route that would bring a flood of migrants into the biggest Amazon tropical forest reserve in Ecuador. Because it turns out that, if the petroleum map is superimposed on Ecuador's map for environmental protection areas, Block 16 is right in the middle of Yasuni National Park.

For environmentalists, Yasuni is a symbol of resistance to the overwhelming influence of oil in Ecuador's Amazonian region. It is claimed that Yasuni is "the most biologically diverse region on our plant's land surface."[22] This is how the park is described by the International Union for the Conservation of Nature (IUCN), which claims that Ecuador's tropical forests contain 10 percent of all the floral species in the world.

> Yasuni National Park has an area of 750,000 hectares that is representative of the Ecuadorean Amazon. . . . it lies within an area thought to have been a Pleistocene refuge for butterflies, forests, lizards and birds. . . . the fauna includes populations of crocodile, alligator, boas, ocelots, jaguars, peccary, tapir, capybara, deer, river dolphins, primates and important sites for migratory species of birds. Since a jaguar requires at least 10,000 ha, only 100 jaguars can exist per million hectares. This emphasizes the need for large areas of virgin rainforest to be protected for all these species to survive in equilibrium.[23]

Is this any place to put an oilfield? ask outraged environmental activists. Anthropologists protest that in and around Block 16, between the Napo and the Pastaza, are the homelands of tribal people, such as the Waiorani, who are famous for their deadly blowpipes and attacks on missionaries. Ecuadorian governments in the past have paid little attention to these recriminations, and Petroecuador does not exclude even national parks from exploration and development, if oil is found. Before Yasuni National Park was created in 1979, the Cuyabene faunal reserve near Shushufindi, covering 254,700 hectares along the Napo river, was thrown open to a consortium of Texaco and Petroecuador, the state oil company. Roads led to invasion by at least 1,000 colonists of the western half of the reserve, where Siona and Secoya Indians had their hunting grounds. These have been occupied and thousands of hectares of forest land have been cleared for small farms.

With environmentalists in full cry over Yasuni, Conoco contracted an ecological consultant, Douglas McMeekin, to help draw up a working plan for Block 16 that would contain the damages. Conoco's exploration work had been carried out with helicopters supplying the camps and drilling sites, avoiding opening roads. But the pipeline was something else, because of the heavy cost of putting it in without a roadway, which also would be needed for maintenance. Conoco submitted an operation plan in mid-1989 that required Petroecuador to discount Conoco's costs for environmental protection, and put up $30 million to pay for the pipeline from Block 16 to the main trunk. The Conoco project came to a halt while Petroecuador pondered.

"We are doing the best we can under difficult circumstances. We are under contract to Petroecuador," said Alexander B. Chapman, director of Conoco's international exploration and production services. But David Neill, assistant curator of the Missouri Botanical Garden, was skeptical after a visit to Yasuni. He wrote: "There are many uncertainties about the fate of Yasuni park—if the ecosystem is not protected, the remaining wilderness of the entire Ecuadorian Amazon will also disappear within a few decades."[24] National Resources Defense Council and Rainforest Action Network unleashed an international barrage. *World Rainforest Report*, an ecological hotline publication, said the road would open the area to colonization, gold extraction, and water pollution that had "proved lethal to the indigenous cultures and ecosystems of Amazonia."[25] This is the all-or-nothing position: all ecology and no oil. But McMeekin observed, "if Conoco decides to pull out there are a lot of other companies waiting to move in that don't have the same concern over environmental protection."

There is a more fundamental problem in Ecuador's Amazonian oil development, however, than the fate of one park. At the time Yasuni became an issue for the environmental network, Ecuador was asking the World Bank for a $100 million loan to finance Petroecuador's oil development program. Under World Bank guidelines for loans affecting tropical forest areas, this application required an environmental impact assessment.

A World Bank team was sent to make such an evaluation, and the report that emerged put the oil loan on hold.

A study published in 1987 on the basis of satellite images covering the most heavily settled portions of Ecuadorean Amazonia had shown that 45,270 square kilometers of forest that was intact in 1977 had been reduced to 39,876 square kilometers in 1985. That was a 12 percent loss in primary forest in eight years. It was clear from the images that the routes of destruction were the highways and roads that made up the transportation infrastructure for the oilfield. The World Bank team led by James F. Hicks, an economist from the energy division, accompanied by Herman Daly, an environmental economist, examined Ecuador's Amazonian development strategy, based fundamentally on all-out oil development and colonization, and concluded:

- Oil reserves would probably be exhausted in 25 years so it was time to think about alternative investments.
- Ecuador's population (10 million in 1989) would double in 25 years if it continued to grow 2.8 percent a year. If even half of this increased population were sent to colonize the Amazon region, this demographic pressure would "irreversibly destroy" the Amazonian ecosystem's renewable resources and rich but fragile biodiversity.
- To avoid an ecological and social disaster, Amazonia's contribution to Ecuador's long-term development required a new strategy in which agriculture and cattle ranching would be limited, and emphasis would shift to "the apparently great potential in forest products, including selective and renewable commercial wood and production of wild fauna and flora that have been little investigated but have an attractive commercial potential."[26]

The report recommended the drawing up of an agro-ecological zoning map for the Amazon region that would serve as the basis for protection areas where the forest should remain intact because of the biological diversity and presence of tribal groups, as is the case at Yasuni. The zoning would also define areas best suited to commercial forestry and agriculture. This pessimistic diagnosis of the future of Ecuador's Amazonia was too drastic, however, for Ecuador's political leadership to swallow. President Rodrigo Borja, a moderate Social-Democrat elected in 1990, called vaguely for a review of Ecuador's environmental legislation, but went looking for private foreign investors to continue the oil development policy, regardless of the World Bank loan. Oil companies remained eager to bid on new concession areas, including Conoco, which signed another $84 million exploration contract in 1990 for a new Amazon block south of Block 16 in a consortium with Murphy Oil and Canadian and Taiwanese investors.

By then, however, the World Bank had its foot in the environmental door. Negotiations on the oil loan had been broadened to include a long-term strategy for development of the *oriente*, as the Amazon region is called, that put heavy emphasis on environmental protection; and a $10 million technical assistance project to upgrade Ecuador's national environmental administration was linked to the oil loan.

Ecuador's four Amazon provinces, including the sub-Andean valleys, cover nearly half of the country's 275,000 square kilometers. By Amazonian standards, Ecuador's share is not large, but for Ecuadoreans it has unusual importance: politically, because Ecuador still hopes to recover part of the territory it lost in the military conflict with Peru, which deprived Ecuador of access to the Amazon river; and economically, because while the oil lasts, the *oriente* holds the key to Ecuador's growth.

Before oil was found, there had been little colonization of the Amazon region. After Texaco struck oil in Napo province in 1972, the rush into the area of oil exploration crews and road builders drew thousands of workers from coastal and highland Ecuador and from Colombia. Pipelines and highways over the Andes soon followed. Towns and airports sprang up where only missionary outposts and Indian villages had been. The virgin forests were invaded by settlers who had only to clear the forest to claim the land as theirs. This took place without any fiscal incentives, or directed colonization, as in Brazil, but the result was the same.

The *oriente* population soared from 46,000 in 1959 to 173,000 in 1974 and 258,000 in 1982 (the last census). It was estimated at 400,000 in 1990. Half of this population was in Napo province, the center of oil production. Facing the invading settlers were 60,000 native Indians. To hold on to their lands, many of the tribal people were forced to give up their traditional way of life in the forest as hunters and nomad farmers and become settlers in land colonies, with individual lots.[27] Some became small cattle ranchers, others grow coffee, and one Indian community, Guayusa, has planted African palms to supply a new vegetable oil industry in the region. But nomadic Indians like the bellicose Waiorani were driven deeper into the remaining forest or reduced to small reservations. There are no legally recognized, demarcated indigenous areas in Ecuador's *oriente*, where all unoccupied land belongs to the state.

A journey from the Andean highland to the center of the oil operations shows how fast the enviromnent gives way under the impact of oil and colonization. A new highway descends in hairpin turns along the Papallacta river, which feeds a beautiful mountain lagoon that supplies water by an aqueduct to Quito, the national capital. Frequent rockslides off the steep slopes showed the geological instability of the area. An earthquake in 1987 tore up big sections of the old highway into the *oriente* and destroyed 45 kilometers of oil and gas pipelines, producing a spill of

150,000 barrels. At the base of the active Sumaco volcano, squatters have burned off what remained of the dense forest after lumber companies removed the timber. Big mudslides came with the rains.

Along 60 kilometers of asphalted road to Loreto, a lowland agricultural center, there were great gaps in the forest occupied by cattle ranches and small farms. An African palm plantation had taken the place of 5,000 hectares of primary forest. At Coca, the rivers are crossed by ferry barges to reach the so-called *Via de los Aucas*, a dirt road that serves a dozen oil companies exploring south and east of the main oil production centers at Lago Agrio and Shushufindi. More than 600 wells have been drilled, and each chews up a 15-hectare clearing. Exploration has been authorized over more than 1 million hectares.

Everything taking place in the region—oil exploration, palm oil plantations, small colonization, cattle ranching—is destructive of the forest, which is reduced, at best, to piles of logs along the roads waiting to be trucked out. Lumber companies buy logs from settlers, who gladly sell the timber for small payments to help pay for land clearing. Under this arrangement, the lumber companies avoid the legal requirement of reforestation when trees are cut. This is unsustainable, quick-profit forestry with vast waste of timber resources. Ecuador's wood exports in 1988 were only $11 million.

The underlying problem is that Ecuador's strategic planners equate economic development with land occupation as if the region were a theater for military operations. This doctrine was expressed by General Carlos Aguirre, the Amazon expert at the Ministry of Defense, who described the region in 1988 as "an enormous empty area that can receive a large population from the interior and from abroad and develop immense agricultural, livestock, industrial, timber, and other profitable activities that could raise our economy to unsuspected heights."[28] Among the possibilities mentioned was a huge Amazonian hydroelectric dam on the Coca river to generate 3,000 megawatts of electricity for Ecuador and Colombia.

This is obviously a different strategy for Ecuador's Amazon region than that recommended by the World Bank or the United States, which has established clear guidelines against lending if tropical forests would be eradicated or tribal people adversely affected.

Without helicopters it is unlikely that Amazon oil exploration would have gone far, so it is fitting that the helicopter is the most important tool at the service of environmental protection in oil production areas. Conoco's airlift operation at Block 16 to avoid opening roads has been repeated by Shell in its exploration program at Camisea and by Brazil's Petrobras in the first commercial oil and gas field found in Brazilian Amazonia.

Brazil's long quest for self-sufficiency in oil has concentrated exploration in offshore fields on the Atlantic coast that have reached daily production of 650,000 barrels. But Petrobras never gave up on Amazonian exploration, and this paid off in 1988 with the discovery of oil at Urucu, south of Tefe, a port on the Amazon 650 kilometers west of Manaus. Earlier, Petrobras drillers had found substantial gas deposits along the Juruá river, west of Tefé, but not enough to justify a pipeline. Urucu, with estimated reserves of 50 million barrels of oil, and additional gas to increase the combined reserve with Juruá to 20 billion cubic meters of gas, changed this.[29]

Initially, Petrobras barged Urucu's oil to an existing refinery in Manaus, which will be expanded to 10,000 barrels a day, supplying all the western Amazon. Later, Petrobras intends to gather the gas from the shut-in Juruá wells and combine them with the Urucu gas, which would justify a pipeline to Manaus, according to Minister of Infrastructure Oziris Silva, a former president of Petrobras, The Juruá–Urucu wells are not a huge gas source, like Camisea, but they would be a big boost for energy-short Manaus and Porto Velho. Bottled domestic gas from Urucu has cut the price in half in these cities.

Urucu's production is important, said Wagner Freire, Petrobras's former director of exploration, not because of the initial production, but because the structure showed that oil and gas are both present in the ancient pre–Cambrian Brazilian shield. The geology of Brazilian Amazonia is complex because of the intrusion of molten basalt layers from the earth's mantle into the paleozoic sedimentary sands and tar layers where oil and gas are found. Geologists interpreted the Juruá gas finds as "traps" where the volcanic heat had volatilized hydrocarbons into gas. Urucu showed that some liquid petroleum survived the magma intrusions. Whether there is enough oil and gas to make this more than a minor commercial success remains to be discovered by further exploration.

Environmentally, the Urucu field and the pipeline system are far less damaging than the impact produced by a major construction project, like a dam. The base camp at Urucu and the drilling sites were all supplied by a fleet of helicopters, eliminating a road from Carauari, on the Juruá, the nearest town with an airport. Pipelines can also be laid by helicopter and buried underground for safety. There is a clear advantage in extending gas consumption in Amazonia, compared with the potential disturbances represented by big hydroelectric dams or charcoal-fired industries. The extent to which gas can become a dominant fuel depends on how much more gas is found by further drilling.

A flight in a Petrobras helicopter from Carauari to Urucu passes over 100 kilometers of intact forest canopy spreading as far as the eye can see in every direction, expect for small clearings where the Juruá wells were drilled. At these sites, several hectares of forest were burned and the land was scraped by a bulldozer. The ground is baked hard as brick and nothing has grown back but a few weeds, despite torrential rains. The sites are examples of what extreme land abuse can do—but it is unusual to find land so disturbed in Amazonia that rank vegetation does not grow back in clearings.

Sensitive to the environmental concerns that have surged in Brazil, Petrobras brought a team of ecologists to Urucu, including two former directors of the Amazon research institute, to advise on how the field should be operated. The decision not to build a road from Carauari was based on the team report. At the well sites, gas is flared and the flames scorch the surrounding wall of green forest. The tree canopy trembles under the rotors of the helicopters hauling pipes, compressors, mobile houses, and pallets loaded with supplies. A drilling camp is a noisy, dirty place, but Urucu is contained destruction. The airborne operation increases costs, but it spares the region the consequences of a road and it improves the corporate image.

Another oil company that learned from Conoco's experience was Occidental Petroleum, which also has a block in Ecuador. When Occidental opened negotiations for a block in Peru next to the Shell find at Camisea, it discovered that the block contained part of the Manu National Park, which is the equivalent in Peru to Yasuni in Ecuador. Occidental refused to work the block as it was offered and asked the Ministry of Mines to redesign the area, excluding the park. Manu has been penetrated by wildcat goldminers, but it will not be violated for oil.

SEVEN

El Dorado Slept Here

Hidden beneath the trees and the rivers, Amazonia's geological crust contains treasures that rank with South Africa, Canada, Australia, and the Soviet Union in the variety and quantity of recoverable minerals. The full potential is uncertain because Amazonia's mineral structures have only begun to be explored. But what is known has been sufficient to trigger the greatest gold rush in history, overturn world tin markets, and make Amazonia a major supplier of aluminum and ferrous ores and metals. El Dorado is a myth whose time has come.

The best inventory of Brazil's Amazonian mineral reserves comes from the National Department of Mineral Production (DNPM), which conducts geological surveys of the region. Although the estimates for primary minerals are in some cases incomplete, the proven reserves were iron, 18 billion tons; cassiterite (tin), 325,000 tons; manganese, 91 million tons; gold, 900 tons; bauxite (aluminum) 2.6 billion tons; chromium, 114,000 tons; copper, 1.2 billion tons; zirconium, 99 million tons; kaolin, 1 billion tons; potash, 493 million tons; limestone, 4.5 billion tons; diamonds 4 million carats.[1] Before 1960, these riches were almost entirely unknown.

The annual earnings from Amazon minerals is a gray area, because much of the trade in gold, tin, and gemstones goes through illegal, unregistered channels. Using the DNPM's low estimate of 90 tons of gold production from Amazonia annually, and converting bauxite and limestone through energy generated in Amazonia into aluminum and cement, the annual product value of the Amazonian mineral sector is at least $4 billion. Minerals are already by far the greatest single source of income in Amazonia, and the prospect is for much more to come.

The geology of Amazonia contains an enormous variety of rock and earth formations that span at least 1.6 billion years in the history of the earth's crust. The Brazilian central shield and the Guiana shield, the two weathered shoulders of the central trench of the Amazon, are the oldest rock structures in South America. They already existed in the pre–Cambrian period when this continent was still joined to Africa, the ancient Pangea. When tectonic plate movements began the separation of South America from Africa, the Andean range did not exist and what is now the Amazon basin was a continental sea open to the Pacific ocean on the west. It was only with the rise of the Andes, driven upward by another tectonic plate movement originating in the Pacific, that the course of the Amazon rivers was forced eastward toward the Atlantic.[2]

The accumulation of sediments from several inland sea periods, and more recently from the massive erosion of the new Andes, has made a layer cake of alternating organic and crystalline rock formations under the Amazon basin going back to the age of the dinosaurs. At different times during the 400 million years of the paleozoic era, outpourings of magma from the earth's mantle intruded huge sections of hot rock between the layers, and volcanic eruptions created rich mineral deposits that sometimes rise in bald domes above the green forest cover but may also be covered by the more recent sediments that continue to fill the basin.

These processes have given rise to several very different forms of mining in Amazonia. On the one hand, Amazonian geology produces huge deposits of minerals that can only be worked by industrial hard-rock mining methods. This is the case of the Carajás hills of eastern Para, where the world's largest deposit of high-grade, sulphur-free iron ore is in close proximity to large manganese, copper, gold, bauxite, and other mineral deposits. The Carajás mining district, owned by the Cia. Vale do Rio Doce (CVRD), has been developed through a $3.5 billion mine-railroad-port complex that is Amazonia's biggest industrial investment and one of the biggest mining operations in the world. Other big mines are the Rio do Norte Company's bauxite operation on the Trombetas river, the Amapa manganese mine of the Brazilian Antunes group, the kaolin mine at Jarí, and the Pitinga tin, titanium, and zirconium mine north of Manaus, the world's richest known tin deposit, owned by Paranapanema Mining, a private Brazilian company.

But eons of weathering and erosion of old rocks by water and heat, and the more recent deposits from the breakdown of unstable Andean structures, have filled Amazon riverbeds with widely scattered alluvial deposits of gold and tin. This is the world of placer miners, who have been scouring the rivers for nuggets and grains since the seventeenth century, when Brazil's colonial mines, worked by slaves, were the world's largest source of gold. From the start, independent placer miners worked outside the law in Brazil. Many were escaped slaves. They searched for gold and diamonds in inaccessible hills, called grimpos, beyond the control of the Portuguese taskmasters and the king's tax collecters. From grimpo, they came to be known as *garimpeiros*, as in "hillbilly."

Although big mines extract and process minerals at fixed locations, with large capital investments and relatively few people living in company towns, the mining system that predominates in Amazonia is based on the gypsy style of the *garimpeiros*. They are extractors of gold and tin who rush from depleted old locations to new discoveries such as a swarm of locusts. In contrast with other extractors of natural products, like rubber, Brazil nuts, or palm hearts, the *garimpeiro* activity is essentially predatory and unsustainable. It removes a nonrenewable resource virtually exempt from tax or royalty payments and devastates the environment without control or reparation.

Recent technology has converted these wildcat miners from pick-and-shovel artisans into airmobile resource raiders who can carry their search into the most remote parts of the Amazon. They destroy rivers and forests with jet-nozzle pumps, diesel-driven dredges, diving gear, and chainsaws, and pollute the rivers where they work with tons of mercury. *Garimpeiros* are present throughout the Amazon basin. They are in the Colombian frontier regions of Guainia and Vaupes, on the border with Brazil. Gold panners have turned up big finds in the Peruvian Madre de Dios province, in Ecuador's Napo region, and in Bolivia's Beni river watershed. But by far the greatest number are in Brazil.

From eastern Para and northern Mato Grosso to Roraima and the upper Rio Negro, there are at least 400,000 (some say 600,000) *garimpeiros* scrambling for gold, tin, and diamonds at 1,200 different locations. They are accompanied by an auxiliary corps of prostitutes, gold buyers, merchants, pilots, gunmen, and pharmacists who make up a universe that some estimates place at over 1 million people. The annual purchases of pumps, motors, hoses, chainsaws, mercury, diesel fuel, aluminum pontoons for dredges, airplanes, and other supplies for the *garimpo* economy are said to be a $500 million business in Brazil. The amount of gold and tin that is smuggled out of Amazonia is far in excess of the amounts on which taxes are paid. In close proximity to the cocaine traffic in Bolivia, Peru, and Colombia, it is believed that clandestine gold is used to "launder" drug money on a large scale.

Garimpeiros operate on the "democratic" principle that nobody owns mineral assets until they are found by someone. These "barefoot geologists" have made some of the largest mineral discoveries in Amazonia. Jose Candido Araujo, the legendary "Ze Arara," was an illiterate peasant from Maranhão panning for gold on the remote Crepori river in the Tapajós watershed when he found the *garimpo* he called "Liberty" in 1958. This discovery kicked off the modern Amazonian gold rush which concentrated initially on the Tapajós, where 400 *garimpos* have produced an estimated 600 tons of gold since 1958, and a steady 7 tons of gold annually in recent years.[3] Ze Arara is said to have personally accumulated 400 kilos of gold, worth $5 million, with which he bought ranches, beachfront apartments in Rio, and a fleet of airplanes—all paid for in cash from a suitcase.

Then came the discovery of gold on the Madiera river in 1980, first bringing hundreds of divers working from rafts with airpumps and later bringing a fleet of 2,000 dredges that scoured 150 kilometers of river bottom on the Amazon's most violent tributary. From Periquitos at Abuna to Belmont below Porto Velho, there are 18 rapids where gold accumulated over centuries in deep pockets. Whenever a new strike is made, divers and dredges flock to the hot spot, called a *fofoca*, and fight for a place like felines feeding on a carcass. Lucky investors made fortunes, and their crews got 30 to 35 percent of the gold. While the bonanza lasted, Porto Velho boomed with gold shops, supply stores, cabarets, jerry-built housing, and boatyards making dredges. But at least $60,000 to $70,000—or 5 kilos of gold—was needed to break even in a season. By 1990, there were more losers than winners, and many *garimpeiros* departed for new horizons. A pall fell over the city.

The boom drew international adventurers, including a tight clan of Taiwan Chinese, who operated 50 dredges and smuggled their gold directly to Taipei. Among the foreigners was Rafael, known as "el argentino" (*garimpeiros* don't use last names), who was a naval architect in Buenos Aires before he came to seek his fortune in Amazonia. He extracted some kilos of gold, built dredges, blew a lot of money on cabaret girls, and wrote poetry. Rafael's "Profane Oration" conveys the *garimpeiro* obsession with gold:

Yellow God,
Who art in the depths of the Madiera,
Hallowed be Thy name.
Thy Kingdom come, Thy will be done,
On the waters, the rafts, and the dredges,
On the men, the birds and the fish.
Give us this day our daily gold, and
Forgive us our debts, as we
Dive,
Navigate,
Dredge,
Work,
Fight,
Whore, and
Die in your adoration.

The semi-industrial miners of the Madiera recovered at least 80 tons of gold before production began to fall. In 1989 it was 6.4 tons, and by 1991 it was less than 3 tons. The dredges they developed were state-of the-art with metal pontoons carrying a two-story superstructure that housed four workers and a cook, a 200-horsepower diesel engine powering a boom with a rotary drill bit and suction tube reaching 30 meters to the river bottom, and the pumps for a sluice-box system to wash gold ore. As the bonanza ran down, the price of dredges fell from 3 kilos of gold to as little as 1 kilo and the desperation of some owners to put dredges into new areas led some to risk everything on a run through the rapids. In 1991, at the Teotonio falls, beer-drinking *garimpeiros* placed bets and cheered from the shore as dredges were driven over the foaming cataracts. Some made it to the Belmont *garimpo* below the falls, where there was still some gold;

others broke up on the rocks and sank. It was the yellow god devouring its young.

The most dramatic illustration of the *garimpeiro* boom-and-bust cycle is Serra Pelada, which was discovered in 1980 on a ranch near the Carajas mining district of eastern Para. CVRD geologists had visited the site, but they did not activate the company's exploration claim because they failed to detect a gold pit of pharaonic proportions. That was unearthed by *garimpeiros* who took the property by storm. They divided the pit into what came to be 13,000 individual registered claims, each 9 meters square. Up to 80,000 men, including claim-partners, pick-and-shovel workers, and bearers carrying 20-kilo sacks of earth out of the pit on their backs, worked side-by-side in the mud in search of nuggets and pay dirt. They opened a crater 110 meters deep and a kilometer wide, bigger than Maracana, the world's largest stadium.

Primitive as it was, the "pit dogs" of Serra Pelada produced 41 tons of gold in the decade from 1980 to 1989, with a peak of nearly 14 tons in 1983. Less than a decade later, it was over. Cave-ins and floodwaters buried the pit floor. Production in 1989 was 200 kilos. A shantytown of 30,000 people that mushroomed around the pit was like a city slum in the tropical forest. Women nursing babies begged on the streets for food. Idle men sat staring at the empty pit, waiting for someone to provide pumps and scrapers to recover the mine. "Every penny I had is down there," said Juvenal Leal da Silva, a heavy machine operator who came to Serra Pelada with $215,000 saved up from eight years of working in oilfields and diamond mines in Gabon and South Africa.

The *garimpeiro* juggernaut did not stop at Serra Pelada, however. It moved on to new sites on the Peixoto de Azevedo–Teles Pires river system in northeast Mato Grosso, to the Rio Fresco in Kayapo Indian country in Para, and then to the Northwest frontier at Roraima on the Couto de Magalhaes and Uraricoera rivers, where thousands of *garimpeiros* invaded Yanomani Indian lands. These were no longer barefoot prospectors acting alone. Fleets of modern dredges, hundreds of airplanes, and tens of thousands of veteran garimpeiros were transferred to each new front with the financial backing of big gold buyers in Sao Paulo who deal in international gold markets from New York to Tel Aviv and Hong Kong.

Corporate miners who have been driven off Amazonian claims by *garimpeiros* blame some of these traders for providing the invaders with money and arms, and the protection of lawyers and populist politicians. Mine managers and geologists, as well as *garimpeiros*, have been killed in conflicts over mine sites. The insecurity created by these violent disputes brought large-scale mining investments in Amazonia to a halt.

Defenders of the *garimpeiros* argue that they are poor, desperate victims of the "social situation" created by lack of access to land for peasants and urban unemployment in the national economies of the Amazonian countries. "They have come to Amazonia because this is what they imagine is the last frontier of survival. They have carried out a mineral reform by occupation because the conservative land owners prevented an agrarian reform," said José Altino Machado, the founder-president of the Association of Garimpeiro Unions of Amazonia (USAGAL).[4] Machado, whose heavy gold necklaces convey the wealth he has accumulated as an airtaxi operator, is a promoter of *garimpos* in the Tapajoz district and in Roraima. His organization lobbies in Brasília for unrestricted access to Amazonian mining sites by *garimpeiros*, and Machado has good military connections; but he failed to win election in 1990 as federal senator from Roraima, where *garimpeiros* are the major political issue in the state.

The Amazon gold rush did more to populate the region, without any fiscal incentives or government subsidies, than any of the "geo-strategic" development schemes designed by economic planners for Amazonia. The *garimpeiro* migration has been spontaneous, and the mode of mineral extraction involves no official credit. Most of the basic documents, such as the Five-Year Plans for Amazonia, don't even mention *garimpeiros* as a major economic factor. Yet, the gold production of Brazilian Amazonia has become the second most important item, after petroleum, in Brazil's mineral sector.

The theory that Amazonia has served as an escape valve for social pressures created by demography and inequality in the national economies has its best illustation in the *garimpeiros*. They have come mainly from rural areas of the Northeast and southern Brazil, where there is little new land available for the sons of small farmers. In the Andean countries they have also been largely younger men from the highland villages for whom the alternative was outmigration to urban slums. Those who have gone to Amazonian frontiers in search of gold are generally the most vigorous and enterprising.

The *garimpeiros* are contracted by *garimpo* investors—Amazon ranchers, road builders, airtaxi operators, and gold dealers—who provide free food, equipment and fuel. In the typical deal, the workers keep 30 percent of the gold or tin and the backers get the rest. The men work at their river camps from dawn to dusk. They live in makeshift shelters covered by sheets of black plastic, with mangy dogs for companions. Posing for a photograph, "to send a picture back home," they look like the farmboys in the sepia daguerrotypes of Civil War volunteer regiments.

If the crew finds 1 kilo of gold in a six-month season, that is enough for each *garimpeiro* to get 10 grams a month, which is worth double the minimum salary, but the financier loses money. If there is a big find—over 25 kilos—the workers become rich enough to buy a home, a taxi, or a week of wild times. The backers invest in cattle ranches, urban real estate, hotels, pharmacies, and bars—or a new *garimpo*.

Mining industry leaders don't dispute that there are "social" causes that make the *garimpeiro* phenomenon irrepressible, but they say that this does not provide an adequate explanation for the violence directed against large mining companies. Noevaldo Araujo Teixiera, former president of the Brazilian Society of Geologists and a Sao Paulo consultant to large mining companies, put it this way:

> *Garimpeiro* activity as it is exercised today, without any technical or legal control, is highly pernicious. The *garimpeiros* are exterminating Indians and are being exploited themselves by crooked businessmen. In this climate of lawlessness, conflicts run wild, geologists are killed, mining resources are wasted, more than 60 per cent of the gold production is contrabanded, and ecological balances are upset. The loser in this is the nation, and the winners are a handful who convert the poverty and desperation of the needy into an instrument of political and economic pressure for the benefit of those who manage this illegality.[5]

The failure to bring the *garimpeiro* activities under legal and technical control is a glaring example of the administrative irresponsibility that has put the sustainable development of Amazonia in jeopardy. This is not just an environmental problem of polluted rivers, but an economic and social disorder that undermines the possibilities of constituting stable communities in the region. One of the clearest illustrations of this is the experience at Serra Pelada.

When the *garimpeiros* occupied Serra Pelada, CVRD demanded that they be evicted, but the military regime then in power considered this politically risky. Instead of an armed confrontation, the *garimpeiros* were given a ten-year "right of use" permit, and the pit was placed under the administration of an army counterinsurgency officer, Sebastião Rodrigues, known as Major Curio. He banned liquor, guns and women; began each day's work with a flag-raising ceremony; and set up a cooperative that kept a record of gold production and a registry of claim owners.

This system worked while gold recovery was easy, but it fell apart when production required investment and technical control. No provision was made for maintaining the pit or for basic services for the human community at Serra Pelada. Gold was taken out, but nothing was put back in. At the end of the military government in 1985, Serra Pelada was a bankrupt cooperative, a slum, and a mutilated mine. As production plummeted, desperate *garimpeiros* occupied the Carajás–São Luís railroad bridge over the Tocantins river in 1987, demanding money for restoration of the pit. State police, flown in by helicopter, opened fire and killed at least six men.

With the mine in terminal crisis, Curio reappeared at Serra Pelada and was elected secretary general of the cooperative. He offered a plan to reactivate the pit with $16 million from mining firms supposedly interested in processing the tailings with industrial methods. In other words, capital and technology would save Serra Pelada from the ruin wrought by the *garimpeiros.*

The *garimpeiro* system appeals to those who see it as a way of distributing wealth through small entrepreneurs and a large labor force in place of concentration of ownership in a few mining corporations using capital-intensive, low-employment technologies. In this sense, *garimpos* have been a magnet drawing many migrants with economic initiative who have invested much of the capital they have acquired from frontier mining in local ventures. The consumption of the *garimpo* population has provided a major market for Amazonian farmers and merchants. It has been a source of part-time employment and nonfarm income for hard-pressed farm families. These advantages are offset, however, by other, more serious drawbacks in the unregulated *garimpo* system, as it has worked during the great gold rush.

In economic terms, the Brazilian government only knows with confidence the destination of about 25 percent of Amazon gold production over the decade of the 1980s and not much more about the surge in tin that put Brazil in the forefront of world production. In both cases, the volume of unregulated *garimpo* production has depressed international metal prices, reducing export income, and has diverted a major income stream into an underground system where asset strippers pay virtually no taxes on billions of dollars in revenues.

A measure of the enormous amount of gold that has been diverted from *garimpos* into illegal channels is the gap between the estimates from reliable trade sources and the official figures. The DNPM reported gold production in 1988 of 34.1 tons; trade sources placed the total production that year at 123 tons, of which 96 tons were estimated to have come from Amazonian alluvial deposits. In 1989, the official figure rose to 48.6 tons; the trade estimate was 102.6 tons.

The Brazilian Central Bank estimated in 1990 that $8 billion in gold had been smuggled out of Amazonia during the previous decade, without paying taxes. The lost exchange would have been enough to pay one year's interest on the foreign debt. The claim seemed to be supported by international trade statistics that showed that neighboring countries such as Uruguay, which has no goldmines, exported 27 tons of gold in 1987. The evasion was due, in part, to short-sighted Brazilian policies. In 1987, when the world price fluctuated around $12 per gram, the Central Bank set a price equivalent to $7 per gram. The result was token sales by gold dealers in Brazil to cover tax appearances and massive contraband abroad. In 1990, the Central Bank began paying the same price internally as the price abroad.

The effect on international mineral trading of *garimpo* production is highly disruptive. Brazilian gold sales totaling about 140 tons in the first half of 1990 contributed to a 15 percent fall in the world price.[6] Although Brazil is still far behind South Africa's annual production level of 600 tons and trails the Soviet Union, the United States, and Australia, recent production of more than 100 tons a year has overtaken Canada, where mines are being closed because of low prices.

The international tin market suffered an even greater fall due to the *garimpeiros.* In this case, destabilization came from the accidental discovery of a major tin site at Bom Futuro in Rondônia in 1987. An illegal logger found a large grayish nugget of what proved to be nearly pure tin wedged in a tractor tire. A mining exploration claim on the site, near Ariquemes, had been filed in 1982 and was later sold to Paranapanema, but news of the discovery of tons of tin waiting to be picked up on the surface brought thousands of *garimpeiros* rushing to the new bonanza.

In the violent struggle in Brazil for control of minerals between the free-wheeling *garimpeiros* and the mining companies, there is an intermediate area of semi-industrial miners who began as *garimpeiros* but graduated into mechanized mining with hired labor. Under Brazilian law, *garimpeiros* are recognized to have a right to extract minerals they find if they work individually and with artisan tools. That was hardly the case at Bom Futuro. The mine was developed by a score of large local investors, including the mayor of Ariquemes, the town nearest the mine, who used *garimpeiro* cooperatives to claim the site and won a court ruling against eviction by Paranapanema.

With the ruling, and strong support from the state governor Jeronimo Santana, later impeached for embezzlement by the state legislature, the *garimpeiros* kept the mine, but the 15,000 pick-and-shovel miners were soon working for the big operators who moved in heavy earth-moving equipment, over 500 industrial jigs, and fleets of planes and trucks to take out ore. A brisk contraband trade in tin sold the output to a ring of large mining companies. The Rondônia state government was satisfied with a new source of mineral taxes and payoffs, and disregarded the enormous pollution caused by Bom Futuro's dumping wastes into two major rivers. The 100-hectare site is a devastated patch in the dense forest that looks, with its trenches and craters, like the battlefield at Dien Bien Phu. It is a striking example, along with the flooded gold pit at Serra Pelada, of the environmentally destructive results of unregulated mining in Amazonia.

By 1989, Bom Futuro was the largest single tin mine in the world, producing 25,000 tons that year. This output, dumped on the world market, drove tin prices down so severely that it undermined the International Tin Agreement and forced Bolivia, once the world's largest producer, to close its expensive mines. Bom Futuro, along with other Amazon tin mines, placed Brazil on a par with Indonesia and Malaysia as the leading tin exporters in the world. But in view of the price collapse and the illegal marketing of Bom Futuro's tin, the Collor government closed down the operation until it was placed under the management of a consortium formed by Paranapanema, other mining companies, and four *garimpeiro* coops. An official investigation said Bom Futuro's wildcat operators had avoided paying taxes on $300 million of tin, as well as causing millions of dollars in environmental damages.

It does not take complicated natural resource and environmental accounting to figure out that the *garimpeiro* system is enormously wasteful. The ancient sluice-box method used by *garimpeiros* fails to extract the very fine gold particles that make up at least 25 percent of the alluvial ore content, so this is lost. The value dumped into the rivers, on the basis of an annual production of 100 tons, is in the range of $300 million. Just as there are inexpensive technical means to reduce mercury pollution, there are chemical processes for sthe recovery of fine gold. A modern mining company would use these and apply the additional income to cover environmental impact costs that are legally required of a mining company. This does not count, of course, in *garimpeiro* accounting.

In ecological and health terms, the most severe danger arising from the *garimpeiro* system is the use of mercury. Gold particles adhere to mercury when it is mixed with washed paydirt. The mercury is burned off with a blowtorch and pure gold remains. *Garimpeiros* rarely use a closed retort, so volatalized mercury goes into the atmosphere, condenses in water particles, and winds up in the rivers. There it enters the organic food chain of algae, minnows, and carnivorous fish, in increasing concentrations at each step. Edible fish samples from the Madiera and Tapajós rivers have been analyzed containing up to ten times the mercury levels considered safe for human consumption by the World Health Organization.[7]

The fear is that massive mercury pollution of the rivers will reproduce in Amazonia the disaster known as Minimata disease. In Japan, mercury waste from caustic soda plants dumped into the ocean at Minimata and Nigata caused 1,200 cases of methylized mercury poisoning from fish consumption, with fatal or crippling damage to the central nervous system, blindness, and births of deformed festuses. There is no evidence yet in Amazonia of public health problems directly attributable to mercury poisoning, but many researchers say it is only a matter of time until bacteria in river sediments converts inocuous metallic mercury into deadly methylized mercury (CH_2 Hg) that will poison the food chain for years.

The scale on which mercury pollution is taking place is alarming. From 1980 until 1989, the annual importation of mercury in Brazil closely paralleled the gold boom, going from 25 tons to 266 tons. Of this amount, about 165 tons cannot be traced to industrial or odontological buyers, so it is assumed to have gone to *garimpos.* Since the rule of thumb is that one gram of mercury is used to recover one gram of gold, this coincides with current estimates that Brazil's *garimpeiro* output of gold is over 100 tons a year. An environmental research team that surveyed the Peixoto de Azevedo watershed in the state of Mato Grosso found that 21 tons of mercury had been used to produce 11.5 tons of gold in 1985. On the Madiera, it has been estimated that 87 tons of mercury were consumed between 1979 and 1985.[8]

In 1989, President Sarney issued a presidential decree banning the use of mercury, with no practical effects because nothing was done to put controls into practice. Brazil has received offers of technical assistance on reducing mercury use from West Germany, Japan, South Africa, the World

Bank, and major companies such as Bechtel Engineering. The problem is not really technical, however, since there are devices, such as retorts, that can reduce mercury dispersion, and there are alternative methods. The problem is cultural, since it is the *garimpeiros* who have to be persuaded to change their ways. The establishment in the *garimpos* of a climate conducive to law enforcement, health care, and technical management would be a first step toward environmental control.

The gold rush has also aggravated the long-standing problems caused by the clash between modern occupants of Amazonia and the indigenous communities. The *garimpeiros* invade whatever lands adjoin the rivers that contain gold, whether the are private property or Indian reservations. If there is resistance, there is usually violence, and the miners almost always impose their way. But the main hazard is to the health of the natives. The *garimpeiros* are vectors for disease wherever they go, from malignant malaria to venereal infections, including AIDS.

In the case of the Yanomani, the Uru-eu-wau-wau of Rondônia, the Panara of Mato Grosso, and other vulnerable tribes, the effects have been devastating. When Orlando Villasboas, founder of the Xingú indigenous park, first contacted the Panara along the Peixoto de Azevedo river of northern Mato Grosso in 1972, there were 600 Indians. The Cuiabá–Santarém highway brought in thousands of *garimpeiros* who occupied the river, and in less than ten years the Panara population fell to 80. The survivors were evacuated to the Xingú park, where they have recovered in numbers to 140. "But it is not the same. On the Peixoto the land was good and there were always fish," said Kreton Panara, a leader of the tribe.[9]

The growing militancy of tribal groups throughout Amazonia in defense of their rights, with the support of human rights and environmental groups, has brought a large increase in areas that are being demarcated as indigenous reserves. In Brazil, the total area earmarked for Indian reserves is 700,000 square kilometers, of which 380,000 kilometers had been demarcated by 1990; if the total area is legalized, Indian lands will occupy 12 percent of the Brazilian Amazon. In Colombia and Bolivia, most of the Amazon region is being recognized as Indian lands. Ecuador is assigning geographically defined reserves to indigenous communities in Amazonia, although the oil and other subsoil resources remain the property of the state.

The expansion of Indian reserves has sharpened the issue of mining rights because access to the mineral resources in the Indian areas, including oil and gold in eastern Ecuador and Peru, and the mineral districts of Para, Roraima, and western Amazonas, are regarded as major prospective areas by mining companies and governments. One of the most crucial points is the Rio Negro triangle formed by Brazil, Colombia, and Venezuela, where Yanomani, Tucano, Tikuna, and other Indians have major reserves. This is an area where the chimera of El Dorado has already played a historic role in the territorial occupation of Amazonia. In 1774, a

Spanish governor in eastern Venezuela sent an expedition to the upper reaches of the Orinoco watershed, which is divided from the Negro river—and therefore the Amazon system—only by a narrow strip of inundated land called the Casiquiare. The Spaniards had heard from an Indian of "a high hill, bare except for a little grass, its surface covered in every direction with cones and pyramids of gold."

The Spaniards found no hills of gold, but when the Portuguese got wind of the expedition, they sent their own military forces up the Branco river and built Fort São Joaquim, at the junction with the Uraricoera river. From that advanced position they controlled the territory of Roraima, bordering Venezuela and Guyana. Not much happened for the next 200 years, until gold was discovered by *garimpeiros* in the rivers near Paapíu, where a Roman Catholic missionary outpost maintained contact with the reclusive Yanomani clans, made up of over 10,000 Indians who wander between Brazil and Venezuela.

The Yanomani territory is an officially identified indigenous area, out-of-bounds for unauthorized outsiders; but as soon as word of the gold find got out, it was invaded quickly by about 25,000 *garimpeiros*, supported by 400 jungle pilots flying supplies in small airplanes and helicopters into over 100 hazardous airstrips. Scores of Indians began to die of diseases and violence, although some Yanomani village chiefs provided labor to clear airstrips and charged landing fees.

The missionary priests launched an international protest against the "genocide" of the Yanomami. The campaign was led by the bishop of Roraima, Aldo Modgiano, who is from the Italian order Consolato de Torino. Anthropological societies and human rights groups joined the outcry, and rainforest alliances discovered that these hunters and forest gatherers were defenders of the environment. Davi Yanomani, a member of the tribe, received a United Nations Environment Program award. Within Brazil, however, President Sarney divide the Yanomani reserve into an archipelago of 19 separate villages within a national forest, and granted *garimpeiros* mining areas in the spaces left open. The federal Indian agency expelled the missionaries from their Yanomani outposts, and the gold rush continued up the Uraricoera river and into the Surucucu hills on the Venezuelan border.

President Collor faced an international uproar over the Yanomani issue when he took office in 1990. He quickly flew to Roraima, ordered military and police to blow up the *garimpeiro* airstrips, and reversed the order creating mining reserves in Yanomani territory. By 1991, the *garimpeiro* front in Roraima had been substantially reduced to about 800 miners who continued working the Uraricoera river with dredges, outside the Indian reserve. In November 1991, Collor signed a decree consolidatintg the reserve covering 9.4 million hectares (see Chap. 8).

The exclusion seemed only a lull, however, until the Brazilian congress adopted a new mining code defining the rules for mining and timber extraction in indigenous areas. The Yanomani reserve is an enormous area which covers one-quarter of Roraima. Anthropologists say this continuous area is necessary for the survival of the Indians in their forest-dependent way of life. But the 300,000 non-Indians in Brazil's newest state consider the subsoil resources in the reserve essential for the state's economic life. Roraima's economy revolves around mining and the commerce associated with *garimpeiro* activity. The state's first elected governor, Ottomar Pinto, refused to allow state police to participate in the federal ouster of the *garimpeiros*, and he paid bail when they were arrested. Roraima's congressional delegation is made up of strong advocates of mining concessions in indigenous areas for big mining companies, such as Paranapanema, and for *garimpeiro* cooperatives. Under the Brazilian constitution of 1988, such mining licenses can only be granted with the consent of congress and of the Indian communities, which are supposed to receive compensation payments.

The explosive effects of *garimpeiro* discoveries cross national borders and create diplomatic problems. Brazil's 4,000-kilometer frontier perimeter with Colombia and Venezuela is a highly mineralized region that is under the military control of an army command center at São Gabriel da Cachoeira on the upper Rio Negro. Jungle troops train local Indians, maintain agricultural schools, provide medical and dental services along the rivers, and airlift basic supplies to local communities. This is part of a special frontier security project called Calha Norte, or Northern Slopes of the Amazon, that the Brazilian armed forces consider one of their major missions. It is a $100 million-a-year item in the military budget.

The main area of operations for Calha Norte is the far north-western Amazon area of Brazil known as "The Dog's Head," from its configuration on a map. This area between the Vaupes and Negro rivers, bordering with both Venezuela and Colombia, is covered by some of the most intact rainforest in Amazonia. There are very few settlers, but there is important mining activity. A subsidiary of Ecopetrol, the Colombian state oil company, is developing a large gold mine near Puerto Inirida, where the mineral body extends across the border with Brazil to a location called Capurro, where a mine is being developed by Paranapanema. The border is well defined and there is technical cooperation between the two corporate mining companies, so harmony reigns.

Where *garimpeiros* are active, however, conflicts erupt along the borders. The Traira river on the Brazil–Colombia border near Pari Cachoeira, an old Salesian mission town, runs into a Tucano Indian reserve that contains large alluvial gold deposits. Paranapanema made a deal with the Tucano leaders to develop the Traira and brought in a private security force that drove off a band of *garimpeiros* on the Brazilian side of the river, with several dead. The *garimpeiros* continued working on the Colombian side, where a camp of 4,000 miners came under the control of Colombian guerrillas. In 1991, the guerrillas clashed with a Brazilian army border patrol on the Traira, killing three soldiers and losing four men. Antigovernment guerrillas operate freely in some areas of Colombian Amazonia, which is full of clandestine airfields used by drug traffickers.

Problems have also arisen along the ill-defined Venezuelan frontier with Roraima. The gold rush that brought 25,000 miners into the Yanomani Indian reserve crossed over into Venezuela, and this was headlined by the Venezuelan press as "invasion of the national territory." Venezuelan military forces expelled the Brazilian *garimpeiros*, confiscating their equipment, and some diplomatic exchanges at the presidential level were necessary to calm public opinion.

The Roman Catholic church's Indian missionary movement in Brazil is critical of the "militarization" of Amazonia, which it sees as an instrument for imposing the interests of the "national society," such as mineral exploitation, on the tribal lands. But the situation of the missions on the ground in Amazonia is weak in relation to the advance of outside forces into the region. The days are over when missionary teams traveling the rivers for days in canoes were the principal contact with the natives. The missions are in decline. The numbers of foreign priests and nuns willing to do jungle service has dropped, and not enough Indians are entering seminaries to fill the gap with a native clergy.

The Amazon military command says its presence is necessary to control *garimpeiros*, drug traffickers, and guerrillas crossing the border from Colombia. The importance that the military attach to the mineral resources is evident. The only road built in the "Dog's Head" by military engineers goes to Sete Lagos, 65 kilometers north of Sao Gabriel, where the Brazilian government's Mineral Resources Research Corporation (CPRM) has drilled 12 holes in an ore body deep in the forests. CPRM geologists say they have found the richest deposit in the world (2.9 billion tons) of niobium, a rare metal used in steel alloys, as well as some uranium.

Amazonia's already estabished mines, and the likelihood that more geological surveying and exploration will turn up new deposits, have convinced many observers that mining should play the leading role in the occupation of the region because it offers the best return on investment at the lowest environmental cost. One of these boosters is João Orestes dos Santos, the chief geologist of the CPRM office for the western Amazon in Manaus. Santos says Amazonian mining production could treble, "bringing viable economic development while preserving 99 percent of the region's ecosystems," in contrast with land colonization and ranching.[10] Another mining advocate is Samuel Hanan, a top executive at Paranapanema, who says that all the mining sites in Amazonia could be developed without altering more than 12,500 square kilometers, or 0.35 percent of the region.[11]

These arguments in favor of a mining priority for Amazonia are on good economic ground. Although the mineral resources are nonrenewable, they are abundant and they are profitable. Their contribution to sustainable development of Amazonia depends on putting a fair share of the proceeds from these assets back into the region in diversified investments. The environmental case for big mining ventures is also good. The major Amazonian mines, including Carajás, Trombetas, and Pitinga, have all invested substantially to control pollution and repair ecological damages where they dig. Big mining companies are not a threat to the environment within the confined enclaves where they are operating.

But two big problems remain to be faced if mining is to provide a solid foundation for sustainable development. One is a basic reform of the *garimpeiro* system, with its wasteful strippage of irreplaceable mineral assets, environmental pollution, tax evasion, and social anarchy. The other, no less important, is the recognition by big mining companies that megaprojects on the frontier produce social and environmental "externalities" that must be foreseen and managed with the same care as the more limited effects of mining within the fences that surround their enclaves. As a first step, the disasters that have already taken place beyond the fences need to be repaired, as much as possible, by corrective actions.

* * *

Marabá, Tucurui, Carajás, Imperatriz—these names are landmarks for the most important route into Amazonia of the modern era. The Carajás corridor, as it is called, has been for eastern Pará what the Cumberland Trail was for American pioneers going to the Ohio Valley; but instead of the slow trek in ox-drawn carts, these modern migrants came faster and in greater numbers by truck and railroad. Apart from land, the attractions at the end of the trail were the world's largest iron ore mine, a gold pit where 80,000 miners once clawed the earth, and a dam that is the Grand Coulee of Amazonia.

The Carajás corridor, combining highway, railroad, and power projects, is the largest concentration of raw economic power and new population anywhere in Amazonia. The corridor's geographical center is Acailândia, where the railroad, the power transmission towers from Tucurui, and the Brasília–Belém highway intersect east of Marabá and north of Imperatriz. The projects brought in thousands of construction workers. The heavy building activity turned isolated areas into poles of commercial attraction. Frantic speculation in real estate and discovery of placer goldmines added to the fever. In contrast with Rondônia or northern Mato Grosso, this is an industrial, as well as an agricultural, occupation.

The first surge of migrants came with the new highways, like the Belém–Brasília and Trans-Maranhão that connected the peasant masses of the Northeast with the enormous wedge of land formed by the watersheds of the Tocantins and Araguaia rivers. This is a transition zone between the dry *cerrado* uplands of northern Goiás, which were split off in 1988 to form the new state of Tocantins, and the pre-*hylea* forest lands of western Maranhão and southeastern Pará. The dynamics of this occupation led westward into the watershed of the Xingú river once the Araguaia valley was occupied. The Carajás corridor lies squarely on this path of penetration of the *hylea*.

The newcomers transformed sleepy old trading towns on the rivers, such as Conceição do Araguaia and Marabá, into centers of frontier violence over land and mines. Two economic systems, two cultures, and two visions of Amazonia clashed here. Well-heeled ranchers and colonization companies from southern Brazil came with big investment plans, often subidized by the government. They displaced old settlers and peasant squatters, who fell back on the old towns. The poor migrants from the Northeast brought nothing but a desire for land, and their poverty soon became an insoluble problem for the new municipalities that emerged without resources for education, health, or housing.

Then came the discovery that the Carajás hills, between the Araguaia and Xingú rivers, hold the largest iron mine in the world. The ore body, in four sparsely covered humps that rise out of the dense rainforest, was discovered accidentally in 1966 by geologists who landed their helicopter in a clearing to refuel.[12] U.S. Steel was initially in on the find, but the Brazilian government decided to take full control of Carajás. CVRD, the state mining company, became the sole developer. It began to drill and discovered that in addition to 18 billion tons of 66 percent pure iron ore, Carajás also contained manganese, bauxite, copper, nickel, and gold in vast quantities, as well as other, lesser minerals, making this one of the greatest mineral deposits in the world.

This discovery did as much to change the image of Amazonia from pauper to prince as oil discoveries had done in other Amazon countries. The collapse of the rubber boom had left Amazonia without solid economic arguments to justify large investments. With Carajás, and the simultaneous discovery of vast bauxite deposits at Trombetas, there was a basis for two megaprojects: the Tucurui hydroelectric dam and the Carajás–São Luís railroad. Between the two, they created a new mining rationale for Amazonian development that was reinforced by the encouraging results of an airborne radar survey of Amazonia, called RADAM, that provided the first complete geophysical overview of the region.

The Carajás deposits can produce iron ore for 400 years at the present extraction rate of 35 million tons a year, which is one-third of Brazil's total exports. Brazil intends to holds its place as the world's number-one iron ore exporter, while increasing domestic steel output. Carajás will provide most of the export ore, releasing production from the traditional mines of Minas Gerais for domestic steel plants, which produce 20 million tons a year for Brazil's industries. But to get iron ore and other Carajás minerals to a deep-water port required the construction of an 890-kilometer railroad that begins in the mist-covered Carajás hills, crosses the Tocantins river over a 2.3-kilometer bridge at Marabá, and plunges through a swath of devastated forest to the Atlantic coast at Itaqui, near São Luís, Maranhão.

In addition to providing the indispensable link from mine to port, the railroad line was seen by "strategic planners" and politicians in Brasilia as providing the transport infrastructure for the occupation and broad economic development of Maranhão and eastern Pará. "The basic concept is to develop an 'export corridor' in the north of Brazil. . . . The future availability of energy at Tucurui promises good prospects for exploitation of the vast mineral deposits in the area. . . . The Carajás railway will cross large expanses of land that are suitable for either forestry or agriculture. The forestry components are closely tied to the proposed industrial projects, primarily the production of pig iron which will require 25 million cubic meters of charcoal per year calling for large forests that will need to be planted in order to reduce fuel costs," said a Brazilian government planning document in 1980.[13]

CVRD's mine-railroad-port project became the key component, therefore, in a grandiose but unrealistic scheme called the Gran (Greater) Carajás Program. On paper, Gran Carajás was a powerful institution. As set up in 1980, it was headed by a federal interministerial council, chaired by the minister of interior, with an executive secretary who later became minister of agriculture. The program called for investments of $60 billion for a variety of economic and social purposes in a Gran Carajás area covering 840,000 square kilometers, or one-fifth of the Brazilian Amazon. The council had powers to grant tax incentives to private investors and allocate public funds for social projects in the region, which gave the program some of the attributes of a development agency.

But the secretariate never developed the technical staff or the beaurocratic power to coordinate the actions of a wealth of federal agencies and state governments in charge of programs as diverse as agrarian reform, malaria control, urban waterworks, road construction, and rural education. Gran Carajás, which was run by politicians, became a slush fund without money. It never got any foreign financing, and after the Brazilian financial crisis of 1982 it withered into an obscure office in Brasilia frequented by politically recommended clients seeking tax breaks and credit subsidies.

Independently of Gran Carajás, the CVRD Carajás iron project went forward on its own with World Bank and government financing from Japan, the United States, and the European Economic Community. As a result, eastern Amazonia got a first-rate new mine and a pioneering railroad; but the federal and state agencies, lacking coordination and resources from the Gran Carajás program, failed to deliver the urban development, agrarian reform, rural credit, and health, education, and environmental protection that were acute needs in the whirlwind of human occupation taking place in the Carajás corridor and its area of influence. SUDAM, the regional agency responsible for planning, did just as little. This great failure in regional development planning has contributed to social and environmental problems in the Carajás corridor that put the major economic advances in the region in jeopardy.

* * *

One measure of this failure is the Indian communities within the Carajás area of influence. The mining and power projects that made Carajás a password for modernity take their names from Indians who were once the tribal lords of these forests. Tucurui, where there is now a 2-mile-long hydroelectric dam, is where the Parakana indians fished. The Carajás hills are named for a tribe of Indians, famed for their prowess with canoes, who were ancient rivals of the Kayapo for control of the Araguaia river.

Anthropologists list 24 indigenous areas that are exposed to danger by the developments radiating from the Carajás mines and railroad. The total Indian population is 12,000 in reserves covering close to 3 million hectares.[14] The tribes include Guajajara, Urubu, and Guaja in Maranhao; Apinaje in Tocantins; and Assurini, Surui, Gaviao, Xikrin-Kayapo, and Parakana in Para. Some of these are in legally demarcated areas, but Indian lands have been invaded by settlers, and logged and mined by outsiders, and demarcation lines have been altered to put through roads or build towns.

A $304 million World Bank loan that helped build the railroad required CVRD to finance a $13 million component for Indian protection, which has been applied through FUNAI, mainly for demarcation and removal of settlers from the 631,000-hectare Xicrin reserve that adjoins the Carajás mining district and for cash compensation to the Gaviao and others whose reserves are crossed by the railroad and powerlines. There was less success in other cases. An advisory group of Brazilian anthropologists quit the project in protest against FUNAI's failure to expel invaders from a Guaja reserve, and CVRD had to suspend disbursements to pressure FUNAI into carrying out demarcations and resettlement of the Parakana, whose reserve was partially flooded by the Tucurui reservoir.

Another way of measuring the planning failure is by the environmental alterations in the area. Over the entrance gates to the Carajás mining district is a sign that says "CVRD, Carajás Ecological Park." Indeed, inside the chainlink fence, CVRD controls a 4,000-square-kilometer tract where the rainforest is largely intact and the headwaters of the Itacaiunas river run clean. Squatters and *garimpeiros* who have tried to invade the area have been pushed out by security guards. The only breaks in the forest cover are the company town for about 15,000 people, an airport with a 2,000-meter runway, and the mine sites where 150-ton trucks crawl antlike out of terraced pits and dump ore on endless conveyor belts that feed into crushers and washers. Tailings and wastewater are stored in retention ponds to avoid pollution. Carajás gets high marks for its environmental protection program.

Outside the CVRD mining district, however, there is vast deforestation. In the 47,000-square-kilometer area between Marabá and Rio Maria, which contains the eastern approaches to the Carajás mines, satellite images show that in 1972 only 300 kilometers had been deforested, By 1985, the year that the railroad was completed, 8,200 square kilometers of forest was gone, and in 1990 it was 10,000 square kilometers. Land clearing was mainly along state highway PA-150, which runs from Marabá to Conceição do Araguaia, and along PA-275, which is the spur going to Carajás. A dirt road off PA-275 leads to Serra Pelada, the biggest *garimpeiro* goldmine in Brazil, where a stream of red slurry tainted with mercury silts the Itacaiunas river, which supplies Marabá with contaminated fish.

The planning failure takes on an even greater order of magnitude if one considers the Carajás corridor from the standpoint of the new populations drawn into the area. Defending the rights of the 12,000 remaining tribal people in the area is easier than coping with the economic and social problems of 1 million newcomers and the political conflicts to which they give rise. The nine townships that form a 400-kilometer belt between Tucurui and Conceição da Araguaia had a combined population in 1970 of 94,976. A decade later it had grown to 251,226. It then exploded, reaching 892,050 in 1987. The estimated population for 1990 was 1.4 million, or 14 times the level of 1970, when the land rush began.[15]

In some cases, the urban population of towns such as Marabá (240,000) and Tucurui (130,000) is growing faster than the rural, and they can be considered cities, like Imperatriz, the main entry point to this region from the south, which has 250,000 people, and is both a commercial boomtown and a center of rural violence. The inability of town administrations to provide education, health and law-enforcement services can be measured by the meagerness of local tax revenues. Until 1988, Marabá had a budget of $5 million and Tucurui had less than $2 million. Both towns had not received payment in full of their share of federal and state tax revenues in 1987, and were deep in the red. Only token amounts were being paid for town support by CVRD and Eletronorte. Gran Carajás, as noted earlier, had failed completely.

Understandably, a World Bank end-of-project critique of the Carajás iron loan in 1989 concluded that the results were a mixture of successes and failures. The project had achieved its objectives in increasing Brazil's share of the world iron market, thereby earning foreign exchange and generating "significant regional employment and income benefits." But the report said the broad socioeconomic and environmental impacts of the project had not been adequately foreseen. It said this placed both CVRD and the bank under a "moral obligation" to help federal, state, and community authorities "provide needed infrastructure and services to the rapidly growing rural and urban populations—as well as to limit environmental degradation—in the region."[16]

For instance, the CVRD project included an investment of $144 million for township development, which was spent mainly relocating occupants of a slum area in Sao Luiz around the new port terminal and for the new town within the Carajas site. But there was also money for waterworks, a hospital, and some houses for railroad workers at a town called Paraupebas, which has mushroomed around the main entrance to the mine. In its loan commitment with the World Bank, CVRD estimated a Paraupebas population of 10,000 in 1988. In fact, the town holds 70,000, including a booming redlight shantytown district called Rio Verde where *garimpeiros* come from miles around. The rural areas around Paraupebas are a combination of large ranches, agricultural colonies created by agrarian reform projects, and squatters, who frequently try to invade the CVRD reserve. Except for this friction, there is little connection between the mine and its surrounding communities.

Faisal Salman, the first elected mayor of Paraupebas, is a medical doctor and an enthusiastic "green" campaigner. A banner across the new city hall building proclaims "Paraupebas, Green Heart of Para." Along rural roads there are billboards with Salman's motto: "Better to plant than to cut." Salman says that he tells farmers to reforest their plots. He has also encouraged a consumer cooperative that has a supermarket in the Carajás company town to buy from local vegetable and fruit farmers, instead of bringing produce from distant sources. "What we really need is a technical assistance program for the small farmers so they learn how to produce better. That's the kind of thing that Vale do Rio Doce should support," said Salman.

The World Bank critique put the same idea in a broader context. The report said:

> While the iron ore project . . . can in many ways be viewed as a model of effective environmental management in connection with the physical installation and day-to-day operation of large mining and transport facilities, at the same time—and ironically—it also clearly exemplifies an inadequate approach to environmental planning and control in the larger region affected by these investments . . . blame for the relative neglect of potential environmental impact in its larger area of influence must be shared between [CVRD and the bank], as well as by the Brazilian government more generally.[17]

The social and environmental problems of eastern Para began long before the railway or the dam were started. The influx of migrants come over highways that were built into the region when nobody knew about Carajás or Tucurui. It is clear, however, that the Carajás project and Eletronorte's Tucurui have brought an industrial and urban dimension to the occupation that was not there with the initial settlers. The solution to the regional social and environmental problems can only come through institutions like CVRD and Eletronorte that have a capacity for organizing economic and social action that is unmatched on the frontier. State and local political institutions are weak and lack the resources to deal along with overwhelming social problems.

* * *

A ride on the Carajás Express provides vivid impressions of the transformations that are under way in this key Amazonian sector. The train leaves promptly at 8 A.M. from the Guardian Angel station in São Luís. The ten-car passenger combination gets priority over the ore trains, which thunder down the main line five or six times a day, each with 15,800 tons of iron ore or manganese in 160 gondolas pulled by diesel locomotives. At the Itaqui port, a conveyor belt system can load ships of up to 250,000 tons in a day. The main customers for Carajás ore are steelmakers in Japan, Germany, Italy, and France, who helped finance the project.

The passenger service carries over 400,000 persons a year, but the train is half empty as it leaves São Luís. After a two-hour leisurely roll through a delta of irrigated ricefields and palm groves, the train stops at Santa Inez, a crossroad town in central Maranhão that is reached by buses from all over the Northeast. After Santa Inez, it is standing room only with more than 1,000 peasants, peddlers, *garimpeiros*, nuns, prostitutes, ranchers, and migrating families with numerous children sitting on suitcases, sacks of flour, or any free perch.

The train climbs to a low mountain ridge at kilometer 478 that separates the tropical plain from the Amazon forest. From there onward, the forest has been cut and reduced to ashes on both sides of the track as far as the eye can see. In some places, plots of mandioc and bananas planted among the black stumps show the presence of small settlers; in other sections, fenced lands with thin cattle feeding on brush are the mark of ranchers who have replaced the forests with short-lived, increasingly worthless pastures. In the dry season, before the rains bring back some green, it looks like a bombed-out battlefield. The only place that has been spared along the railway is the Guaja Indian reservation at Caru where the clear waters of the Pindaré river run through dense forest cover.

On the train, nobody seems concerned about the landscape. Passengers speak with confidence about what they are doing on the frontier. Albino de Jesus Martinez and Jose Fernando Ferreira da Silva both work at the Carajás mine and are returning from a visit to relatives in Maranhão. "Tucurui and Carajás changed my life," says Martinez, who was a 23-year-old store clerk in São Luís when he heard in 1981 that workmen were being hired by contractors for Tucurui. He signed up, worked at the dam for four years, and used his off-hours at the remote construction camp to take a course in mechanics. With Tucurui finished, he went to Carajás, where he now operates an 18-meter perforator, a complex machine with nine diesel motors that drills holes for explosive charges in the ore body. He passes around photos of himself at the controls of the monster machine and with his wife and child at his home in the Carajás company town. It's hard to tell which makes him prouder.

Ferreira da Silva came from an even humbler background. He said he was one of 16 children in a rural family at Barreirinhas on the coast of Maranhão, where mandioc and fish are the only food. To escape a lifetime as a man with a hoe, he went to live with a relative in São Luís and "with great sacrifice" he finished a technical high school course. He was 19 years old when he got a job in 1980 with a geological team doing exploration work at Carajás. Now he works in the automatic loading operations that fill the trains with ore. He is also an active union member. "We have a strong union and we were ready to stop the mine when the company rejected our demands for an adjustment against inflation, but we won in court and they paid. Carajás has been good for me, but workers have to organize and fight for their rights," he says. Ferreira da Silva has come a long way from rural Barreirinhas, where there are no unions and those who have stayed behind "all live in misery."

Most of the passengers don't work at Carajás or Tucurui, however. Many are on their way to the frontier to find land, and others are looking for jobs. That is what brings them to the Acailândia station, where a sawmill town that is on the Belém–Brasília highway is entering the age of metallurgy. Two industrial plants have been installed to convert Carajás iron ore into pig iron, the first dirty step toward making steel. The smelters are fired by charcoal, a primitive but cheap way to produce iron when wood is abundant. The surrounding area, shorn of trees, is dotted with igloo-like clay kilns that give off foul vapors as they bake wood from the surrounding forests into charcoal. Two more pig iron smelters are in operation up the line at Maraba, and two more are being put in on the way to São Luís. The installed pig iron capacity in the Carajás corridor is approaching 400,000 tons a year.

The pig iron producers are all from Minas Gerais, the southern mining state, where dwindling forests, rising costs for charcoal, and growing environmental pressures have made life difficult for pig iron makers. At Acailândia's "Industrial Center" they have a cost-free area for factories beside the railroad tracks, with transportation to and from the port at Itaqui, high-grade ore at a cheap price, and an abundant supply of charcoal from sawmill wastes and trees felled by land clearing. They have also received tax exemptions and Amazon investment subsidies from the federal government.

The first pig iron plant at Acailândia was Viena Iron Works of Maranhao, the Amazonia branch of the traditional Valadares family of Minas Gerais which has been in the pig iron business and national politics for years. Wellington Valadares, Viena's engineering director, gave his view of the coming of pig iron to Amazonia as we sat in a ramshackle office by the light of a gas lamp because of a power failure, a common event in Acailândia. He explained that Brazil produced 3 million tons a year of pig-iron for export to the United States, Japan, and Europe "because foundries there see it as a dirty, low-tech business with high labor costs." In contrast, he said Brazil has low labor costs, excellent iron ore, and, in Amazonia, plenty of charcoal.

"I say, each parrot to his own perch. We have a comparative advantage in pig iron and we should be encouraged to develop our capabilities," said Valadares. "If the ore is not turned into pig iron here, where will the jobs be? The Carajas mine has created very few jobs. One day it will just be a big hole in the ground and the railroad line will be a place for kids selling oranges and peanuts along the tracks," Valadares added. He said that his plant was a $3 million investment that had created 200 direct-hire jobs and would give work to another 800 persons as charcoal makers and truckers. Some day, he mused, there will be a steel mill at São Luís to transform pig iron into higher value steel, and metal processing industries will turn out finished products—in other words, a new version of Minas Gerais.

Valadares said the pig iron plants were not going to destroy the Amazonian forests, as critics of this subsidized activity claim, because the charcoal comes from trees that would be burned anyway by ranchers clearing land or from the waste product at sawmills. "It is not economical to bring charcoal from deep in the forest when there are so few roads. When local supplies run out we will have to make our charcoal from plantation forests," he said.

IBAMA has been putting pressure on the pig iron plants to meet the requirement that they create renewable sources of charcoal or shut down. Valladares said Viena had begun preparing for that day by buying a logged-out ranch property near Acailândia where 50 hectares have been planted in fast-growing eucalyptus supplied by CVRD's tropical forestry research center at Acailândia. He said that in seven years his plantation would supply 50 percent of Viena's annual charcoal needs. The Viena plant capacity with two furnaces operating full time is 100,000 tons a year, and Vale do Pindare, the other pig iron operator, is at the same level. Each ton of pig iron consumes 750 kilos of charcoal, so a production level of 200,000 tons annually would need 150,000 tons of charcoal.

Many environmentalists hotly oppose pig iron plants in Amazonia. They attribute severe deforestation of Minas Gerais to the pig iron producers, and say the same will happen to the Carajás corridor. Philip Fearnside, a resident American ecologist at INPA in Manaus, has calculated that the annual production target of 1.2 million tons of pig iron announced by the government when the Gran Carajás program handed out subsidies to 13 firms would consume so much charcoal that 20,000 square kilometers of native forest would be cut each year to meet the demand. IBAMA subsequently canceled the fiscal incentives granted to five pig iron smelters that had not begun investments and suspended any new ones.

The prohibition of pig iron production in Amazonia for environmental reasons is not a satisfactory solution, however, for the thousands of workers in the smelters, charcoal kilns, and forests, or for many more who are looking for work in the Carajás corridor. There are environmentally benign ways of producing charcoal that do not depend on forest clearing and could bring reforestation to degraded forest areas. Reforestation and forest management would create even more jobs if adopted by the pig iron industry.

CVRD foresters have been experimenting for 15 years with eucalyptus varieties at a tropical forestry center at Linhares, Espirito Santo, where cloned species are being mass-produced for plantations that supply Brazil's booming short-fiber cellulose industries. These varieties are being tested in the Carajás corridor at three CVRD experimental forestry centers between Marabá and Pindare-Merim in Maranhão. Some varieties have been developed for high charcoal yields, and they can be reproduced on a large scale from tissue culture.

At Acailândia, solid ranks of eucalyptus grow off the Brasília–Belém highway at a CVRD forestry station. The rapid growth of these specimens has attracted local attention, and demand for plants is high, not only from pig iron producers and ceramic makers, but from ranchers. "We know now what varieties adapt well to local conditions, and the best varieties we have produce 300 cubic meters per hectare. With that volume, 2,000 hectares a year produce enough charcoal for 200,000 tons of pig iron," said Salim Jordy Filho, director of the Acailândia forestry station.

If these figures are right, a 200,000-ton pig iron producer would need to plant only 20,000 hectares to obtain a 30-year supply of charcoal. Eucalyptus varieties tested by CVRD grow to cutting size in eight years in Brazil's high-rainfall areas, and then grow back for a second and third cutting in a 25-year cycle. It follows that an industry production level of 1.2 million tons could be sustained from plantations occupying 1,000 square kilometers, instead of 20 times as much native forest, as calculated by Fearnside. The economic question remains whether pig iron producers will decide to make the necessary investment. A plantation could cost $400 per hectare for which there is no special credit now. But if IBAMA applies the law on sustainable charcoal procurement, the pig iron producers may have no alternative but to turn to a reforestation system, such as CVRD has developed.

The importance of CVRD's forestry program for Amazonia goes well beyond the problem of a sustainable supply of charcoal for pig iron production. It has opened up the prospects of converting degraded land of the Carajás corridor into a "carbon sink" where tree plantations soak up significant amounts of CO_2 from the atmosphere through photosynthesis. CVRD has a direct commercial interest, of course, in continuing to sell Carajás ore to the pig iron smelters and in promoting expansion of metallurgy based on charcoal, including iron-based alloys, so energy wood plantations are part of this strategy. CVRD is also a diversified company that has major cellulose investments, including a joint venture with Japanese partners in a 600,000-ton plant in Minas Gerais. CVRD's eucalyptus research made that cellulose project possible, and the "know-how" has been transferred to experimental stations in the Carajás corridor in preparation for new cellulose projects that will use the railroad's cargo capacity.

This diversification from mining into forestry is a strategy that combines environmental considerations and a response by CVRD to demands that it assume a broader role as a catalyst of development in Amazonia. When the Carajás mining project was being installed, CVRD's major concern was to bring the project onstream at the lowest financial cost. The initial budget of $4.5 billion was stripped of superfluous expenses, and the actual cost of the mine-railroad-port complex was $3.5 billion. During that stage, whenever CVRD was called upon by politicians and social pleaders to assume "external responsibilities," they got a stock answer from Eliezer Batista da Silva, the president of the company through the Carajás construction period: "We are a mining company, not a regional development agency." CVRD prefered returning $74 million economized from a $304 million World Bank loan that helped finance the project rather than spend it, as the bank suggested, on community projects along the railway.

This frugality protected CVRD's investment, but it referred everything having to do with land occupation, urban development, and environmental control outside the immediate areas of the mine-rail-port project to federal and state agencies that had conflicting interests, inadequate programs, and scarce funds. As pointed out earlier, the Gran Carajás program failed to deal with the crisis that developed beyond CVRD's direct responsibility. But CVRD, in the face of the criticisms, has not disengaged from the problems created by the Carajás corridor and has come up with an ingenious reforestation plan that could significantly increase employment while reducing global atmospheric pollution.

The CVRD forestry plan is a long-term move toward recovery of degraded pastures and protection of native forest that opens up potential employment for thousands of workers. Batista, after leaving the CVRD presidency, remained the head of international operations and is the company's elder statesman. Global concerns over Brazil's environmental performance drew Batista's attention to the possibility of repairing damages to the Carajás corridor through a major reforestation effort that would be economically attractive to investors, including CVRD. The proposal, based on work by the University of Sao Paulo's Institute of Advanced Studies, combines the contribution new forests can make to slowing global climate warming, by capturing CO_2, a greenhouse gas, from the atmosphere, with the creation of a new source of cellulose for pulp and paper industries in the Carajás corridor.

The proposal calls for joint-venture investments between CVRD and foreign cellulose companies for industries that would be supplied from up to 1 million hectares in eucalyptus plantations in degraded areas along the railroad line. The plantations would be operated by an "eastern Amazon forest management center" created as a private foundation with an international scientific advisory board. Each reforestation area would be linked to protective management of adjacent native rainforest that remains in the Carajás corridor.[18] "This would reverse the environmental degradation process and at the same time provide another source of cargo for the railroad and lots of new jobs," said Batista.

According to CVRD, the Forest Management Center would be able to support ten pulp plants, each producing 420,000 tons of cellulose a year, which would employ 40,000 in the forest and industrial sectors and create 80,000 more service and other indirect jobs. The industrial investment alone would require $6.5 billion, basically from private investors. For the reforestation efforts, the plan depends on low-cost financing from a global "green" fund that would provide at least $1 billion for reforestation of 1 million hectares as part of the campaign against climate warming.

The Collor administration embraced the Carajás corridor reforesation plan as an Amazon priority. Perhaps the most important factor was the employment opportunities that would be created by a private-sector pulp and paper industry without need for much public financing. But the increase of stable jobs in the Carajás corridor would also address environmental problems created by social instability. "Generating permanent jobs in better-qualified positions will avoid men drifting into migrant predatory farming or pick-and-pan gold mining. This will consequently cause a drastic reduction in the anthropic pressures on natural resources, destruction of virgin forests, wide-spread use of slash-and-burn land clearing and river pollution from improvised mining," Batista said.

If the Forest Management Center lives up to these notices, CVRD would cease to be the villain of the Carajás corridor and would become the white knight of environmental protection. It might also make a lot of money. CVRD, which had a net profit of $750 million from its mining and metallurgy activities in 1990, has great expectations for its pulp and paper plans. The management is confident that the world market for cellulose will absorb the Amazon production, which could reach 4 million tons annually by the year 2010. CVRD estimates that the production of wood from 1 million hectares of plantation forests during a 15-year period would obtain $2.2 billion from cellulose sales to the pulp and paper industries; this calculation was based on the very low price of $3 per cubic meter, with production per hectare of 300 cubic meters. Armed with these arguments, Batista began a worldwide search for partners in Scandinavia and Japan for two big cellulose projects, each providing for reforestation of 150,000 hectares. But cost estimates for reforestation at close to $1,000 per hectare made prospects uncertain without "soft" financing on ecological grounds. This was not available in 1991.

The inclusion of forestry projects in a chapter devoted to Amazonian mining should not be seen as a digression because the role of mining in sustainable development is the conversion of nonrenewable mine income into diversified investments that are economically and environmentally sustainable. When mining companies such as CVRD do the job of conducting forestry research and promote forestry enterprises that reforest degraded areas, they are contributing to Amazonian economic development in ways that repair ecological damages and generate badly needed jobs, training in forestry and industrial technology, local and state tax revenues, and, in the long run, community stability, the keystone of environmental protection. This is the basic difference between large industrial mines and the *garimpo* mining system.

PART III

Generations of little men have nibbled, like mice, at the edges of Amazonia. . . . Those who would solve the problem of its development must be possessed of superhuman imagination and boldness. . . . They must be willing to forget much that is familiar and start anew as if on another planet.

William Lytle Schurz, *Brazil, The Infinite Country* (New York: Dutton, 1961), p. 64.

EIGHT

The "New Frontier"

Getting things right in Amazonia is not as daunting in theory or in practice as implanting life on Mars. It is a social problem that involves a rational ordering of human activity that will make the use of natural resources compatible with the stability of existing ecosystems of the region. The relevant question for this study is how to bring about human behavior in modern Amazon communities that will satisfy economic and social needs by means that are consistent with a healthy, sustainable environment. A new mentality has to develop, based on experience, inventiveness, and socially transmitted values. The process by which this evolves is what I call the "new frontier."

Environmental militants picture the 1980s as a decade of destruction in Amazonia. Those concerned with economic growth see it as a decade of wasted opportunities. Both are justly critical of a human occupation that was reckless, rootless, and in many cases lawless. But for all its errors, the decade was not an unmitigated disaster, as radical critics of "developmentalism" claim. It produced many remarkable discoveries in natural resources and a wealth of human experience in tropical ecosystems that place Amazonia in sight of new frontiers.

At the local level, settlers who were attracted to Amazonia by the lure of land and easy access to frontier assets, such as gold and timber, have discovered that the exuberant tropical world where they now live is fragile, deceptive, and often dangerous. They are learning the hard way, through agricultural failures, exhaustion of nonrenewable resources, epidemic diseases and social violence, that special care must be taken in the use of land, water, forests, and minerals if economic activity is to be sustainable.

During this period of experimentation, predatory human behavior, inappropriate technologies, and marketing errors have brought much of the economic thrust in Amazonia to an impasse. The "old frontier" is in decline because there are basic gaps in critical areas of knowledge on how Amazonia should be developed. There is also severe social fragmentation. The personal initiative that brought the settlers to the frontier has often been wasted. But the settlers are still looking for better ways of organizing economic and social activities. The critical need is for a support system of essential community services, applied research, rural extension, credit, and market mechanisms.

At the international level, there should be strong interest in helping the "new frontier" emerge. The "discovery" that tropical deforestation can alter the global climate has changed Amazonia's relations with the outside world. Makers of world opinion have created a frightful scenario of forest fires that could turn Amazonia into a wasteland, leading to floods and famine for all humanity. As a result, this unknown region, of no concern in the past to any but a handful of scientists, is on the agenda of mankind's "critical issues." The forests, climate, tribal people, natural resources, drug smuggling, land conflicts, and human rights in Amazonia have all become subjects of international concern. This is a highly sensitive issue in North-South relations.

The global clamor over the Amazon rainforest reached an emotional peak in late 1988 when the assassination of Francisco "Chico" Mendes by order of a rancher made this appealing leader of the Brazilian rubbertappers into a "martyr symbol" of resistance to forest destruction. The intensity of ecological criticism this frontier murder turned on Amazonia baffled and irritated the military and foreign service professionals who dominated Amazon policy during the insecure Sarney government. These sectors cultivated the same arrogant nationalist attitudes against "foreign interference" in Amazonia that had once tried to keep this international river closed to foreign-flag ships. They opposed whatever seemed in their eyes to "internationalize" Amazonia, including offers to convert ("debt-swap") parts of Brazil's foreign debt into funding for nature programs to protect Amazon forests. But the international uproar over the Mendes killing could not be ignored because Brazil's negative posture on the environment had contributed to a confrontation over money with multilateral development banks (MDBs) and Brazil's foreign creditors.

Trouble began when Sarney's government suspended payment on the foreign debt in 1986. This line was later abandoned and Brazil sought a renewal of financing; but by then the MDBs, under severe pressure from donor countries and international "green" movements, had imposed new requirements for lending that added environmental "conditionality" to the usual economic criteria. This infuriated Sarney. Breaking the rule of silence that usually surrounds negotiations with international lenders, Brazil denounced the World Bank in early 1989 for withholding approval of a $500 million loan for power development, including several Amazon dams, partly on environmental grounds.

"The World Bank is meddling in our internal political affairs. . . . We Brazilians run things in Brazil, and we don't accept interference," said an indignant Sarney.[1] He had just returned from the funeral of Emperor Hirohito in Japan, where world leaders, including President Bush and President François Mitterand, had lectured him on Brazil's environmental record, making it clear that refinancing of Brazil's foreign debt required a new attitude. "In my diplomatic experience, I never saw such all-out pressure on Brazil," commented Paulo Tarso Flecha de Lima, then secretary general of the foreign ministry.

Two months later, Brazil swallowed its prickly pride of "sovereignty" and adopted a more pragmatic position. Sarney announced an environmental management program he called "Our Nature," which was made up of a package of decrees and legislation that included suspension of tax incentives for land clearing in Amazonia, a campaign to reduce burning of forests, and the consolidation of the national environmental administration into a new federal agency. Sarney took the occasion to denounce international ecologists for "a crude campaign that affects our products, our people, and our institutions, creating problems for the stability of our democracy."[3] All the same, Sarney came a long way toward meeting the international concerns over environment that had contributed to Brazil's financial isolation.

Fernando Cesar Mesquita, the first president of the new Institute of Environment and Natural Resources (IBAMA), admitted later that "it was done, regrettably, under international pressure and not everyone in the government was honestly behind the new policy."[3] But the shift toward a more active environmental policy was well received by public opinion, where international criticism and local environmental groups had implanted a fast-growing interest in "green" issues in Brazil. There was soon frustration, however, because environmental programs lacked budget support. When an expected immediate inflow of international funds to support an emergency environmental effort did not materialize, there was bitter criticism. Mesquita, an uninhibited journalist who had been Sarney's press secretary, detected dishonesty in the international "green" campaign against Brazil.

"In Italy, millions of readers of the magazine L'Espresso received a form petition, to be signed and mailed postage-paid to the prime minister, saying: 'Don't buy Brazilian iron ore. Don't let the European Community finance development projects in Amazonia.' The cost of the campaign was supposedly sponsored by the World Wildlife Fund, but we learned in Italy that mining companies that compete with Vale do Rio Doce iron exports were involved in the financing," said Mesquita. He disparaged opposition from President Bush and U.S. senators to a trans-Amazon highway through Acre to the Pacific: "It has nothing to do with ecology. . . . [what] they don't want is our soybeans, lumber and other goods competing with theirs at lower prices in the Pacific markets." Mesquita also questioned the "moral authority" of President Mitterand to criticize Brazil on global environmental dangers while France continued testing nuclear arms at the Mururoa atoll in the Pacific, "where Mitterand ordered the sinking of a Greenpeace ship."[4]

Nonetheless, during the last year of the Sarney government, Mesquita injected dynamic personal leadership into the makeshift unfinanced "Our

Nature" program. He led airborne inspection teams that arrested illegal loggers redhanded. There was a reduction in forest clearing and burning. With Brazil in a more cooperative mood, the World Bank concluded a $137 million loan, with German co-financing, for a national environmental program that was the first of its kind by the bank with any of its 155 member countries. Disbursements did not begin until a year later, after the new government of President Collor took office, but 1989 was a turning point toward the "new frontier," where sterile confrontations must be replaced by international cooperation for environmental protection and scientific knowledge of the tropical forests.

Deeper changes that could have dramatic importance for the future of Amazonia then came both from Brazil and on the international front. The hesitant steps taken by Sarney became a rush toward international cooperation under Collor. The new president understood better than Sarney the importance of the environmental issue for Brazil's foreign relations. Collor, who was 40 years old on taking office, said he belonged to the John Lennon generation that wanted mankind at peace with nature. In addition to his generational vibrations, Collor saw that the rainforest "mania," as Sarney called the campaign to "save the Amazon," could be turned to Brazil's advantage if environmental protection of the Amazon elicited large-scale international financial and scientific cooperation.

Collor signaled a new posture by appointing José Lutzenberger, a militant ecologist, to a prestigious job as presidential secretary for environmental affairs. Like many of Collor's early appointments, the choice seemed like a stroke of clever marketing. Lutzenberger gave the Collor administration "instant credibility" on the international environmental circuit where he was well known for his tirades against deforestation in Amazonia and as a feisty critic of Brazil's past development policies. However, with no experience in government and mediocre scientific credentials, he was put in charge of IBAMA, a complicated agency in need of reform, and the policy-setting National Environmental Council (CONAMA). A more technical input to environmental policy came from Jose Goldemberg, a physicist and former rector of the University of Sao Paulo, whom Collor named secretary of science and technology. Goldemberg, a specialist in energy conservation, represented Brazil ably in international negotiations on global climate change and in other scientific forums. His pragmatic problem-solving approach balanced, and sometimes came into conflict with, the visionary, impractical style of Lutzenberger, for whom the enemy was modern industrial civilization.

On the diplomatic front, Collor ordered the foreign office to round up the necessary votes in the U.N. General Assembly to make Brazil the host for the United Nations Conference on Environment and Development (UNCED), better known as ECO-92. In a trade off with Canada, Brazil got the conference and Canada got the secretary generalship of the meeting for Maurice Strong, former head of the Canadian International Development Agency, who had been the coordinator of the first U.N. environmental conference at Stockholm in 1972. That meeting gave birth to the United Nations Environment Program (UNEP), but it left the rich First World and the developing Third World countries far apart on poverty, development, and pollution. Brazil, along with India, argued that the root cause of environmental problems was poverty, and they were the strongest holdouts at Stockholm against subordinating economic growth in the Third World to environmental restrictions and population control.

Collor's objective in bringing ECO-92 to Rio de Janeiro was to change Brazil's image as an environmental hooligan into that of a responsible custodian of Amazonia, worthy of international support. This called for a radical break with the nationalist obduracy that had aborted earlier attempts to bring international scientific research to Amazonia. The classic example of Brazil's porcupine stance on Amazonia was the rejection in 1951, on the grounds of "defense of sovereignty," of an agreement to create an International Institute of the Amazonian Hylea under the auspices of the United Nations Educational, Scientific and Cultural Organization (UNESCO). Brazilian scientists—then, as now, starved for funds—had promoted the initial studies for this institute and the Brazilian government had signed the draft agreement during the second UNESCO conference at Mexico City in 1947. The institute was to have been set up in Manaus, under an international board, for the purpose of promoting and coordinating scientific studies of the Amazon region. The other Amazonian countries were supportive of the plan. So were Britain, France, and Holland, which were to have been members of the institute through their then existing Guyana colonies. But a nationalist outburst in Brazil sank the Amazonian Institute.

When Rio de Janeiro was still Brazil's capital, the Military Club there was a major political forum. Frequented by senior officers, the club was where the campaign for nationalization of Brazil's oil industry got its vital military support. This elite audience was treated on June 27, 1951, to a "geopolitical" address by Artur Bernardes, a former Brazilian president and leader of the *o petroleo e nosso* (the oil is ours) campaign, who attacked the proposed Amazon treaty with nationalist emotion.

Bernardes said the Brazilian Congress should reject the agreement because it exposed Brazil to losing control of an "internationalized" Amazonia. He argued that under such an agreement, Brazil, "disposing of one hundred percent dominion over Amazonia," would put itself "on a footing of equality with other countries that . . . don't own one foot of land in the region." He said that ratification of the proposed convention would amount to "alienation of our dominion over all that region through a document in which Unesco simulates that it is giving something to the Amazonian states, but in fact is taking something from them."[5] In the nationalistic political mood of that time, the Amazonian Institute treaty was buried.

The next decade saw another spasm of nationalist indignation over a new "threat" to Amazonia. This time it was a bizarre scheme proposed by the Hudson Institute, a "think tank" created by the late Herman Kahn, a futurologist who used to help the Pentagon think out strategic questions. Kahn's associate on Latin American affairs, Robert Panero, a wealthy Colombian-American businessman, conceived the idea that Amazonia's future would be well served by putting a great dam across the main river at Monte Alegre, 800 kilometers upriver from Belém. The project envisaged flooding 600,000 square kilometers of forest, thereby creating an artificial Amazon Great Lakes system.

The benefits of this billion-dollar project were alleged to be water transport and power. The lakes, like the Great Lakes of the United States and Canada, were to provide an all-year regional water transport system, involving Brazil, Peru, Bolivia, and Colombia, with a canal at Casiquiare eventually connecting the Amazon heartland with the Orinoco basin of Venezuela. This was supposed to provide bulk transportation for great inland mineral deposits, detected by aerial surveys, and generate a colossal 70 million kilowatts of electricity at the dam site. The inundated area required relocation of most of the cities of the middle Amazon, including Manaus, but this was not regarded as important. The ecological effects, including the flooding of 20 percent of the Amazon rainforest, were left out of the calculation.

The Great Lakes project was formally submitted to the governments of Brazil, Bolivia, Colombia, and Peru in 1967. The response in Brazil was a violent press campaign against the "imperialist" project and a congressional inquiry against the military government, led by Bernardo Cabral, then a young federal deputy from Manaus and later Collor's first minister of justice. He was dropped by Collor when an indiscreet affair with a fellow cabinet member, Minister of Economy Zelia Cardoso de Mello, became a political embarrassment. In the Great Lakes affair, the project was dropped. Whatever the merits or deficiencies of the proposal, they were not considered in the torrent of nationalist rhetoric that killed the idea.[6] The same emphasis on "sovereignty" in Amazonia was evident in the Brazilian delegation's positions at the Stockholm conference on environment in 1972 and in the diplomatic initiatives that led to the Amazon Cooperation Treaty in 1979 as an "umbrella" against external interference in the region.

The change under Collor in Brazil's posture on international cooperation in Amazonia was a radical break, therefore, with a long-standing political position. Collor was gambling on an improvement in the possibilities for North–South cooperation in economic and environmental matters of mutual interest. The gamble received a quick payoff. The annual summit meeting of First World leaders at Houston, Texas, in July 1990 made a political decision to offer Brazil financial support for an Amazon forest protection and resource management program that could test ways of saving tropical forests worldwide.

Declaring that "the destruction of tropical forests has reached alarming proportions," President Bush and his fellow Group of 7 (G-7) leaders from Germany, France, Britain, Italy, Canada, and Japan welcomed "the commitment of the new Government of Brazil to help arrest this destruction and to provide sustainable forest management." The G-7 said they wanted a program in Brazil that would "counteract the threat to tropical rain forests in that country" and serve as an experience that could be "shared with other countries faced with the tropical forest destruction."[7]

This proposal, at the initiative of Germany's Helmut Kohl, was a breakthrough for international environmental cooperation in Amazonia. The G-7 decision addressed, for the first time, global tropical deforestation in the place that matters most—the Brazilian Amazon heartland. If rainforest countries had an OPEC, Brazil would be the Saudi Arabia. Tropical rainforests are found in southeast Asia, West Africa, and Central America, but the Amazon basin of Brazil alone contains one-third of this type of forest. Not only is this the largest continuous tropical humid forest in the world, but compared with Asia's decimated primary forests, it is still largely intact. So, if the problem of vanishing tropical forests is going to be tackled head-on, Brazil is the place to begin.

Until recently, the tropical forest question in the western hemisphere has only been nibbled at in experimental stations in Costa Rica, Puerto Rico, or the Panama Canal zone, or through study missions at ecological niches in unpopulated parts of Amazonia. This work has been of scientific value as basic research on flora and fauna, climate, soils, and water in tropical ecosystems. But in terms of biodiversity, the great storehouse is the Amazonian rainforest where there is nothing approaching a complete genetic inventory and biochemical analysis of the enormous variety of living species, many still unclassified. Until this challenge to scientific knowledge is addressed on an Amazonic scale, which means a major mobilization of scientific and financial resources, all the jeremiads about genetic extinction and vanishing species will be hollow lamentations.

With attention properly focused on the Brazilian Amazon, the G-7 commissioned the World Bank and the Commission of the European Communities to help Brazil prepare a specific proposal for the Amazon pilot program. This was a major opportunity for putting into practice the integration of economic and environmental aspects of development that became one of the World Bank's major concerns in the late 1980s under the presidency of Barber Conable. The problem for the drafters of the G-7 Amazon program was how to develop an operational plan that would meet the global environmental concerns of the First World, focused mainly on halting deforestation, putting off climate change, and conserving biodiversity, and, at the same time, serve Brazil's interest in developing a new strategy for sustainable economic growth in Amazonia.

Work on the proposal began in late 1990 under the coordination of

Nancy Birdsall, a senior economist in charge of environment in the World Bank's Latin American operations division. Brazil set up an interministerial committee charged with producing an overall statement of Amazon policy and specific projects. Missions and papers flowed back and forth between Washington, Brussels, and Brasilia. It is politically significant that the G-7 proposal went forward in Brazil without the ultranationalist reactions of the past. Birdsall's skillful leadership of the World Bank–EEC team gave the Brazilians the primary initiative in presenting the ideas contained in the proposal. As visualized by Brazil, the proposal had four guiding principles.

1. *Environmental burden sharing.* International benefits deriving from better management of Amazon ecosystems, such as reduction of atmospheric CO_2 and preservation of biodiversity, justified foreign financial transfers on highly concessional terms. Foreign funds would be applied mainly to cover costs of forest protection programs and access to new environmental technologies, such as solar and biomass energy, developed by the G-7 countries.
2. *Additionality.* Participating financial donors would provide concessional loans or grants for the Amazon program in addition to any other existing development lending, including World Bank loans.
3. *Social forestry.* Efforts at forest preservation would fail unless they contributed to a viable standard of living for residents of the region, without whose support preservation could not succeed. Legitimate development objectives of private individuals and local communities should be embodied in projects testing sustainable development with minimal deforestation.
4. *Long-term commitment.* There is no quick technological or policy fix for the complex problems involved in tropical forest protection, so long-term investment was needed to improve scientific knowledge and public understanding. Therefore, a five-year program period should be followed by sustained financing through endowments or trusts.

The Collor government was eager to nail down G-7 support for an Amazon program before the ECO-92 meeting, but turning the general principles into a specific proposal proved difficult. The preparatory sessions revealed weaknesses in the government's ability to draw the blueprint for such an ambitious program. Coordination of the proposal fell to Lutzenberger's environmental secretariate, but his leadership was weak. He was spending most of his time flying around the world to address international ecology groups, or taking tea in London with Prince Charles, who made an "ecological visit" to Brazil with Lady Diana in 1991. When Lutzenberg was in Brasília, he shunned his office in the glass towers of the capitol and kept a reclusive residence among the trees of a national park. His inadequate and inexperienced staff had to grapple with problems for which he offered only oracular guidance. Lutzenberger's ineffectual managerial style weakened the environmental sector at all levels, and the G-7 proposal was no exception.

The negotiation with the World Bank and EEC was also hampered by the stiff procedural formality of the meetings. The Brazilian "team" was under the supervision of Clodoaldo Hegueney, a senior foreign service officer assigned to the Ministry of Economy, where he was in charge of international affairs. Brazil's foreign service mandarins regard negotiations as a martial art, where points are scored by outwitting an adversary. This is perhaps an advantage in trade talks or arms agreements, but it is a style that stifles the flexible, accommodating approach that is needed for successful international cooperation at the many levels of government and society involved in the Amazon program.

There was no time for delays, however, because of the pressure to get a proposal ready for the G-7 summit meeting, held in London in July 1991, and the Brazilian bureauocracy cobbled together a presentation that provided a general framework for a new Amazon policy. It contained many new ideas on sustainable economic development and environmental protection; but it was incomplete and more like a wish list for a green Christmas than a balanced strategy. The process that gave rise to this framework gave greater weight to the environmental and scientific sectors in the federal government and less to economic and social areas. Agriculture, mining, health, and education had no direct role. The governors of the nine states of Amazonia, and their federal congressional delegations, were not much involved. NGO environmental groups were consulted and many submitted proposals for "demonstration projects" that accompanied the presentation; but no effort was made to involve Amazon populations directly through public hearings or a system of consultations with communities.

Nonetheless, the Brazilian Amazon proposal represented such an advance over previous positions and contained so many of the elements of a "new frontier" for Amazonia that it won qualified support at the G-7 summit in London in 1991. Brazil proposed a $1.25 billion five-year program, with a "first-phase" outlay of $242.5 million. Japan and France balked at making such a large commitment and the initial request was whittled down to $50 million, which was considered the most that Brazil could use effectively without further studies. During the start-up period, the proposal was to be strengthened by hiring outside consultants on administration and for study of large investment areas, such as reforestation. After months of haggling, the G-7 agreed to an initial $250 million commitment.

The fundamental lines of action in the Brazilian Amazon proposal included:[8]

- Creation and administration of conservation units, with primary emphasis on prompt demarcation of Indian reserves
- Territorial organization of the entire Amazon region through ecological-economic zoning as a legal-technical basis for licensing resource utilization, including land and water
- Upgrading monitoring and surveillance of forest resources through remote sensing and specialized environmental police, in combination with local communities
- Recovery of degraded areas—through reforestation—and incentives for sustainable forest management, natural extractive reserves, and protection areas
- Scientific research, with emphasis on strengthening Amazon institutions and maintaining scientists in the region
- Budget support and personnel training for a national system of federal, state, and municipal environment offices
- Funding for "demonstration projects," including some proposed by environmental NGOs and "grassroots" community organizations

Match this agenda for action with the Amazon situation that has been described in this study and what emerges is an implicit strategy to contain and reverse forest destruction. This strategy consists mainly of reducing the pressure for clearing land by existing settlers through better land-and-forest-use technology, while at the same time excluding a very large portion of the Amazon territory that is still intact from socioeconomic "interferences" that have produced environmental damages in the recent past. Eneas Salati, the climatologist whom Goldemberg reappointed director of the Amazon research institute (INPA), participated in drafting the proposal and argued that conservation units and zoning should preserve as much as 70 percent of the Brazilian Amazon hylea forest.

Salati's claim is ambitious, but not too far-fetched to be used as a "new frontier" target. It bears repeating that 90 percent of the forests of the Brazilian Amazon are intact or essentially undisturbed, despite the more than 400,000 square kilometers that have been cleared. In relation to the full "legal" Amazon, or 5 million square kilometers, the areas that have been stripped of forest are primarily located in the transition zones of eastern Pará and Maranhão, northern Mato Grosso, Rondônia, and southeastern Acre. Beyond that perimeter there is a natural line of defense for the hylea, which occupies the interfluvial lowlands and stretches unbroken over western Amazonia, the north bank of the Amazon, and the Rio Negro valley.

The front-line strongpoints for such a defensive strategy are the indigenous reserve areas that occupy environmentally sensitive positions in Amazonia. In Brazil, more than 500 Indian areas appear on the anthropological maps drawn by the National Indian Foundation (FUNAI); in the sub-

Andean region, Colombia's Amazon areas are already largely assigned to Indian reserves, and in Ecuador and Bolivia there are strong local indigenous movements that have obtained recognition of communal rights in large forested areas.

The identification of these Indian homelands is based on historical-anthropological criteria of tribal occupancy that are independent of agro-ecological zoning. But many of the indigenous units are simply "paper reserves" that have never been fully surveyed, demarcated, legally registered, and placed under effective administration. The Brazilian constitution of 1988, reinforcing its strong chapter on Indian affairs, ordered that all Indian lands be demarcated and duly registered by November 1993, the last year of Collor's government.

If Collor fulfills his public pledge to meet the deadline, the increase in Indian areas would be 405,813 square kilometers,[9] almost entirely in the Amazon region. With the 300,000 square kilometers already registered in all Brazil, this would place more than 10 percent of the Amazon region in the category of protected indigenous areas. Altogether, the sum of Indian areas, extractive reserves, conservation units, and national and state forests could reach over 1 million square kilometers by the year 2000, an enormous area equal to Germany, France, and Italy combined.

There is an understandable disbelief in the efficacy of reserves and conservation units in Brazil based on experience with existing national parks, Indian areas, and ecological reserves. Without adequate administration, such areas are invaded, looted, and put to uses that degrade the environment. Identification, demarcation, and administration of Indian reserves, which are federal public lands, is the responsibility of FUNAI. This is one of Brazil's most demoralized public agencies. Under Sarney, it was shaken by scandals in illegal sales of timber from Indian reservations that led to the indictment of Romero Jucá Filho, who was president of FUNAI before unsuccessfully running for governor of Roraima, on charges of illicit enrichment. Under Collor, FUNAI's budget was slashed 75 percent as part of a general austerity plan. With only $2 million to demarcate 91 identified areas in 1991, Cantidio Guimaraes, a former airforce colonel placed in command of FUNAI, said he faced "an impossible mission."[10] A few months later Guimaraes was fired because, said the Brazilian press, someone had to take the blame for the failure to demarcate the reserves.

A better explanation was that Collor had been embarrassed during an official visit to the United States in June 1991, by a letter from eight U.S. senators questioning Brazil's commitment to protection of Indian rights and Amazonia's environment. On his return, Collor fired Guimaraes before leaving the Brasilia airport and appointed Sidney Possuelo, a respected Indian scout, to the presidency of FUNAI. Possuelo's personal commitment to Indian rights was well known and immediately after taking office he submitted a plan to Collor for the demarcation of a unified Yanomani reserve covering over 9 million hectares. He also signed a contract with a private foundation that had raised $1.2 million for the demarcation of the central sector of the divided Kayapo reserve.

The credibility of any Amazon program to protect the forests and ecosystems depends on a strong Indian reserve program. Collor said his political decision to complete the demarcations depended only on having the funds needed to do the job. In the proposal to the G-7, the demarcation of Indian areas was the largest single item, totaling $66.6 million in the "first phase." In the five-year budget, the total amount for all conservation units, including national and state parks, natural extractive reserves, national production forests, and biological-ecological stations, as well as Indian reserves, was $315.7 million.

There are certain areas in the immensity of the Brazilian Amazon that deserve priority. The creation or consolidation of Indian reserves and other conservation units is ecologically critical because these areas are under immediate pressures from outsiders. Strategically, the most obvious are the state of Acre; the Yanomani Indian lands in Roraima, which suffered a massive invasion by wildcat miners; the Uru-eu-wau-wau reserve in the misty Pacaas Novas hills of Rondônia, where the main watersheds of that agricultural state originate; and the Mekragnoti reserve on the Xingú river, which is the missing link for a "Kayapo corridor" between Mato Grosso and Para that would be a natural barrier to forest destruction west of the Carajás corridor.

In Acre, the land conflict between new settlers and forest people produced an alliance of rubbertappers and Indians, whose resistance to deforestation has kept land clearing to only 5 percent of the state. Two enormous natural extractive reserves were decreed as a result of the Chico Mendes movement. One is a solid block of 10,840 square kilometers that prevents further large-scale land clearing in the southeastern end of the state, and another is a 5,500-square-kilometer reserve in western Acre that adjoins a national park of equal size and two Indian reserves. Acre's identified Indian reserves, together with the extractive reserves, national and state forests, national parks, and ecological station, cover 50 percent of the state's forests, which provide the upland origins for the great Purus and the Jurua rivers that dominate Acre's sylvan ecosystems.

The day all these conservation units and protected areas are functioning, the territorial organization of Acre will be secure. Without them, the landownership situation is chaotic. Registered private land claims in Acre exceeded by 35 percent the total area of the state.[11] This is a legal timebomb that grows out of land speculation in old, unsurveyed rubber estates, often covering a million or more hectares, where the only thing considered valuable was the rubber trees, not the land. The decrepit estates were bought, for a few dollars a hectare, by absentee owners in Sao Paulo. They pay neglegible land taxes, invest nothing, and wait for the extension of the BR-364 highway to Acre's border with Peru to send land and timber values skyrocketing.

The Yanomani are primitive hunter-gatherers in the state of Roraima, north of the Amazon, who occupy forests where the rivers and hills were found to be rich in gold and tin. A rush of wildcat miners into the area brought a commercial boom that elevated that frontier territory to statehood in 1988. That is a status that depends heavily on the continuation of mining as an organized economic activity. The Yanomani live in scattered, fueding clans and are not organized to defend their lands, so they are vulnerable and need protection. Collor expelled most of the miners, but there was strong resistance in Roraima to the complete loss of access to rich mining areas that have produced $100 million a year. State officials proposed a smaller Yanomani reserve as a conservation unit for 9,000 Indians, but operated by a technical administration that could license access to limited mining areas by registered firms or cooperatives. The mining would pay for administration of the reserve and would maintain an activity that provides most of the jobs for Roraima's 400,000 people and tax revenues for the state's economy.

The Yanomani issue came to a head in November, 1991, when the army high command joined the Roraima state politicians in open opposition to formalizing the huge reserve area along the Venezuelan border. General Carlos Tinoco, the army minister, argued that this would jeopardize "national security." Collor called a meeting of the Brazilian National Security Council, where Foreign Minister Francisco Rezek led a cabinet majority in support of the Yanomani reserve. On November 15, Collor signed the decree establishing a reserve area of 9.419 million square kilometers for the Yanomani, *the largest indigenous reserve in the world.* "Our sovereignty remains intact and, in practice, strengthened because Indian territories, under the constitution, are the property of the Union," said Collor. Tinoco, who was present, did not applaud.

Defenders of Indian rights and the environmental community, however, did applaud the decision, which not only met long-standing demands for a Yanomani park but dispelled left-wing opposition claims that, behind the scenes, the military were in control of Amazon policy in Brazil. Possuelo said FUNAI could complete demarcation of the Yanomani reserve in six months, if it received the necessary funds, which were awaited from international sources. This left unresolved the problem of future administration costs for a reserve that is larger than some European countries, and the rights to mineral exploitation in an area on which the state of Roraima depends for most of its income.

The Uru-eu-wau-wau are also primitive hunters, but they number only about 1,200, having suffered severe mortality after Rondônia's virgin lands were occupied by colonization projects, ranchers, and mine prospectors. The furtive Uru-eu-wau-wau withdrew into dense forests of enchanting beauty that cover hills from which pure streams descend to the Candeias, the Jamari, the Ouro Preto, and other rivers of western Rondonia. This natural sanctuary has been filmed at treetop level from an ultralight aircraft by

Vicente Rios, a daring Brazilian cameraman who shot all the footage for Adrian Cowell's "Decade of Desruction" documentaries on Amazonia. The Uru-eu-wau-wau are surrounded by environmental destruction in five adjoining townships, such as the Bom Futuro tinmine, the half-abandoned colonization project at Burrareiro and unauthorized land clearings along the highway from BR-364 to Costa Marquez on the Guapore river.

An Uru-eu-wau-wau reserve of 18,671 square kilometers was identified in 1978 and was demarcated and ready for registration in 1985, but political groups identified with settlers and miners in Rondonia mounted an opposition campaign. Governor Jeronimo Santana, in an angry letter to Sarney, said the reserve had been improperly drawn to include an agrarian reform site where settlers had been given land, and he attacked anthropological consultants of the World Bank whose reports on violations of Indian reserves led to a suspension of disbursement of funds for agricultural development to Rondonia in 1986. "Don't sacrifice the thousands of families who have come to Rondonia to find a better life," pleaded Santana, himself a migrant.[12] Sarney's Minister of Justice Saul Ramos tried to reduce the size of the reservation in a political deal with a Rondonia federal deputy, Moises Benesby, who owned a rubber estate adjoining the reserve and was interested in developing a tin mine behind a facade of putting new settlers on the Indian land.[13] This threatened not only the remaining Uru-eu-wau-wau but a forest area that has high ecosystem conservation value (see Rondônia zoning map in the front matter).

Collor reversed the dismemberment decree, but FUNAI administrators continued to report invasions of the Uru-eu-wau-wau reserve by lumbermen and miners, allegedly in collusion with a local IBAMA inspector, who was fired. The environmentally correct course was to create a combination of contiguous conservation units, including the Indian reserve, the overlapping Pacaas Novas National Park, two natural extractive reserves for rubbertappers on the Ouro Preto and Cautario rivers, and the Jamari national forest. The conversion of the wildcat mining camp at Bom Futuro into a licensed operation under environmental controls was necessary to reduce pressure on the Indian reserve as well as river pollution. Some of these measures were cautiously advanced, but there was a lack of administrative organization at IBAMA and weak enforcement of laws that would inhibit violations.

The Kayapo corridor is a defense line for central Para that runs the length of the Xingú river, south of the Transamazon highway, with the Mekragnoti reserve serving as the link between the existing Xingú Indian park in Mato Grosso and an already constituted Kayapo reserve in Para called Gorotire. The total of the three areas would cover over 9,400 square kilometers in a critical zone that is surrounded by cattle ranches, colonization projects, mining camps, and lumber mills.

Megaron Txucarramae, a nephew of Raoni, the best known Kayapo chief, has run the Xingú park as an employe of FUNAI for five years.

Megaron said a continuous Kayapo area was necessary to unite the Kayapo clans in defense of the entire Xingú homeland. There are 3,100 Indians of various tribes in the park, which was created as a refuge for threatened indigenous people in 1960 by the Villasboas brothers, famous Indian scouts for the old Indian Protection Service that preceded FUNAI. But migrants from the "national society" have surrounded the once-isolated park, and the Indians have grown more assertive of their rights. They want homelands, not a refuge, said Megaron.

> We hear the bulldozers and chainsaws of the lumbermen cutting trees just outside the park for the sawmills at Marcelândia to the west, but they don't enter the park because we are vigilant. To the east and south, there are big soybean and cattle farms around Canarana. Going north, there are *garimpeiros* who keep trying to get into the park but our guards put them out. With help from our friends, we can protect ourselves and the forests of the Xingú.[14]

Help for the Xingú park has come from Sting, the British rock star, who accompanied Raoni on a world tour, from the Vatican to Tokyo, and raised over $1 million for the Amazon Indian cause through benefit concerts. The Brazilian government was notified that the money would be available whenever the authorizing decree for the Mekragnoti reserve was issued. The amount was sufficient to demarcate—at about $500 per kilometer—the entire reserve, which lies on both sides of the Xingu river. While the Kayapo waited for the decree, interest earned from the money placed on deposit helped the Xingú park set up a patrol system for the 3,400-square-kilometer area, with 11 watchtowers around the park perimeter communicated by radio with a command center. Sting's connection with the Kayapo, which was criticized initially by some as a promotional stunt, was placed under the management of a nonprofit foundation, Fundacao Mata Virgem, with an advisory board of respected anthropologists and physicians who organized health programs for Indian tribes.

Indian reserves are a measure of the commitment by an Amazon country to do its part in conserving Amazon forests. Indigenous rights are a politically sensitive issue that tests the resolve of a government to implement environmental policies when they come under pressure from interest groups. From an ecological standpoint, Indian reserves combine, at least in theory, the protective features of a conservation unit with social equity for threatened ethnic minorities. FUNAI's Possuelo, one of a disappearing breed of Brazilian Indian scouts, has said of the Kayapo reserve: "At a minimum, in addition to protecting the Indians, it will preserve from destruction an extended area of forest where their history took place, where their heroes created their universe and where their ancestors are buried."[15] These traditions, as well as intimate knowledge of the sylvan fauna and flora, are supposed to qualify Indians as benign "friends of the forest."

Yet, Indian reserves, like other conservation units, are not a self-fulfilling panacea unless they are administered firmly, like the Xingú park. When Indians acquire a level of acculturation they become involved in interethnic relations, including the commercial attractions of the marketplace. Among the Kayapo, some local leaders, such as Toto Pombo of the Gorotire clan, trade in timber and sell gold rights within the reserve to outside miners for cash payments. Others hand over mahogany trees, worth a fortune to loggers, for a pickup truck, a generator and TV set, or food and medicine. The Gaviao of Para, who get a right-of-way fee from the Carajás railroad, don't harvest Brazil nuts any more in their reserve; they hire workers to do it. Timber deals in the Rondônia Indian reserves, attributed to corrupt FUNAI officials and loggers, involved payoffs to tribal chiefs, such as Jose Itabira of the Surui or Oita Mina of the Cintas Largas, who have homes in towns and deposit their money in banks.

The reality, not the theory, of Indian reserves and other conservation units raises two central questions that a "new frontier" Amazon forest conservation program must answer to be sustainable. One is administrative, and the other is economic. Who is going to manage the reserve, and what sources of income will cover the costs? The two are closely related. As much as possible, the reserves should be under indigenous administration with outside assistance for health, education, and legal services that can be provided by NGOs, if they have funding. Fundacao Mata Virgem and the Kayapo are an example. Economically, reserves for Indians and extractivists should be self-sustaining, for if forest dwellers don't generate enough income from their resources to support the occupants, their forests are unlikely to survive. Experience has shown the folly of depending on government offices to maintain reserves; they have neither the resources nor the motivation.

The "old frontier" is littered with the wreckage of official programs that failed in Amazonia because of severe administrative limitations and insufficient funds. This includes major projects, with the World Bank and the Interamerican Development Bank (IDB), which began as "model" programs for protection of the environment and Indian communities in areas opened to colonization by new highways. On paper they were impressive programs, but in practice the results served mainly to point out the pitfalls to be avoided by the G-7 program and other new ventures.

The World Bank's great project in the Amazon rainforest was the $450 million Northwest Development ("Polonoroeste") loan in 1981 that financed paving of the BR-364 highway from Cuiabá in Mato Grosso to Porto Velho in Rondônia. The loan included $193 million for a rural economic-social program to bring orderly agricultural settlement to the region, while protecting the ecosystems and the indigenous people of the area. The program was for settlement of 28,000 families, which later became 43,000 families, on good-quality soils, with credit for perennial tree crops, such as coffee and cacao, that would make them stable homesteaders.

After five years, the Polonoroeste program was declared a "disaster" by its critics and an economic casualty by its friends. Many reasons have been advanced to explain the problems of what Barber Conable has called "the most troublesome project the World Bank ever took on." There was a serious underestimation of the migratory influx that would come with the paving of the highway. Families seeking land in Rondonia alone exceeded 50,000 between 1983 and 1986; and later studies have raised the figure to over 100,000. This mistake was compounded by the failure by state land officials to classify the areas to be occupied in terms of suitability for agriculture, forestry, or conservation. Lack of control over squatters led to vast clearing of forests on infertile land, unsuitable for sustained annual cropping. Satellite images for 1990 indicate that 27 percent of Rondonia's forests have been removed.

Another fundamental cause of the Polonoroeste "disaster" was the low priority the Brazilian federal government assigned to its obligations under the agreement, both in allocation of credit and in provision of funds for environmental and Indian protection. Invasions of Indian communities and ecological reserves led to an international uproar after the bank had disbursed its funds for the environmental subprogram to Brazilian agencies that failed to do their job. The federal development agency, in charge of Polonoroeste (SUDECO) failed to provide the counterpart funds budgeted in the agreement to finance small farmers who were supposed to plant coffee and cacao and become stable occupants of suitable land. As Dennis Mahar, one of the World Bank designers of Polonoroeste, observed, the virtual sabotage of the small-farmer strategy took place while cheap official credit was subsidizing investments by ranchers for deforestation, fencing, and pasture formation.[16] A technical review in 1987 described the consequences:

> Deforestation has accelerated sharply and uptake of sustainable farming systems has been less than expected. . . . much of the better land is denied to small farmers because it remains under-used in the hands of speculators. . . . for [small farmers] the temptation to sell and move on after a brief period of slash-and-burn cultivation has often proved irresistible.[17]

The IDB went through a similar painful experience. Not to be outdone on the Amazon frontier by the World Bank, the IDB extended a $146 million loan for the paving of BR-364 from Porto Velho in Rondonia to Rio Branco in Acre. As part of the deal, the contract included a $10 million Environmental Protection and Indian Community Program (PMACI), for which IDB put up 40 percent and Brazil was to provide 60 percent. PMACI specified that 28 Indian reserves in the highway's area of influence were to be promptly demarcated and given healthcare and that ecological safeguards be taken by creating state forests, ecological stations, and extractive reserves against an expected influx of settlers. In its first stage, PMACI was administered by an urban architect with no experience in Amazonia from an office in the federal Ministry of Planning in Brasilia, which was 2,000 kilometers from the scene. After disbursing $100 million for the pavement of half of the 550-kilometer highway, IDB was confronted by evidence from environmental NGOs, led by the Environmental Defense Fund, that the road to Acre was contributing to the destruction of forests where rubbertappers lived and that Indian reserves were being dismembered, not demarcated.

PMACI provided the platform for the first international appearance of Chico Mendes at the IDB annual meeting held in Miami in 1987. He denounced the Brazilian government's failure to live up to its commitments under PMACI, and he lobbied for his proposal that Brazil create natural extractive reserves to save the forests. Six months passed, while contractors continued paving the road, before IDB blew the whistle. It suspended loan disbursements in late 1987 when it was notified by the U.S. Treasury and influential U.S. senators that a $22 billion capital replenishment being sought by the bank would not get through the U.S. Congress without reforms in the bank's management, including closer control over the environmental consequences of its loans. The IDB presidency was then assumed by Enrique Iglesias, a former foreign minister of Uruguay and long-time United Nations official, who was in tune with the environmental lobbyists in Washington. He decided that it was important that PMACI be salvaged, for the reputation of IDB as well as for Brazil, the bank's biggest client.

Negotiations on a "definitive action plan" (PAD) for PMACI began in May 1988. Clodoaldo Hegueney, the same diplomat who was later in charge of the negotiations on the G-7 proposal, led the interagency group in Brasilia. Behind Hugeuney, with a right of veto, sat a colonel from the National Security Council. When the IDB asked about demarcation of Indian reserves, the answer was that Acre was a sensitive "frontier area" that could not be handed over to the Indians in large units, so the FUNAI demarcations had to be reduced. In this scheme, Indian villages were to be placed within national forests, where concessions to cut timber can be issued. The Indian representatives, during a meeting in Rio Branco with representatives of IDB, confronted air force colonel Carlos Nascimento, assigned to PMACI as a national security monitor, with an angry refusal to accept truncated reserves. Nascimento stormed out of the meeting, insulting missionaries and environmentalists who were present.

The Indian stand was backed by IDB, but implementation of demarcation depended on the Brazilian government. An impasse was avoided by deferring the demarcation issue until after PMACI delivered promised health, education, and economic services to the Indian communities. The IDB sidestepped the reserve controversy by saying that its money was to be used for "socioeconomic" needs, but not for demarcation. After nine months of nit-picking talks, the PAD emerged in a 262-page document that contained an $8 million work plan to be completed in 20 months and a solemn "declaration of intentions" that said:

> The elaboration of the Definitive Action Plan of PMACI and the commitment to carry it out represent for the Government of Brazil much more than simply compliance with the clauses of a contract. It is, in truth, the expression, above all, of a national preoccupation with the environmental and indigenous questions of the country, and constitutes a pioneering effort to define a new form of occupation and model for development of Amazonia.[18]

Fine words, and the IDB believed that there finally would be action; but what followed between 1989 and mid-1991 was almost complete paralysis of PMACI. A small amount of PMACI funds were released by Brazil for an "emergency program" during the dry season of 1989 to combat forest burning and illegal logging in Acre as part of the Sarney government's belated "Our Nature" program. Jose Rente Nascimento, a forestry economist who did his Ph.D. dissertation at the University of Minnesota on Amazon development, was by then the PMACI coordinator. During an inspection trip to Acre he was beaten up by employees of Jorge Moura, a prominent Acre businessman, whose lumberyard was fined when a raid found it full of illegal stocks of wood. Federal police investigated the assault, which hospitalized Rente with a head wound, but no charges were ever brought against Moura, whose civic services to Rio Branco include a park bench displaying his name in the city's main square. But PMACI's main problem was not frontier violence; it was bureaucratic strangulation in Brasilia. During the last year of the Sarney government and the first year of the Collor government, the federal funds for PMACI were first tied up in red tape and later frozen by the Ministry of Economy during an austerity drive against inflation. David Atkinson, IDB's energetic representative in Brazil, visited Acre three times; met with rubbertappers, Indians, and state officials; and made promises that could not be kept because there was no money from Brasilia.

Finally, in June 1991, when PMACI was about to expire, the Brazilian government released $3 million in counterpart funds that kept the program going. With this, IDB extended the loan agreement for another year and resumed disbursements, but with an important change in emphasis. Under persistent pressure from Atkinson and Rafael Negret, a Colombian ecologist on IDB's staff in Brazil, IBAMA agreed to decentralize PMACI, which was broken up into 86 small projects administered directly by interested communities and local NGOs. This meant, for instance, that funds for health services in Indian communities went directly to the field, instead of through clumsy federal or state agencies. This was more efficient and strengthen the institutional roles of the NGOs on the ground. Another innovation provided a $500,000 grant from PMACI funds to demarcate part of the Chico Mendes Reserve, for which IBAMA had no budget.

Ironically, lack of Brazilian funds for highway construction delayed IDB disbursement of $10 million to complete the paving of the Porto Velho-Rio Branco highway. In practice, PMACI's environmental and indigenous protection role took priority over completion of the highway, which was as it should be. Demarcation of the Indian areas and conservation units in the area of influence of the highway should be completed ahead of the extension of BR-364. This was the original idea when PMACI was put forward as a "model of Amazonian development."

For the federal bureaucrats and environmental consultants who drafted the G-7 proposal, however, the problems in implementing Amazon programs are not in Brasilia, but lie in the attitudes of the frontier people and the weakness of government controls over the things they do. An early draft of the document outlining Brazil's proposal contained this warning to the G-7 donors:

> When it comes to implementing plans it must be recognized that there are limits to government`s capacity to intervene in the Amazon region which are not found in other regions. In an often socially chaotic pioneering atmosphere, government institutions are generally weak. Migrants know little and care less about government plans or wishes. With an intense, single-minded motivation to improve their own lot as best and as fast as they can, they often regard attempts to channel development or to preserve and protect the environment as interferences with progress.[19]

From what has been seen in this study, that statement was an accurate description of the precarious relationship between public authorities and Amazon settlers. But it failed to diagnose the causes and draw the right conclusions. Government institutions are weak because the federal and state agencies responsible for executing policies and enforcing laws have lacked leadership, motivated personnel, and stable funding. That is largely a problem of political priorities, ingrained beaurocratic practices in the federal and state governments, and incompetent management. If funds are insufficient, it is because they are going to other areas with more political clout. If migrants "know little and care less" about government plans, it is because they are not consulted by remote beaurocrats or given a voice or participation in drawing up and implementing the plans that affect them. The cause for concern should be not the initiative of the settlers, but the failure of the government to define clearly, and then protect, the areas of public interest that are peculiar to Amazonia's conditions as a mosaic of unique ecosystems and as a resource frontier that is still largely in the public domain.

The Brazilian constitution of 1988 declared the tropical forests of the "legal" Amazon—a specific geographic area of 5 million square kilometers that enjoys special tax statutes—to be a "national patrimony," which should mean that the forests, and the ecosystems of which they form part, are a public asset. Therefore, all Amazon forests, including those on privately owned land, should be considered, in the words of Anglo–American jurisprudence, to be "clothed with a public interest." Felling trees cannot be an unlimited private right when the consequences can affect the environment in which the community lives. "When . . . one devotes his property to a use in which the public has an interest, he, in effect, grants to the public an interest in that use, and must submit to be controlled by the public for the common good," ruled the U.S. Supreme Court in *Munn* v. *Illinois*, a landmark decision in 1877. That concept should be the legal basis for regulating forest use in Amazonia through a system of agro-ecological zoning that uses technical studies of soils, topography, climate, and biosystems to define how land should be used and conserved.

Important measures in that direction were taken under the Collor government. Congress approved a new Agricultural Policy Directives law that reversed the criteria that made deforestation a "land improvement" condition for obtaining land titles, and gave preferential status for lines of credit and tax incentives to forest management projects, reforestation, and conservation of native forests; the law also assigned agricultural credits to "nonpredatory" extractivism by Indians and other forest producers.[20] A new tax law regulating fiscal incentives that had subsidized land clearing by large ranchers and colonization projects in Amazonia restricted the granting of such incentives to areas "of recognized agro-livestock aptitude," as determined by an interministerial group in charge of ecological-economic zoning of Amazonia, involving the Secretariate for Strategic Affairs, IBAMA (environment), FUNAI (Indians), and INCRA (agrarian reform institute).[21]

It is the political-legislative role of government to define the rules of the game and lay out the playing field in Amazonia. The issues of social equity, economic utility, and environmental ethics can be argued from varying philosophical positions. The choices will predictably generate resistance from major economic and political interests. But choices have to be made, and then need to be carried out with strong administrative leadership. Implementation encounters many difficulties. Most, if not all, Amazonian national governments are administratively weakened by chronic fiscal crisis. In the frontier regions, social conflicts, drug trafficking, and armed insurgencies threaten governance at all levels. But these are obstacles that leadership has always had to face in the raw conditions of Amazonia.

In Brazil, the administrative needs of the environmental sector were not met by Lutzenberger's personal approach to the problems facing Amazonia.[22] Lutzenberger descended from German rural immigrants who settled in Rio Grande do Sul, Brazil's southernmost state, which shares the temperate pampa grasslands with Uruguay and Argentina. Vehemence is regarded as a virtue among the gauchos of the border states, and Lutzenberg vented strong emotions in his environmental convictions. With a mystical attachment to Nature reminiscent of German Romanticism, he embraced the Gaia creed, which holds that the natural world functions as a single living organism, a Mother Earth ecosystem that will inflict a terrible retribution on industrial man for his "unnatural" aggressions against the biosphere. In contrast with these modern "barbarians," Lutzenberg sees the Indians of the forests as the paradigm of a higher civilization in harmony with Gaia.

With this prophetic approach, Lutzenberger took a scornful view of environmental "palliatives." He said that cleaning up polluted air or recycling industrial waste was "just bandaids" and did not address the "philosophical" questions of modern man's "alienated" behavior toward the natural world. "We can't halt the devastation with palliatives, like putting exhaust filters on cars or opening more national parks. What the nations of the world have to discuss are what has brought man to such a predatory relationship with the living organisms of the planet, and how we can unite to save the world from destruction," said Lutzenberger, a former pesticide salesman who speaks with the fervor of a convert and the voice of revelation.[23]

Lutzenberger raised questions that may be suitable for "postmodern" philosophical speculation but that are beyond the horizon of environmental concerns of most of the people in Amazonia. Frontier people in Brazil and the other Amazon countries are concerned with specific problem situations, such as the living conditions of slash-and-burn peasants, wildcat miners, and predatory loggers who are the invaders of Indian lands and destroyers of ecological stability. Amazon populations are victims of social conditions that undermine life on the frontier, for example epidemic malaria, child mortality, and lack of schools. This is fundamentally an economic problem. The delivery of health and educational services on the frontier can only be financed when the economic activities of settlers puts some money into their pockets and the treasuries of local townships. Only with this economic foundation will there be sustainable communities where the human ecology of Amazonia will improve. Saving the forests from destruction may be the primary declared objective of the Amazon rainforest conservation program backed by the G-7, but the operational means to achieve this goal must offer tropical forest populations practical economic and social alternatives to their destructive practices.

Under Lutzenberger, the environmental sector seemed adrift on specific problems. IBAMA's patchwork bureaucracy needed a drastic overhaul to do an accumulation of complex tasks for which it was ill prepared. These included the responsibilities of a national forest service, control of fisheries, administration of national conservation units, regulation of production and trade in natural rubber, and policing of an enormously lucrative clandestine traffic in crocodile hides, turtle oils, rare feathers, parrots, monkeys, and lesser petshop species. Typically, it had too many people where they were least needed, and too few where the need was great; ten times as many people worked at the Botanical Garden in Rio de Janeiro as were assigned to all the Amazon. But decisions on retraining, relocating, or replacing staff proceeded at a snail's pace, despite a general administrative reform taking place throughout the federal government. This was not a problem that drew any close attention from Lutzenberger.

The National Environmental Program, backed by the World Bank, was only kept afloat by the able leadership of Maria Lourdes Davies de Freitas, who had designed and directed the model environmental protection plan for the Carajás mining district. The counterpart Brazilian funds needed to release the World Bank's much larger contribution only became partially available a year after the original agreement was signed. It took months to get the routine approval of the loan through the Brazilian Congress because Lutzenberger didn't like talking to politicians.

The political contradictions and administrative weaknesses in the government's environmental sector produced a congressional investigation, during which Lutzenberger admitted that he did not know how to reform IBAMA. Enemies of environmental controls in Amazonia, such as Governor Gilberto Mestrinho, took the offensive. They demanded the removal of the president of IBAMA, Tania Munhoz, an architect and ecological activist, who took the job after Lutzenberger fired his three previous appointees. On October 2, 1991 Munhoz resigned under pressure, and Lutzenberger named Eduardo Martins, a young primatologist who had found favor with the irascible chief, as head of the agency. In practice, IBAMA's main activities were run from then on by Davies de Freitas, who was simultaneously in charge of the National Environmental Program, the G-7 Amazon Rainforest program, and the planning and finance department of IBAMA. But ambiguities on policy continued and the reorganization of IBAMA awaited strong leadership.

In the critical area of forestry policy, a National Program for Forest Conservation and Development that should have been the cornerstone for Amazon forest management was drafted in 1990 by Jose Carlos Carvalho, one of Brazil's most experienced forestry engineers. Carvalho had been secretary general of the national forestry institute (IBDF), and when that agency was merged into IBAMA, which he helped to create, Carvalho continued as secretary general and then head of the natural resources division. The comprehensive forestry plan prepared by Carvalho, after wide consultations in Brazil and abroad, identified 250,000 square kilometers of high-priority conservation units and natural extractive reserves, proposed reforestation of 20,000 square kilometers of degraded areas, and set up 12 national forests for technically regulated timber production.

Lutzenberger went along with reforestation, but sidestepped making a decision on the plan because it involved timber cutting in the Amazon rainforest, which he considered an "aggression." The alternative to managed production is, of course, uncontrolled predatory logging, which results not only in ecological damages but in the loss of tax revenues and stable employment from a major Amazon resource. In practice, a public asset is plundered for personal profit because IBAMA did not organize an effective national forestry service, which was part of the forestry plan. With the plan in limbo, Carvalho left IBAMA, an agency that could ill afford to lose competent people.

The greatest of Lutzenberger's contradictions was produced by the collapse, during the first year of the Collor government, of the internal price for natural rubber. Rubbertappers, like Indians, are held to be natural defenders of the forest ecosystems since they extract latex from standing trees, as maple syrup is tapped in Vermont. As such, IBAMA administers a subsidy program that was created in 1970 to maintain this traditional activity by ensuring a market for all native production at a price equivalent to the national minimum wage. The rubbertappers would be wiped out by competition from imports and synthetic rubber if the subsidy scheme did not protect native rubber.

Brazil's domestic production of natural rubber, of about 30,000 tons, supplies only 30 percent of national consumption. The rest comes from imports from southeast Asia, where the cost is far lower than in the Brazilian Amazon. Half of Brazil's natural rubber is produced on plantations, mainly outside the Amazon; the rest comes from 75,000 families in Amazonia for whom rubber is the main source of cash income. Plantations get the full subsidy price because they market on a large commercial scale and have a product that meets industrial-quality standards. The rubbertappers don't meet these standards and they have a primitive marketing system. In 1990–91, prices were depressed by a recession that cut demand from tiremakers and other industrial consumers, who prefer imported rubber. The price paid to tappers by itinerant buyers in the Brazilian Amazon was the equivalent of 45 U.S. cents a kilo, which was about half the IBAMA subsidy price. The price for rubber in the forests was the lowest in 25 years in relation to the national minimum wage. The average tapper harvests 500 kilos a year, which provided a rubber income of $225, or $19 a month; the national minimum wage was about $45 a month. The price collapse forced many to give up and migrate to urban slums.

The natural rubber subsidy was financed by an equalization tax on imports by tiremakers and other industries, which pass on the added cost to consumers, so ultimately the public pays. The tax, known as TORMB, provides IBAMA coffers with $50 million a year, which is supposed to support rubbertappers. But the money has been used mainly to support the IBAMA bureaucracy. Despite desparate pleas from "people of the forest" organizations that TORMB revenue be shifted directly to the rubbertappers, IBAMA resisted and Lutzenberger failed the Amazon producers. The tapper leaders and many of Brazil's environmental militants, who had rallied behind "Lutz" as a comrade, began to criticize him as a "decoy duck" whose plumage gave a false ecological respectability to the Collor government.

The lame explanation offered in defense of the Secretariate of the Environment and IBAMA was that Lutzenberger had not been able to obtain the money for his sector from the tight-fisted Ministry of Economy. There was a bit of truth to that, because other ministries made the same gripe, but some, like the Ministry of Health and the Ministry of Agriculture, were successful in putting through major administrative reforms, increasing benefits to the neediest, despite the budgetary squeeze. What the public spending crisis made clear was that Brazil's public sector was overextended and was financially unsustainable. This was as true for the environmental sector as for all state enterprises. One inescapable conclusion was that the presidential environment secretariate and IBAMA had to increase their efficiency and allocate limited resources where they were most needed. Another was that NGOs had to be given a greater role in the administration of environmental projects and that private enterprise had to be encouraged to create forms of production and marketing that were appropriate for the Amazon environment.

A necessary prerequisite for the involvement of the private sector in an environmental protection strategy was the issuance of clear ecological-economic zoning directives. The idea of zoning has been supported for more than a decade by prominent Amazon businessmen, such as Joao Carlos Meirelles, president of a national livestock association and a founder of colonization projects in northwest Mato Grosso. The legislature of the state of Rondônia has already approved such a division of its territory into six zones, which would leave 30 percent of the state in permanent conservation units, another 30 percent under forest cover in extractive reserves and production forests, and only the remaining 40 percent to be occupied by forest-altering activities.[24] The eastern part of Acre was also zoned in some detail by the Brazilian Institute of Geography and Statistics (IBGE) for PMACI, with a map showing which areas were suitable for pasture and agriculture and which should be reserved exclusively for forest.[25]

The law ordering the ecological-economic zoning of Amazonia, passed by the Brazilian Congress in 1990, established three broad categories of land use based on aptitude: (1) productive areas, divided into agriculture, livestock, forestry, mining, or multiple uses; (2) reserve areas, unsuitable for short-term use, because under existing technologies the cost of environmental damages would exceed the economic benefits from occupation; and (3) permanent conservation areas, including Indian reserves.

All ecological-economic zoning is a technocratic exercise that can only be implemented with the political agreement of proprietors and local communities that have to support enforcement mechanisms. Zoning regulates land, forest, and water use down to the township level, where such determinations are bound to be controversial. When it is implanted, however, territorial organization based on zoning gives occupants of land the security to make investment decisions based on the technical aptitudes of their property. Rubbertappers retain their extractive reserves, riverbank populations have their fishing rights, ranchers can develop cattle in areas suitable for pastures, lumber operators can cut timber under licensed management plans in production forests, wildcat mining is placed under technical and environmental controls, and agricultural settlers are directed toward fertile soils. In most Amazon states, the area devoted to ecosystem conservation under ecological-economic zoning will exceed the area devoted to economically productive activities, which will be able to generate tax payments that will sustain the public costs of environmental protection.

A new Amazon economic system can then develop in an orderly way, and the creative potential of private enterprise will move, at all levels, toward sustainable activities in place of the predatory practices of the lawless "old frontier." Examples of this are already available in the mining and forestry sectors, which are the most economically viable activities in Amazonia and therefore the most able to carry out environmental protection.

Everyone agrees that reforestation of degraded pastures in Amazonia is a good thing, so there is no zoning controversy when it comes to planting trees where they used to be. The crossroad formed by the Brasília–Belém highway and the 900-kilometer Carajás–São Luís railway is the epicenter for the most devastated area of the Amazon. Within the 250,000 square kilometers of the railway's "area of influence," deforestation reached 38 percent in 1985 (see map in front matter), even before the freight cars loaded with iron and manganese began to roll from the Carajás hills in the Araguaia–Tocantins watershed to coastal Maranhão. As was seen in Chapter 5, highways had opened the area to settlement long before the CVRD railroad was built, but this state mining company and the World Bank, which financed the mining project, receive the blame because they left the broad consequences of creating a major "pole of attraction" in a frontier region to other federal and state agencies that failed to organize regional development. This was an error of omission, in which CVRD and the bank recognize a share of responsibility, but the question now is not the past, but what to do about the continuing deterioration produced by landless settlers and ranching expansion.

Many environmentalists look at the desolate landscape of charred trees in the Carajas corridor and see only a natural disaster, which it is, and they try to shut down any activity that leads to further forest clearing, such as ranches and pig iron furnaces fueled by charcoal. An entrepreneur with environmental sensibility like Eliezer Batista, who was president of CVRD when the Carajas railway was built, sees the cleared land as an opportunity for massive reforestation, which has an economic value. Thousands of hectares that have been reduced to low-grade pasture could be better employed, economically and ecologically, as industrial plantations for cellulose, rubber, and fruits, and as energy forests.

Such artificial forests would relieve pressure on the native forests from wildcat timber companies and industrial fuelwood consumers, such as pig iron smelters that are installed in the region because of the high-grade iron ore from the Carajas mines, cheap charcoal fuel, and the railroad. The plantations would also provide stability for thousands of small farmers growing trees and would provide jobs for workers in cellulose pulp factories. These activities would generate increased freight for the railroad and expand tax revenues for the townships.

"Sustainable development is accepted by everyone, but the elusive question remains of how it can be financed. We propose to finance environmental protection of the remaining forests by attracting private capital for reforestation and forest management that can be profitable undertakings," said Eliezer Batista, who became director of CVRD's international operations and a sort of roving diplomat, after handing on the presidency to younger management.[26] Batista rounded up support from Brazilian private corporations—including big paper manufacturers such as Aracruz, Simao, Ripasa, and Jari; the national airline, Varig; and a steelmaker, Mannesmann S.A.—for the creation of a Brazilian Amazon Rainforest Foundation (BAF). This was set up in June 1991 as a private, nonprofit foundation with the international participation of the Rockefeller Foundation and of the Business Council for Sustainable Development of Geneva, led by Stephan Schmidheiny, a Swiss businessman who was the top private-sector adviser to Maurice Strong, secretary general of ECO-92.

The purpose of the Brazilian Amazon Rainforest Foundation is to act like a private development fund, promoting research and design of projects that "reconcile the environment with development" and obtaining the necessary financing to carry them out. The target in the Carajas corridor is reforestation of 1 million hectares. It costs about $1,000 to plant a hectare of eucalyptus so reforestation on the scale foreseen by the program would cost over $1 billion. But there would be other projects on a smaller scale that would also provide environmental gains, such as rubber plantations, fruit orchards, quality woods for furniture factories, and other forest-based industries, financed by the foundation from donor funds.

Batista convinced Collor and Lutzenberger to include the BAF forestry program in the G-7 proposal because of the "global value" of reforestation and forest protection as a "carbon sink." Donors were told that funds provided to BAF that were used for commercial plantations would be repaid on banking terms, but on the condition that the money would remain at the disposal of the foundation to finance other environmental projects. This recycling of funds is what would make BAF a private ecological-economic development fund, under an international board of directors.

The technical viability of the Carajás reforestation plan rests on 20 years of tropical forestry research by CVRD that preceded the entry of the mining company into a major cellulose project in Minas Gerais with a Japanese partner. CVRD's research center at Linhares, Espirito Santo, not only preserves one of the last major sectors of Atlantic forest, but is the genetic laboratory-nursery where superior varieties of eucalyptus and other fast-growing species have been developed for cellulose production. The cloning of high-yield mother trees is a technique that has given Brazil world leadership in short-fiber cellulose production from eucalyptus. Based on this technology, Aracruz, CVRD, and others are installing plant capacity for over 1.6 million tons by 1993 with an investment of over $5 billion.

This advanced biotechnology and tropical forestry know-how has been transferred to the Carajás corridor, providing an enormous headstart over earlier reforestation experiences, particularly the famous Jarí project where Daniel Ludwig, a U.S. shipping magnate, cleared 100,000 hectares of native forest to implant a cellulose Xanadu on the north side of the Amazon in the 1960s. The hasty Ludwig made no forestry tests and planted gmelina, a fast-growing Asian exotic, along with Honduras pine and the daglupta species of eucalyptus, which did not adapt to the rainfall conditions. The gmelina failed and the wood yields fell so short of expectations that Jari was seen as a forestry failure when Ludwig sold out to a Brazilian consortium in 1980. But CVRD and Jarí foresters have since concluded, after observing results with other eucalyptus varieties more suitable for Amazon conditions, that Jarí's failure was lack of research, since the cellulose yields from the selected varieties was 30 percent higher.[27] Jarí is being replanted with the new varieties.

Since everything in Amazonia is experimental, the input of researchers and laboratories of a major corporation, such as CVRD, can make an extraordinary contribution if they are turned to the problems of the region. CVRD has maintained ecological research posts along the railroad corridor that have studied the phyto-ecology of three distinct areas since 1982 and forestry centers since 1985. The principal center near Acailândia has experimented with a wide range of native species and can provide producers and nurseries with high-quality fruit and wood trees. The center also has the modern facilities for rapid reproduction of planting material by tissue-culture cloning of selected commercial species at low cost.

If energy wood consumers, particularly the pig iron smelters along the Carajás railway, are forced to meet their needs for charcoal from new high-yield plantations, that will not only reduce pressure on the native forest, but will also provide a certain number of forestry jobs directly and guarantee other jobs in the metallurgical sector. This is a positive economic-ecological combination since the price of charcoal from plantations (about $65 a ton) should allow pig iron production to continue. CVRD foresters believe that 20,000 hectares of eucalyptus planted on degraded pasture lands could provide the charcoal needed for 250,000 tons of pig iron per year, which is the size of the larger smelters.[28]

But the reversal of environmental degradation in the Carajás corridor is not just a question of high-tech reforestation or big corporate investments. The main problem in the region is the presence of a very large floating population of landless peasants who are destroying remaining stands of forest. This is an agrarian reform problem, and the question is, to what extent can the forest management centers provide for this unskilled population and turn them into productive producers?

Batista said new plantations of eucalyptus, and perhaps acacia and pine, can provide a productive solution for poor farmers in the area because they will be incorporated into the forest centers as mini-producers. "On fifty hectare plots they can plant part in commercial trees, for which there will be a guaranteed market, and the rest of the land can be planted in fruit trees, like *cupuacu*, or rubber, for which there will be industrial processing. They will also have some land for their household needs," said Batista. The number of plot holders will depend on the number of centers that are installed. Plans call for ten.

Renato Moraes de Jesus, director of the CVRD forestry center at Linhares, has promoted this kind of small-landholder forestry in Espírito Santo, working through municipal nurseries supplied from the center with planting materials. He believes that large-scale reforestation is possible only through a program that makes the best technology available to family-size rural innovators who are prepared to adapt new ways. This is the essence of the "new frontier." Speaking at a conference on reforestation at the University of Oregon, de Jesus said:

> It is difficult to convince the farmer to utilize an arboreal element in the management of his property and it is difficult for him to understand the implications of the increasing levels of CO_2. . . . The viability of programs of this nature can only be assured if the planting unit is on the small scale of the rural landowners, where the problem becomes a question of forestry extension and environmental education. . . . it is indispensable to make available all of the technology involved in these programs and to orient the landowner as to how, where and why to plant trees.[29]

Communities of small foresters in the Carajás corridor, functioning as satellites to the forest-based industries, would benefit from CVRD's planting material and technical assistance. Their labor would reduce planting costs, although there would have to be an income flow to the small producers during the period of tree planting and early growth. Later, there would be a potential agroforestry income from supplying food to the forestry and industrial workers who will be drawn to the area. CVRD's contribution to social forestry would improve the company's standing in the region, with added dividends in providing a social answer to ecological criticisms of plantation forestry in the tropics. The main environmental benefits would be the stabilization of part of the floating population through a form of land tenure that would associate reforestation with protection of the native forest that would form part of such communities.

It is noteworthy that the major reforestation component of the Brazilian Amazon program came from the economic enterprise sector, with a method of financing and an operational plan that addressed economic and social problems undermining the human ecology of the region. The program did not demand public funds, which the Brazilian government can't provide, and it contributed scientific research and proven technology at no cost. If the centers are profitable, an important part of the Amazon region will acquire its own source of funds for environmental protection from a sustainable economic activity. Paraphrasing Clemanceau's comment on war and generals, the environment is too important a matter to be left to (just) ecologists.

NINE

Friends of the Forest

I love the wild no less than the good.
Henry Thoreau (1847)

The municipal auditorium in Rio Branco, the capital of Acre, was packed the night a public hearing was held on a life-and-death issue on the Amazon frontier—the right of a rancher to clear the land of forest. Two rubbertapper leaders took the floor in defense of their Amazon world of trees. "For many years, out in the forests, we have been putting our bodies between the chainsaw gangs sent by the ranchers and the trees that provide our way of life. Now, we are doing the same thing here in the city, at this hearing, with the law on our side," said Julio Barboza, president of the National Rubbertappers Council (CNS), who grew up in a forest community on the Acre river. Gumercindo Garcia, wearing a red tee-shirt stamped in white letters with the PT of the Catholic left Workers party, put the plea in a political nutshell. "We want a moratorium on forest clearing because we know that for us, the local people, the sustained use of forest resources is better than cattle ranching. This is our political proposal for all of Amazonia," said Garcia, who studied agronomy before becoming an organizer in the rubbertapper movement.

The public hearing in May 1990 was a historic occasion. It was the first environmental impact hearing ever held in Amazonia under a law that gave the state environmental agency (IMAC) authority to protect the rainforests that cover the state of Acre from demolition by big ranchers. The rubbertappers were joined in presenting their arguments for forest preservation by ecologists and geographers who came from São Paulo and Rio de Janeiro to testify. The ranchers were represented by politicians, lawyers, agronomists, and economic consultants who defended the right to clear more than 5,000 hectares for cattle pastures on a ranch where over half the original forest cover was still untouched.

For more than 400 people in the auditorium, the memory was still fresh of the assassination two years earlier of Chico Mendes, the rubbertapper leader from Xapuri, who organized the nonviolent resistance movement against the occupation of forest land by ranchers. At the time, the killers of Chico Mendes were in jail, but had not yet been tried. Tensions were high because of the uncertainty over the trial and its outcome. Political hostilities were developing over the election of a new state governor. But the hearing moved the conflict over land and forests in Acre from the violence of midnight ambushes to a civic debate.

The frequent cheers for the rubbertappers during the three-hour hearing showed that the sympathies of the crowd, made up mainly of young professionals, university professors, students, and militants of the PT, a national party Chico Mendes had helped found. These partisans were not disappointed. Two weeks after the hearing, Marco Antonio Mendes, superintendent of IMAC and environmental adviser to the governor, denied the owners of the 21,143-hectare Bela Alianca ranch permission to clear 5,800 hectares to provide pasture for cows. The decision was one of the reasons that deforestation in Acre by ranchers in 1990 fell to a third of the rate two years earlier. Mendes (no relation to Chico) was an economist from the University of Sao Paulo who went to Acre for a frontier adventure and stayed on to create the environmental agency. The hearing was based on a much neglected resolution adopted in 1986 by the National Environmental Council that required that any landowner seeking to clear an area of Amazon forest producing a "significant environmental impact" must first apply for a license from the state based on a technical inventory of the forest and an environmental impact statement. IMAC ruled that a challenge to the statement required the application to go to a public hearing.

That procedure was adopted by Governor Flaviano Melo as a mechanism for peaceful solution of land conflicts that had given rise to the violence that killed Chico Mendes. The Acre death toll included at least 100 rubbertappers, as well as many settlers, Indians, and ranch workers. The hearing showed that there are legal ways to protect the forest without violence when local populations, such as rubbertappers and Indian alliances, get organized with support from environmental activists from home and abroad.

In the case of Bela Alianca, the CNS presented a challenge carefully prepared by an advisory group of geographers, anthropologists, lawyers, and ecologists. This combination of grassroots organization, scientific knowledge, and "green" political activism can mobilize communities, raise funds, prepare legal briefs and technical opinions, and use political leverage to regulate cattlemen, land developers, loggers, and miners who cut, burn, and plunder native forests that should be protected.

A year later, at the eastern end of Amazonia, the Acre public hearing procedure was repeated at Laranjal do Jarí, a river town on the border between Amapá and Pará. Nearly 500 persons, including federal and Amapá state environment officials, rural union leaders, ecologists, and lawyers, but mainly townspeople and caboclos from the Cajarí, Maracá and

Jarí rivers, debated at a public hearing the proposed construction of a 125 kilometer highway through a rainforest area designated as an extractive reserve for rubbertappers. Emotions ran high because merchants in Laranjal and landowners in the area who favored the road project had stirred up animosity toward the rubbertappers, accusing them of trying to prevent "progress." Surprisingly, for a meeting where violence was feared, the result was constructive.

Before traveling to Laranjal by riverboat—the only available means of transportation in that region—Pedro Ramos, vice president of the National Rubbertappers Council (CNS), had been clubbed by assailants in a night attack in Macapá, the capital of Amapá. The assault was reminiscent of the murder of Chico Mendes; but Ramos went to the meeting with a bandaged head and took part in the debate on the environmental impact statement on the road prepared by a consultant for the highway contractor. When it was over, the rubbertappers had made it clear that they were not against the road, linking Laranjal to Macapá, but insisted only that before the road was opened the reserve area be legally demarcated and land claims by absentee ranchers be resolved.

The Governor of Amapá, Annibal Barcellos, the main sponsor of the road, then issued a decree setting up a special mixed commission, including rubbertappers and environmental NGOs, to supervise demarcation of four extractive reserves, with the removal of "invaders," including *garimpeiros* working the streams in the area for gold. The agreement also provided state financing for a Brazil nut shelling plant to be run by a rubbertapper cooperative. Contrary to what had been feared, there was no polarization, and the agreement offered tangible benefits for an isolated area without the displacement of traditional populations. "The debate showed that a road, in itself, is not progress unless the rights of the people are protected. Now that there is an agreement on the pre-conditions, the road that everyone wants can go forward," said Carlos Walter Porto, an anthropologist from the Federal University of Rio de Janeiro, who spoke at the hearing.

At Amapá, as in Acre, the environmental hearing procedure showed that when a community can be organized to use the legal means available for imposing conditions on public works and use of natural resources, political and social compromises are possible, even in Amazonia. To do so, economic, social, and environmental interests must all be reconciled.

The ruling in Acre on Bela Alianca came as a rude shock to the ranchers, most of whom were attracted to Acre from more developed areas of eastern Brazil by the eagerness of the old-time operators of rubber estates to sell the land cheap. In the 1970s, the state government rolled out a red carpet for the cattlemen and facilitated their acquisitions by allowing the conversion of flimsy "use rights" into titled property. The rubbertappers, having no recognized rights, had no say, and the only way to resist eviction was by *empates*—human barriers formed by rubbertappers, with their women and children, against the chainsaw gangs, usually backed by armed guards.

By the time impact statements were required for forest clearance, ranchers and small farmers had deforested more than 500,000 hectares in the eastern Acre river valley, or about 15 percent of the land, although forests in the rest of the state were largely undisturbed. Large stands of forest remain on most ranches, since half must be maintained by law, but ranchers say they must be free to clear their land up to the limit because cattle breeding in the state requires more pasture.

Dirceu Sanchez de Zamora, then president of the Acre Cattlemen's Association, was one of the few large landowners who went to the Bela Alianca hearing. He walked out halfway through the debate. "This is all politics," said Sanchez. "We are as much ecologists as anybody because we also live off the land, but on the basis of reality, not ideology. We don't destroy, we produce food."

Dirceu and his brother Sidnei own five ranches in Acre, with a total area of over 60,000 hectares and 15,000 cows. They sold family lands in western São Paulo developed by their father, a Spanish immigrant, and with the proceeds they bought ten times more land in Acre. Many other out-of-state investors did the same thing, including some of the biggest meatpackers in Brazil, such as Gerardo Bordon. But Bordon pulled out of Acre and left a 60,000-hectare property undeveloped when the rubbertapper conflict around his property became violent. Many other investors have lost interest because of delays in paving BR-364 from Porto Velho to Rio Branco, and the almost complete lack of road maintainance in Acre.

A visit to Sidnei's Cipoal ranch, ten miles from Rio Branco, found hundreds of pure-bred gray Nelore steers feeding among fire-charred stumps on lush pastures divided into well-fenced lots. There were 550 hectares of new pasture on land cleared in 1989 and sown with a combination of grasses and legumes for year-round fodder for steers brought from other ranches or trucked from neighboring Bolivia. The brothers sell 1,600 steers a year, worth over $500,000, and employ 180 people. Asked if his activities were profitable, Sidnei replied:

> The conditions for cattle raising in the Acre valley are the best in Brazil. There is no frost, as in the south, or dry season to reduce pasture, so animals fatten all year and are ready for market in three years. The calving rate we get here is ninety per cent. I have visited ranches all over Brazil and in Texas, so I know what I'm talking about. We didn't come here with any government subsidy. Do you think we be investing our own money here if it didn't make economic sense? We are businessmen.[1]

* * *

Conversion of run-down rubber estates into cattle ranches may be good business, but it was bad politics. The massive expulsion of rubbertappers, and their desperate resistance to these evictions, is what created the rubbertapper movement, headed by the CNS, based in Rio Branco. This has become a force strong enough to challenge further expansion of ranching not only through legal hearings, but through the creation of "extractive reserves."

Extractive reserves are communal forest production units, created by federal decree, where associations of rubbertappers and other gatherers make a living from gathering natural products, such as latex and Brazil nuts, with little alteration of the forests. The rubbertapper movement's main objective is to obtain from the federal and state governments in Amazonia the formation of the largest possible number of these units. There are incipient extractive reserves in Amapá, Pará, Rondônia, and Amazonas, but the center of this movement is Acre, where it was first launched. Responding to pressure from the CNS and international environmentalists, in March 1989 Brazil's national government created the "Chico Mendes Extractive Reserve." This is an immense block of undisturbed native forest that sets aside 970,570 hectares for the exclusive use of 1,200 rubbertappers and their families. The reserve stands as a physical barrier to further advances up the Acre river valley by cattle ranchers and land developers. Another huge reserve of 506,000 hectares was created at the western end of the state in the upper Juruá river valley.

The concept of the "extractive reserve" is a powerful political legacy left by Chico Mendes to the rubbertapper movement. The idea grew out of discussions between Mendes, then an unknown rural union organizer in the backwoods of Xapuri, and Mary Helena Allegretti, an anthropologist from the University of Brasília. Allegretti went to Acre to do research on the exploitation of the *seringueiros* (as rubbertappers are called in Brazil) by operators of rubber estates. She was working in 1980 in a rural education program in Xapuri sponsored by Oxfam/UK, a Christian-aid NGO, when she met Mendes. From then on her career was devoted to the organization of the rubbertappers into a national movement. By 1985, when the first national rubbertapper congress met in Brasília, the newly created CNS proclaimed in its platform the ideas for extractive reserves.[2]

The theory of the "extractive reserve" has been developed by Allegretti and Stephan Schwartzman, an anthropologist and ecologist who is strongly involved with the rubbertappers. He defines it in a political context that makes these reserves an instrument of agrarian reform, as well as the environmental movement: "Extractive reserves are in essence a proposal for an ecologically sound and socially appropriate land reform in the Amazon."[3] An extractive reserve is, by definition, a citadel for "friends of the forest." In the lexicon of the environmental movement, the "defenders" of nature are Indians, extractivists, and riverbank people, refered to generically in Brazil as *caboclos*. Allegretti has proposed that all these groups be considered candidates for extractive reserves, with an immediate goal of 250,000 square kilometers. This is equal to the area of the state of Rondônia.

Chico Mendes became an international symbol of the "friends of the forest," and not just one more anonymous victim of Amazon land conflicts, because his movement to resist the eviction of rubbertappers from the forests was broadcast as an "ecological" issue by global environmental communicators. Mendes was a natural for the part. His affability, emotive dark eyes, verbal simplicity, and a touch of melancholy conveyed a sincerity that made Mendes a persuasive advocate for a good cause. With these attributes, Mendes's presentation to the world as the face of "environmental friendliness" became simply a case of good political marketing.

The international outreach for Mendes came through Adrian Cowell, an upperclass Englishman who produced environmental documentaries. During the filming in Acre of his "Decade of Destruction," Cowell developed a close relationship with Allegretti and Mendes. He gave prominence to the rubbertappers in his series and loaned Mendes money for an unsuccessful bid for election as mayor of Xapuri. Schwartzman, who had research experience in Brazil's rural conflicts, met Mendes through Allegretti in 1985, and through his Washington connections with the Environmental Defense Fund (EDF) he brought Mendes to the attention of the international NGO community. The making of Mendes as an international figure advanced when he went to England to receive a Global 500 award—the "green" Nobel Prize—from Mostafa Tolba, director of the United Nations Environmental Program, in 1987. Ted Turner's Better World Society award came the following year when Mendes visited U.S. senators and congressmen in Washington. Posthumously, he was awarded the Japanese Sasakawa International Environmental Prize in 1990, which put $250,000 in the treasury of the rubbertapper council.[4]

But Mendes was not an "ecologist," except in a tactical sense. His motivations were social and political. His goal was a forest version of land reform that would give rubbertappers permanent rights to use the forests where they have made their frugal living for generations. Mendes said repeatedly that he was not against highways or other "developmental" improvements in Acre, as long as they were preceded by the establishment of "extractive reserves" for the *seringueiros*—the people of the *seringa*, or rubbertree—and not just as routes for cattle ranchers and colonists to take their lands. It is not the environment, as such, but a fair distributution of the benefits that can be obtained by the development of Acre's society and economy that is the issue for the rubbertapper movement and in all similar Amazon conflicts over land and resources.

* * *

The election of a new governor in Acre in 1990 made it clear that the political situation in that highly politicized state was more complex than a simple confrontation between "evil ranchers" and "friends of the forest." That was not the way a majority of people in Acre wanted the lines drawn.

The creation of the Chico Mendes reserve, the IMAC ruling on Bela

Alianca, and the murder trial in Xapuri of Mendes's killers had created an appearance of polarization as the campaign for governor developed. On the one hand, the CNS threw its support behind an attractive young forestry engineer from an old Acre family, Jorge Viana, who ran as a "progressive" Workers Party (PT) candidate. Viana had been a friend of Chico Mendes and he was a declared opponent of the out-of-state ranchers. Viana had environmental credentials, having worked at the state's technological foundation (FUNTAC) on extractive reserves and the Antimari forest management project supported by ITTO.

On the other side, most of the ranchers and lumbermen lined up behind the candidacy of Rubem Branquinho, a federal deputy whose flashy campaign was supported by wealthy businessmen. These included João Branco, the owner of a Rio Branco TV station and newspaper, who is accused by CNS leaders of having plotted the murder of Chico Mendes. For many ranchers, the control of the statehouse seemed a question of survival. As one rancher put it at the time: "We are under siege. They want to drive us out of Acre, but we are not quitters."

To everyone's surprise, the campaign was peaceful and neither of the extremes won. When the votes were counted in November 1990, the winner was a moderate politician, Edmundo Pinto, who edged out Viana and came in far ahead of Branquinho. The outgoing governor, Flaviano Melo, was elected to the federal senate, and two prominent rubbertappers, Osmarino Amancio Rodriguez and Gumercindo Garcia, running on the PT ticket for the state legislature, were soundly defeated. The voters had delivered a clear verdict in favor of moderation. Or, put another way, for most people in Acre the forest was not the immediate issue.

A runoff election was still necessary, however, between Pinto and Viana because no candidate had won an absolute (50 percent) majority. In the interim between the first and second election, the killers of Chico Mendes were brought to trial at the little courthouse in Xapuri. Darcy Alves da Silva, a scrawny, spectacled rancher from Xapuri, wanted in two other states for homicide, stood accused of ordering the assassination, and Darli Alves Pereira, one of Darcy's numerous offspring from a harem of four wives, had confessed to being the shotgun killer.

Mendes had successfully resisted attempts by Alves and other ranchers to take control of the Cachoeira rubber estate, and he had obtained a warrant from a judge in Parana for the arrest of Alves on a murder charge in that state. A jury of seven men decided these were sufficient motives for Alves to have ordered the killing, and he and his son were found guilty and sentenced to long prison terms by circuit judge Adair Longhini. Both the jury and the judge were threatened with death by the Alves clan if the Alveses were not acquited, but the CNS mobilized 2,000 rubbertappers to Xapuri where they kept a vigil at the courthouse, along with Chico's widow Ilzamar, 400 journalists, and ecotroopers from all over the world, until the verdict was announced. For a moment, the light of justice shone on Xapuri as law came to the frontier. The new victory of the rubbertappers in the Chico Mendes trial had no effect, however, on the runoff election for governor. Pinto won handily, and the PT and CNS had to swallow a bitter defeat. The wealthy landowners had been unable to "buy" the election, but it was also clear that the "peoples of the forest" did not wield the dominant political force in Acre that they had hoped to mobilize, They lost even in districts like Xapuri and Brasileia, where the rubbertappers control the rural workers' unions. As a consequence, the PT–CNS alliance failed to win control of the manna of public jobs and revenues that are the main sources of political influence in Acre.

The election outcome also removed from the state government some key officials who were close allies of the rubbertappers, particularly Marco Antonio Mendes at IMAC and Gilberto Siqueiros, the founder-director of FUNTAC, the state's main point of contact for international financing. Several projects of interest to the Acre rubbertappers had been under discussion with FUNTAC before the election, including an $8 million grant from Canada's International Development Agency (CIDA) for "sustainable forest management," mainly aimed at development of extractive reserves; the second phase of the $3 million Antimari forest management project with ITTO; and various grants and loans promised by IDB, including demarcation of the Chico Mendes reserve with PMACI money.

The election outcome raised difficulties for the donors since it was unclear what the new governor's political relations would be with the rubbertappers, on whom all the donors had built their forest protection strategy. Pinto's profile was not that of a "friend of the forest." He had worked as an administrator at the federal university in Rio Branco, from where he entered politics as a state deputy. During his campaign for governor, Pinto stumped the rural townships promising local roads and completion of the BR-364 highway to the western end of the state, a promise that the former governor had made but failed to keep. Pinto also promised more education and health services, and above all more jobs through encouragement of new private investments. His line was economic development, and he had little to say on the environment. So, what Acre voted for was development.

None of Pinto's goals were attainable, however, without large amounts of federal money and foreign financing. He showed a keen awareness of his fiscal problems in his first interviews with donors. They made it clear to Pinto that his needs would be favorably considered only if he satisfied the environmental interests of these financial sources. This message was reflected in his initial appointments. He chose as his secretary of planning, the key job on economic and social coordination, a professor of economics at the university, Raimundo Angelim, whom Viana had intended to name to the same job if he had been elected. Pinto invited Gilberto Siqueiros, the former head of FUNTAC and a champion of extractive reserves, to be a personal adviser. Marco Antonio Mendes stayed on in Acre as coordinator of PMACI for the IDB, under a United Nations grant, thereby maintaining a liaison between FUNTAC and the rubbertappers and Indians.

At the same time, Pinto's pragmatic political approach took account of the plain fact that in Acre's rural economy there are more small farmers, lumber workers, and ranchers, with their employees—most of whom voted for him—than there are rubbertappers and other extractivists. Therefore, he put FUNTAC, the state's development agency, under the direction of a politically neutral, technically respected agronomist, Judson Ferreira Valentim, who was not seen as an enemy by either the organized rubbertappers or the ranchers, agriculturalists, and lumbermen. An agronomist from the federal agricultural research center in Acre was put in charge of IMAC, which faded in importance.

At FUNTAC, Valentim combined experiences and qualities that are rarely found together on the Amazon frontier: personal knowledge of the problems of the colonist and first-class scientific training. He grew up in a pioneer settlement where he experienced the hardships of the frontier. Valentim's father, a farmer from Minas Gerais, opened a ranch in the virgin rainforests of Paragominas, a township on the Belém–Brasília highway that became famous in case studies of how deforestation leads to degraded tropical pastures. Valentim saw the decline of the family ranch, which his father eventually sold, but he was able to finish his university studies. After graduating from the Federal University of Pará, in Belém, Valentim worked for the federal agricultural research agency (EMBRAPA), and obtained a scholarship for graduate studies at the University of Florida, where he studied the enrichment of tropical pastures.

This was the background and training Valentim brought to EMBRAPA's research station in Acre, where he began experiments and field trials in 1980 on improved pasture management. He developed a pasture technology, combining *kudzu* (an African legume) and tropical grasses, that offers significant environmental benefits as well as economic rewards. This is an example of how problem-solving research can discover appropriate practices for Amazonian conditions that combine economic and environmental objectives.

Valentim's pasture proposal is more than a technocratic fix. It is a practical alternative to deforestation that could modify the destructive behavior of ranchers and colonists, who are the primary agents of forest clearing. If pastures were more durable, there would be less pressure from ranchers for land clearing. More intensive use of already deforested areas could reduce social conflict over land and contribute to political pacification. As a consequence, threats to Acre's forests would have been reduced by the conversion of "foes" to a more sustainable use of land through better environmental management, thanks to a new technology.

Kudzu (*pueraria phaleolis*) is not a miracle plant; it is quite common in tropical countries and it can be a pest when it grows out of control. But when it is properly used, it offers a way to make Amazon pastures more sustainable by overcoming the main problem faced by cattle ranchers in Amazonia: the resurgent growth of shrubs and weeds that degrade pastures formed on deforested land. When Amazon forest canopies are removed, the entry of light after the trees fall produces vigorous regeneration of dormant pioneer species. The traditional method of control, practiced by the Indians long before the colonists arrived, is to burn off the regrowth in the dry season. In a few years, however, this primitive system reduces soil fertility and grasses lose nutritional value. Degraded pastures are then abandoned to the invaders, which rapidly form a secondary forest (just the opposite of desertification). But ranchers then cut down new forest to plant new pasture in a never-ending cycle of destruction of primary forest. This is what has to be stopped.

Kudzu deals with both the weed control and soil fertility problems. This broadleaf creeper grows so vigorously during the wet season that it smothers all competitors, providing a palatable, nutritious fodder and abundant organic mulch when cattle trample the leaves. As a legume, *kudzu* has rhizobium nodules on its roots that fix nitrogen in the soil, fertilizing grasses that provide pasture during the dry season. Sustainable management of this type of pasture requires fencing divisions for careful rotation of cattle feeding, but it is not otherwise expensive. Keeping pastures clean by manual cutting or herbicides is much costlier.

More than 150 ranchers in Acre now employ the *kudzu* technology on 20,000 hectares. Flavio Maia Cardoso, owner of the 1,200-hectare Fazenda Niteroi, a ranch 35 kilometers east of Rio Branco, is one of Valentim's converts to *kudzu*. He owns ranches in Santa Catarina in southern Brazil, and in Mato Grosso do Sul, and by comparison, he considers his pastures in Acre to be the best. "*Kudzu* has been our salvation. It keeps down the weeds, resists the spittle bug, fertilizes the soil and we don't have to burn, which protects the organic matter. The cattle feed all year on a protein source as good as alfalfa. The bottom line is more cattle per hectare and more beef on each carcass," said Cardozo, who keeps close control on his herd and pasture with a portable computer.

The *kudzu* pasture technology has wide application. It is not just for big ranchers, since anyone who has animals to feed needs stable pasture. This includes horses and donkeys that are used by rubbertappers to transport heavy balls of latex and sacks of Brazil nuts from the forest. It also helps solve the pasture problems of small farmers for whom cattle and milk are an important part of family income. Valentim is a natural rural extensionist who likes personal contact with farmers. He spread the news on *kudzu* in visits to agrarian reform colonization sites and through small farmer associations. Demand for *kudzu* seed was so strong from small farmers that the supply of seeds ran out. Production of *kudzu* seed became a profitable business.

The signals from the economic marketplace for the *kudzu* technology are all good. But what about the environment? Some "friends of the forest" are afraid that if *kudzu* makes pastures so good, ranchers will be motivated to burn down the whole of the Amazon. Not so, says Valentim. He argues that the *kudzu* technology should lead to a reduction of pressure on the forest by ranchers who make rational decisions, and he persuaded IMAC to require that ranchers recover degraded pasture as one of the conditions for licensing new forest clearing. Valentim's reasoning is economic, but the results are protective of the environment.

> The utilization of this [*kudzu*] technology in pasture establishment contributes to increase stocking rates from .5 animal/ha to 1.5 animal/ha, during the dry season and from 1.0 animal/ha to 2.5 animal/ha during the rainy season, respectively, in grass and grass-legume pastures. Liveweight gains increase from 90 to 220 kg/animal/year. Meat production increases from 70 to 440 kg/ha/year.
>
> In Acre there are about 600,000 ha of pastures and about 600,000 animals. The adoption of this technology in these areas would allow a doubling or tripling of the number of animals grazing these pastures without the need of additional forest clearing and burning.
>
> Since an increase of this magnitude in the cattle herd will be possible only after many years, we suggest that forest clearing and burning for pasture establishment be suspended in the near future. This will be possible through the reclamation and improvement of the existing pastures using legumes and adequate management practices, allowing a more efficient utilization of the land and of the pastures, thus increasing productivity and profitability.[5]

Valentim's reasoning is political as well as economic. During the IMAC environmental impact hearing, where he appeared as a witness in support of the application, Valentim said it was time for the Acre community to decide whether or not forests should be cleared for agriculture and livestock:

> If a consensus is reached in Acre that cattle ranching should not be allowed to expand further, the state should make that clear so that farm owners are not misled into thinking it is possible. . . . If agriculture and livestock are acceptable economic alternatives, then the state has to define clearly, on the basis of technical studies, which do not exist now, what land should be occupied, what activities should take place, and which areas should be preserved. . . . There are good and bad land owners, as in all activities. Competent farm entrepreneurs are not speculators who exploit nature. If there is a profitable alternative that reduces the environmental impact, that is what they will choose.[6]

The zoning proposal was not on the ballot in the 1990 election, but it had been presented by FUNTAC at township meetings where zoning was discussed as part of a state development plan, prepared by Siqueiros, that gave priority to a combination of extractive reserves and wood industries supplied from technically managed production forests. Valentim pushed forward with technical studies on how to zone the state based on a natural resources map of Acre prepared by FUNTAC from satellite images that provide a rough identification of soils, sources of commercial timber, and watersheds. Mapping has to be refined in greater detail and checked on the ground against property claims to be able to introduce effective micro-zoning. That takes time and money, but when it is done, political decisions on land uses and territorial organization will be possible.

If Acre's citizens can reach a political agreement on the basis of zoning, there should be room for coexistence between ranchers, small farmers, traditional extractivists, Indians, and lumbermen in a state where there is 1 square kilometer of land for every rural inhabitant and 95 percent of the forest is still standing. It is essential for the survival of these forests, however, that the zoning be technically defined, legally enforceable, and generally accepted by the population before BR-364 is extended to western Acre. Otherwise, Acre will be a new Rondônia.

* * *

The economic, social, and environmental crisis brought on the unregulated occupation of Rondônia's virgin lands was the legacy received by a new governor, Oswaldo Piana, when he took office in 1991. Not only were 27 percent of the primary forests demolished; so were the states finances, which his predecessor had left in ruins. For the same reasons as his neighbor in Acre, Piana began showing interest in the environmental issues in Rondônia when he was told by federal officials in Brasília and by international lenders that reduction in deforestation and attention to Indian reserves were necessary if he hoped to obtain financial help.

Before becoming governor, Piana, who has a medical degree, served a term as president of the state legislature where he was known more as a good social mixer in Porto Velho's bohemian nightlife than as a lawmaker. The governorship was thrust upon him by the voters after Senator Olavo Pires, the early favorite, was assassinated in a crime linked to Rondônia's flourishing cocaine traffic. Piana formed a "transition team" of political and technical advisers to decide what to do. As the master plan for the new administration, they dusted off a proposal called PLANAFLORO that had been presented to the World Bank in 1989 but was on the shelf for lack of financing.

PLANAFLORO, or the Rondônia Agro-Livestock and Forestry Plan, was the first major agro-ecological zoning blueprint developed for the Amazon region. The state legislature approved a zoning plan that divided Rondônia's 243,000 square kilometers into six zones, or land use categories. The zoning placed 52 percent of the state in some form of forest reserve or conservation unit, if Indian reserves are included. The World Bank gave the $228 million project a favorable appraisal, saying that it would bring "a significant reduction in the rate of destruction of . . . the remaining natural rainforest" along with benefits for 6,000 Amerindians, 2,400 rubbertappers, 900 families of fishermen, and 55,000 low-income farm families. Small farmers, who outnumber the "forest people" by more than five to one, were to be converted to "agro-forestry," or the association of annual food crops and livestock with perennial trees, capable of providing fruits, fertilization, fodder, fuel, and wood. Cattlemen would be allowed to ranch only where zoning studies authorized pasture (see Rondônia zoning map).

PLANAFLORO received a new name at the World Bank as the Rondônia Natural Resource Management Project. It might as well have been called "Son of Polonoroeste," however, since it was designed to correct the errors associated with the implementation of the bank's earlier $450 million loan that paved BR-364 into Rondonia. The new project was imaginative, but the NGO community found it had disturbing similarities to Polonoroeste, which was enshrined as a prime example of the World Bank's propensity for "bankrolling environmental disasters." As such, the bank management's inclination to lend another $167 million to Rondônia for the new project came under severe scrutiny.

A devastating NGO critique of the new Rondônia project reached the World Bank in January 1990 as the loan was being prepared for submission to the executive board. The 15-page NGO document was signed by more than 30 "friends of the forest" organizations, from Australia to the German Green party and from Sweden to Tokyo's Rainforest Network. The critique was prepared by Bruce Rich, EDF's international program director, and Stephan Schwartzman, EDF's staff anthropologist. This collaboration was a classic example of how NGO "networking," which has become a powerful form of control over MDB lending in sensitive tropical rainforest regions. This control is based on very close contact between the international NGOs and the local organizations that represent forest people. These local NGOs, be it in Brazil, India, or the Cameroons, are political groups, usually in opposition to the local governments, so the NGO formula turns the MDB into an instrument of pressure on governments that don't do right by tribal people or that violate the standards set by the NGOs for protection of the "global ecology."[7]

Rich, a brilliant environmental lawyer, traveled the world for two years—working in an auto factory in São Paulo, Brazil, debating ecological politics with German "greens," and visiting villages in India—before returning to Yale's law school to complete his degree. He then went to Washington, and set about discovering the pressure points in the U.S. Treasury and in U.S. congressional committees with oversight on foreign aid that could force the World Bank and other MDBs to pay attention to the NGOs' environmental agenda. The leverage proved to be the recurrent need of the MDBs to obtain replenishment funds from the donor countries whose legislatures, made up of elective officials, have become sensitive to "green" concerns.[8] The NGO-versus-MDB campaign has contributed to reforms in the banks that have greatly strengthened environmental policies and staffing, but the NGOs maintain the pressure.

For loans involving Brazil, Schwartzman had a wealth of experience and personal contacts in that country. He went to the Brazilian Amazon from the University of Chicago in 1974 as a student anthropologist to research a doctoral thesis. His subject was the Panara Indians of Mato Grosso, who were being decimated by the advance of ranchers and miners as the Cuiabá–Santarém highway reached the dense forests along the Peixoto de Azevedo river. This experience with frontier violence made a lasting impression on Schwartzman, a paleface from Washington, D.C., who studied at a Quaker school and played guitar in a rock band before discovering ethnography. He lived in the Xingú park with Kayapo Indians for two years before returning to Washington, where he joined EDF. His discovery in 1985 of Chico Mendes and the rubbertapper movement through Allegretti shifted his perspective to the broader environmental questions, and the politics they involve. Schwartzman sees "environmental protection" less as management of natural resources and more as "a function of distribution of wealth, forms of land use, and the participation of local constituencies concerned with natural resources destruction"—in other words, political power to the "forest people."

The critique prepared by Rich and Schwartzman drew on information from Allegretti's Amazon Studies Institute (IEA), the CNS, the Union of Indigenous Nations (UNI), Sao Paulo anthropologists, and sources in Rondonia that saw the loan as an election platform for the discredited Santana government. The document was sent with a covering memo to key directors of the World Bank, including E. Patrick Coady, the U.S. executive director and Treasury's man at the bank, asking that the bank staff reply to the criticisms before sending the project to the board.

The NGO critique said Indian reserves and conservation units were being invaded and plundered because Brazil had failed to comply with its Polonoroeste contract obligations on "legal regularization and complete demarcation" of these units. The memo further complained that "NGOs, local unions, indigenous and community groups in Rondônia" had not been consulted by the bank during the preparation of the project. A letter was attached from Osmarino Amancio Rodriguez, secretary of CNS, and Ailton Krenak, national coordinator of UNI, asking for a suspension of negotiations "until local communities and their organizations . . . are informed of the content and purpose of the project." As such, the NGO memo questioned the bank's political judgment more than the technical aspects of the project.

> Past experience in other Bank Brazil projects in the 1980s indicates that the project's prospects for success are slight without explicit, specific requirements for the accomplishment of carefully timed and monitored measures, linked to loan disbursements. This is essential to ensure protection of Amerindian and natural reserves as well as the other land use zones in the project's agro-ecological plan.[9]

The bank management responded to this challenge in an angry letter from Shahid Husain, a Pakistani economist who was regional vice-president for Latin America. Husain said the EDF document was "based on considerable misinformation" and was "not conducive to a . . . useful dialogue" between the bank and Brazil on the environmental issues in Rondônia.[10] But the bank backed off when EDF's well-grounded attack was followed by a letter to the bank from Lutzenberger requesting that the project be held up until it could be reviewed. The Brazilian Ministry of Economy dropped PLANAFLORO from its priority list for foreign loans, which stranded Governor Santana, who had been counting on fresh funds from the project to elect his successor. Instead, Santana was indicted by the state legislature for malfeasance and ended his term under a heavy cloud. PLANAFLORO seemed to be dead.

As soon as Piana took office, however, the political war between Santana and the "friends of the forest" gave way to a honeymoon with Piana. As in Acre, environmental questions were crucial for the state's finances. The new governor realized that the only way to administer his bankrupt state was by obtaining financial help from the Collor government, which was demanding an administrative housecleaning and environmental reforms. These reforms depended on reviving World Bank support for PLANAFLORO, and this required an end of hostilities with the NGOs. In June 1991, World Bank officials met in Porto Velho with the leaders of CNS, UNI, and 11 other Amazon NGOs to discuss their criticisms, and Piana signed a protocol that gave the NGOs consultative status in the design and implementation of Rondônia's agro-ecological program. With that stroke, Piana became a "friend of the forest" and the Brazilian and international NGOs began to urge the World Bank to bankroll PLANAFLORO.

The change in political behavior from Santana to Piana, and the moves made by the new governor of Acre, shows how environmental politics are changing relations in Amazonia. Everyone uses "green" labels for whatever serves their interests, but the need for external financing makes it necessary for governments to act on the environmental agenda. This presents a rare opportunity for cooperation in areas that had been virtually taboo.

The Collor government's commitment to an Amazon forest conservation program with international participation compels every state in Brazil to get in step on environmental policy if they hope to obtain federal money. Other Amazon national governments are moving in the same direction. Revolutionary guerrilla insurgencies in rural areas and drug wars with powerful traffickers in the Amazon region have not deterred the presidents of Colombia, Peru, and Bolivia from adopting strong environmental policies. President Fujimori's environmental code requires costly technical impact studies for the development of mines and oil exploration concessions in the Amazon region.[11]

But miners and drillers are not the most serious threat to the environment of the eastern Andes. This comes mainly from the deforestation connected with peasant coca plantations on steep slopes, where erosion is severe, and from the toxic wastes dumped into rivers by clandestine laboratories that process "paste" for the global cocaine markets. U.S. pressure for military action in the sub-Andean Amazon against illegal cocaine operations has not been accompanied by enough money for rural development programs that seek alternatives to environmentally damaging coca plantations. Acceptance by peasants has been limited.

The role of coca production in relations to sustainable development of Amazonia is contradictory. In one respect, it provides cash income for thousands of Andean farmers who have migrated to the eastern valleys that can't be matched by any other crop- or resource-based activity. This is basic for family survival strategies, and it is hardly surprising that most peasants in the upper Hullaga valley of Peru or in the Chapare and north Yungas areas of Bolivia make this their principal activity. But the amount of the ultimate cocaine income in foreign markets that reaches the peasants is paltry, and the illegal nature of this export means that there are no tax revenues.

The conversion of cocaine money into local assets, known as "laundering," provides a market for gold smuggled out of Amazonia, and it shows up in land purchases and construction of urban villas. The "smell" of cocaine money clings to a prolific smuggling economy in Bolivia, Rondônia, Amazonas, and the eastern areas of Colombia and Peru, where foreign whiskey, cigarettes, electronics, and automobiles are abundant. But estimates that put the annual cocaine income coming into Amazonia in the billions of dollars are not verifiable by any method of accounting and this guesswork can exaggerate the importance to the peasant of maintaining the coca economy. No strategy for sustainable development of Amazonia can use cocaine as a reliable base for calculation of capital inflows for long-term investment. Only a legal ordering of society and a sound economic foundation can provide a sustainable "New Frontier."

TEN

NGOs: A New Agenda

The new climate of intergovernmental cooperation on the global environment presents new challenges and opportunities to the NGOs that have been very successful in raising the level of public concern over ecological and social questions. International NGOs, and their local counterparts, have played a decisive part in opening up Amazonia to environmental reforms. They will continue to play a vital role in forums where international policies on the environment are shaped and in monitoring agencies that carry out development lending. But sustainable development is more than this. It is not an ecological sit-in or a counterculture happening. It is hands on work, over a long period, by people with practical skills and a good touch for common people. The need is for NGOs that are project-oriented, with the ability to work on the Amazon frontiers, face-to-face with stubborn problems for which there are no ready answers. This calls for strong ties between the international NGO community and local groups that are on the firing lines.

The opportunity, and the need, for NGO project involvement on a wider scale arises because the Amazonian governments have decided, from calculation or conviction, that international cooperation is in their interest. But the Amazonian governments, at the national, state, and local levels, are handicapped by very weak environmental agencies and a severe shortage of counterpart funds to match donor financing. At the same time, the major donor countries are anxious to channel environmental financing through NGO channals, as in the G-7 program for the Brazilian Amazon. Increased use of debt-for-nature swaps could also make more funds available through private channels.

Only governments can define the general policy framework for international cooperation, but once these political decisions are made, the implementation of many programs can be done more efficiently by environmental NGOs, voluntary service organizations, and private enterprise, if they are given the opportunity to participate. The number of local NGOs in Amazonia that are project-oriented is still small, but there are some encouraging experiences that show what can be done with small amounts of money when there is local initiative.

Some foundations have been doing creative work in Amazonia for years. The Ford Foundation, which has had an office in Brazil for 30 years, had a 1991–92 project budget of $1.2 million, divided among Acre, eastern Para, and the dry Brazilian Northeast. "Ford has a policy that supports projects from the grassroots to the national policy level," said Anthony Anderson, Ford's director of projects, who is a Yale ecologist with long experience in Brazil. Ford's backing for the rubbertapper movement has been significant. About $280,000 was earmarked in 1991–92 for CNS and IEA to finance the development of extractive reserves, and a rubbertapper cooperative in Xapuri got $50,000 in working capital. An agroforestry project with small farmers near Paragominas and a peasant agriculture project near Marabá were also being funded. Ford, World Wildlife Fund–USA, and the Macarthur Foundation are backing a forest resources research center at Belém created by Christopher Uhl of Pennsylvania State University. Uhl is an authority on eastern Pará, where he has worked for many years in cooperation with EMBRAPA, Brazil's federal agricultural research agency, on recovery of forests in degraded pastures.

IMAZON (Institute of Man and the Amazonian Environment) is the name Uhl and his mixed team of Brazilian and American scientists have given their ambitious program in eastern Pará. They emphasize environmental research on how natural resources are being used by small farmers, lumbermen and riverbank populations to identify and promote environmentally sustainable alternatives to prevailing systems. This provides a platform for scientific research in Amazonia by 18 graduate level scientists, including agronomists, biologists, economists and ecologists, generating information on sustainable resource uses that can be applied at regional, municipal and community levels.

Rural extension to encourage producers to adopt sustainable systems is a basic element of the program. Local experiences have been identified that show that organized peasants can be receptive to technological change when they have land security. The only surviving colonization project in Paragominas, out of five that were implanted along the Belém–Brasília highway, is at Uraim, where the peasants not only fought successfully against encroaching ranchers, but received technical assistance from 1973 to 1978 from a volunteer group of agronomists from the Italian University of Piacenza, sponsored by the German Roman Catholic aid group, MISEREOR. About 50 families at Uraim maintain stable agriculture and livestock production based on black pepper and supply to local town markets of poultry, eggs, bananas, vegetables and fruit.

The IMAZON group is oriented strongly toward practical management systems for land use, fisheries, and forest production. José Adalberto Verissimo, IMAZON's director and a forestry engineer, takes the pragmatic view that "settlement of much, if not most, of the [Amazon] basin is inevitable" because of the "enormous social, economic, and political pressures in Amazonian countries" for occupation. Therefore, preservation alone is not enough and environmental protection requires the involvement of people at the local, productive level in land use systems "that maintain the resilience and productive potential of Amazonian ecosystems."

World Wildlife Fund–USA, Conservation International, Cultural Survival, Greenpeace, and Ashoka are among the NGOS that have opened permanent offices in Brazil, and the National Resources Defense Council put a team in Ecuador. Garo Batmanian, who did his Ph.D. research at Marabá on peasant conflicts, is in charge of the numerous WWF projects, which are mainly scientific but which include technical participation in the Antimari forest management plan. Carlos Miller, a Brazilian ecologist who worked with Conservation International, joined a private foundation in Manaus from where he can monitor projects protecting biodiversity "hot spots" in Amazonia that have become the major conservation target of CI. The European presence included a $1 million grant from the EEC to an NGO project in Pará with small farmers and migrant peasants conducted through the Nucleus for Higher Amazon Studies (NAEA) at the Federal University of Pará. This important project on the crucial problem of shifting agriculturalists is discussed further later in this chapter.

Identification and monitoring of small projects for foreign donors that do not have a representative in Amazonia can be facilitated by groups like the Brazilian Agroforestry Network (REBRAF). This group of Amazon forestry experts was initially put together by Michael Small, an imaginative Canadian diplomat, to advise the embassy on small environmental projects financed from a $500,000-a-year grant fund. This evolved into a permanent office in Rio that is a clearinghouse for up-to-date agroforestry information and for proposals from Amazonian NGOs or community groups with which it maintains contact. REBRAF is run by Jean Dubois, an internationally known Belgian forester with many years in Amazonia.

A good example of REBRAF's potential role as a clearinghouse is an agroforestry project that brought together funding from the German state of Hesse with the skills and enterprise of a Dutch forester, Wim Groeneveld, who has lived in Rondonia for ten years. Groeneveld studied forestry at Wageningen University in the Netherlands, a world center for tropical forest research since the days of the Dutch East Asian colonies. He went to Rondônia to do Ph.D. research, for which he set up an ecological park that is still largely intact outside Ouro Preto, a major colonization site on BR-364. He married in Brazil and settled in Porto Velho where he runs an ecological NGO, Instituto de Pre-Historia, Antropologia, e Ecologia (IPHAE), associated with the federal university there.

Through various programs, including a modern environmental communications center, IPHAE has become a major ecological base in the western Amazon for foreign and Brazilian researchers. Groeneveld also works as a forestry consultant, and he designed a $500,000 recuperation plan for the degraded forest around the Jacunda tin mine in Rondônia, formerly owned by British Petroleum and now by Brascan, a Canadian firm, which is paying for the cleanup.

The state of Hesse elects only a few German Green party candidates to its Landstaat, but citizens are concerned enough about tropical rainforests for the state government to have banned the use of tropical timber in public projects. More importantly, the state made a grant of $250,000 to Groeneveld to finance a project for "the rehabilitation of degraded rain forest lands in Rondônia." The Hesse project is an important experiment in social forestry in which thousands of small landholders in Rondônia will be offered agroforestry alternatives to recover misused land by planting trees. This attempt to reduce creeping deforestation by small farmers is of far greater practical effect than a largely symbolic timber boycott.

Groeneveld is an imposing figure, a six-foot five former volleyball star whose cuts his blond hair in a punky style and puffs tobacco smoke from a goosenecked briar pipe. He speaks Portuguese well enough for peasants to understand his talks on the virtues of alley cropping, which keeps trees that fertilize the soil between rows of annual crops, or the market opportunities for *cupuacu*, a fruit of the cacao family that has a growing market as a frozen concentrate. What makes these spiels of real interest to farmers is that Groeneveld has set up three large, strategically located nurseries in Rondônia from which selected species are distributed at low cost to small farmers on trucks owned by the project. The demand for *cupuacu* plants exceeded the supply in the second year of the project, and 150 farmers with demonstration plots were encouraged by project technicians to become tree producers. The state forestry institute spent millions of dollars provided by the World Bank's Polonoroeste project and put 800 people on its payroll, but it was never able to organize a state-wide nursery system, which crippled reforestation programs.

"These things work when there is an economic basis for agroforestry innovations. We still have a long way to go to be able to present a full technological package, but this will come with experience and research," said Groeneveld.

The Hesse project runs for three years, by which time a local staff of technicians will have been fully trained and a significant number of Rondonia's 100,000 small farmers will have been exposed to agroforestry. Groeneveld hopes that by then the Rural Worker's Unions (STR) participating in the project will seek support for a wider program, financed by an international donor or PLANAFLORO. IPHAE foresees a state-wide small farmer agroforestry program.

A still more efficient line of financial support for agroforestry projects that have an economic potential would be the creation of private investment funds for small projects combining venture capital and research. Project participants, such as small farmers, could receive credits, repayable on "soft" terms but with a fee to the project administrator to assure management continuity and long-term capital growth. This would shift part of the burden for administering environmental projects from governments, which are subject to political conflicts and beaurocratic instability, to NGOs with grassroots developmental experience and, in some cases, private forestry enterprises.

* * *

On the enormous scale of Amazonia, coordination between in-country NGOs is an efficient way to manage information that avoids waste of resources through unnecessary duplication and dispersion of efforts. "Networking" between NGOs with natural affinities and common interests makes it easier to bring together necessary resources from various donors and makes better use of scarce talents. This is the purpose of a sophisticated electronic system put together by Langston "Kimo" Goree III, who is an expert in satellite communications. Working from IPHAE, Goree put computers with modems in a dozen Amazonian NGOs, from Acre to Belem, that can communicate among themselves and with Brasília, Rio, or points abroad on an "eco-network."

But coordination is not always easy in the motley NGO community because of differences in interests, ideology, and personalities. The NGOs of the genus *amici florestae* are differentiated in numerous species that compete for turf, prestige, and money. During the run up to the ECO-92 conference, the international NGOs with an interest in the rainforest redoubled their efforts to lobby government, and as the amount of external financing available for Amazonia increased, there was an explosion of local NGOs. The stridency of the debate over Amazonia's ecological problems rose to the pitch of a Tower of Babel.

Roughly speaking, the NGOs can be divided into two variants, the "plumbers" and the "bashers." They have the same general objectives of giving priority to the environment, but they differ in style and substance over how to go about achieving their objectives. There are political differences and ideological disputes that sow dissent among the self-declared "friends of the forest." This complicates the definition of what should be done to achieve sustainable development in Amazonia.

The "plumbers" want to clean up the environment and preserve nature, but without pulling down the First World's economy. They tend to occupy upscale office space in Washington or New York, where scholars produce papers that are supposed to influence the policy "process" and scientists devote their attention to saving endangered species and habitats. Much of their millionaire funding comes from government contracts, secure philanthropic endowments and environmentally enlightened corporations. Representative "plumber" institutions are World Resources Institute, World Wildlife Fund, Nature Conservancy, Conservation International, and the Sierra Club in the United States, and the Worldwide Fund for Nature and the International Union for the Conservation of Nature, both based in Geneva, with a strong European orientation toward patronage from royal families and the former colonies in Africa and Asia.

One of the most illustrious "plumbers" is Thomas E. Lovejoy, former president of the World Wildlife Fund–USA, who in 1989 moved to the prestigious position of scientific director of the Smithsonian Institution, with a seat on the White House Scientific Advisory Council. Lovejoy, a Yale-educated biologist from New York, is a bowtied eco-diplomat who invented the debt-for-nature swap, a device for turning bad Third World debts into cheap financing for environmental projects. Lovejoy's European counterpart is Ghillean Prance, a whiskered botanist with a seadog's rolling gait, who is director of Kew Gardens in London. Both have a high degree of personal rapport with the Amazonian scientific community from years of research in the field, which in Lovejoy's case includes a long-term study with INPA near Manaus on ecosystem dynamics and changes in species populations in forest blocks of different sizes. Lovejoy's lips go taut with emotion when he speaks of biodiversity—"the most important single issue facing mankind"—but his scientific perspective is cool and his aims are practical.

The "bashers" are of another plumage altogether. They are counterculture political activists, who are mainly interested in Third World "people" issues. They militate on behalf of Amazon Indians or Penan tribesmen in the dwindling forests of Malaysia. They are young but not yuppie, and they are more comfortable organizing campus rallies or fund-raising rock concerts than lobbying in the corridors of power. They work in disorderly offices in urban renewal areas where they generate "class action" lawsuits against environmental offenders and emit streams of testimony for use by "public interest" advocates at environmental hearings. They are good at mobilizing public protests by direct action tactics, such as sit-ins and boycotts, borrowed from the antinuclear and anti–Vietnam War movements. To these tactics they have added the information power of electronic "networking" by which the White House, influential members of Congress, the World Bank, or a foreign office can be inundated instantly with protest mail from scores of NGOs that command tens of thousands of contributors, each a customer for tee-shirts, bumper stickers, and literature on the "cause."

Randy Hayes, the leader of the San Francisco–based Rainforest Action Network, is one of the more colorful "bashers." He began as a documentary filmmaker in the antinuclear movement and switched to the rainforest as the "green wave" began to roll. He takes a scornful view of "plumbers," with whom he competes for prestige and money. "They are the environmental jet set, in cahoots with the men in grey suits at the World Bank who think they run the world. They lack passion and the instinct for the jugular," said Hayes during a visit in 1990 to the Brazilian Amazon in search of environmental outrages and rainforest militants.[1] Hayes provides his network's 40,000 members, with units on 300 campuses, with a manual listing "7 things to do to Save the Rainforests." The recommendations include a boycott on tropical wood products and hamburgers from ground up tropical cows, solidarity with tribal people, outlawing of "agrotoxic" chemicals, and no more MDB money for Amazon roads. This is the typical agenda of "bashers," which include the Natural Resources Defense Council, Greenpeace, Friends of the Earth chapters in many countries, and a profusion of rainforest alliances and networks.

The international NGOs, with their different programs and styles, pull in millions of dollars, mainly from tax deductible voluntary contributions, and in some cases by the marketing of products from the rainforest that benefit indigenous people. This is big business. Just the top ten environmental organizations in the United States had budgets totaling over $200 million in 1989.[2] Although this went mainly for domestic programs, it is a measure of the cash flow that environmental NGOs can generate. This marketing is supported by a cast of Third World rainforest stars who travel frequently, in the wake of Chico Mendes, on a circuit that includes fund-raising concerts at Madison Square Garden, dinners in Paris with Danielle Mitterand, Vatican audiences with Pope John Paul II, and meetings in Washington at Congress and the World Bank, as well as an endless run of rainforest symposiums, roundtables, and forums from Germany to Japan.

Leaders of the self-styled "peoples of the forest" in Amazonia who have participated frequently as actors in this high-powered promotion have not failed to perceive the importance to them of the NGO marketing. One of the globe-trotting leaders is Ailton Alves Lacerda Krenak, an urbanized descendant of the almost extinct Krenak clan of the Botocudo Indians. Krenak grew up in São Paulo, where he worked in a print shop, read radical political literature, and became a friend of anthropologists at the University of São Paulo who brought him into the Comissão Pro-Indio, a militant political group for Indian rights. He worked briefly with FUNAI, visiting many Indian tribes, and married a French woman before blossoming as a communicator and political strategist for an indigenous movement called Union of Indigenous Nations (UNI), of which he is the self-styled "ambassador." UNI has had an alliance with the National Council of Rubbertappers (CNS) since 1989 called the "Uniao dos Povos da Floresta" (UPF), which is recognized in Brasilia as a major force for Indian rights and extractivism. This group has had only limited success in attempts to negotiate a broader Amazon front, including small settlers represented by the STRs. They all are opposed to large landowners, but the main demands of the rural unions is more land for settlement, whereas the UPF alliance wants forest reserves.[3]

In Krenak's sophisticated analysis, cooperation between the "peoples of the forest" and international NGOs has succeeded in making the First World "feel responsible for the destruction of the Amazon and for the social catastrophe resulting from this deforestation." This guilt can be eased, according to Krenak, by giving money to UNI and other tropical NGOs, and by consuming Rainforest Crunch ice cream or breakfast cereals made with Brazil nuts and cashews from the rainforests. When these natural products are marketed by NGOs, like Cultural Survival, profits from such sales go back to Indians and rubbertappers who are going to "save" the Amazon forests. Krenak makes clear that the basic reason for going to the market in "capitalist society" is not commercial, but to win international support for the permanent indigenous political objective of land rights, "to guarantee our place in our traditional territories."[4] In Amazonia, this can mean control over millions of square kilometers of potentially valuable land and resources.

UNI is a member of COICA, an Amazon "indigenous coordinating organization" based in Lima that represents Indian movements in Peru, Ecuador, Bolivia, and Colombia. Evaristo Nagkuag Ikanan, the Peruvian president of COICA, told a seminar on global development and the environment during the IDB's annual meeting in Montreal in 1990 that the Indians were against illegal coca planters, cattlemen, miners, dam builders, and oil companies in Amazonia. But Nagkuag told the MDBs that if Indian rights to self-determination and permanent possession of their territories were recognized, indigenous people might consent to allow "Amazonian development projects which would be compatible with the indigenous people's principles of respect and care for the world around them. . . . and their concern for the survival and well-being of their future generations." The Indian leader, who is from a tribal community on the Maranon river, added that this would "guarantee the future of the Amazon basin . . . for all of humanity."[5]

These are sensitive political questions of involving territorial rights and national cultural policies. Krenak and Nagkuag are seeking international support for their negotiating position by an active commercial presence for rainforest products in external markets and by appeals to the ecological interests of "humanity." These are shrewd tactics because they bring outside forces, including First World governments concerned with deforestation and climate change, into the Amazon dispute on the side of the Indians and extractivists. The large amount of time that the leaders of the "peoples of the forest" spend in foreign capitals is justified in the name of cementing alliances with NGOs that provide funds and mobilize campaigns in support of these demands.

Thanks to the environmental concerns over Amazonia, these campaigns have awakened a belated recognition in the First World that the human rights of indigenous peoples have been trampled during the modern occupation of Amazonia, as exemplified by the rubbertappers of Acre and the destruction of the Cuyabene Indian park in Ecuador by oil companies and settlers.

Recognition and correction of these injustices should not lead to the mistaken conclusion, however, that the indigenous or traditional forest people are the only legitimate "voice of Amazonia." They have rights and interests to defend, but so do millions of other people in Amazonia who aspire to another form of life that is based on different cultures. To lose sight of this would be environmental reductionism. It risks a severe political backlash from people in Amazonia who feel that their interests are not being protected.

Amazonia is a multiform society, where diverse segments have to be included in a strategy for sustainable economic development in a stable environment. However the environmental agenda is defined, the social and economic reality of Amazonia today shows that traditional extractivists and indigenous people are a small minority, outnumbered 15 to 1 by modern settlers. The newcomers have brought to the region an urban structure, consumer economy, information system, and production technologies that are little different from those in the more developed areas of their "national society." It is not realistic to expect that the modern sector will adopt the ways of the traditional "peoples of the forest," although it can and will assimilate much that is useful from them. In turn, the forest people are being exposed to modern ways that are transforming their outlook and habits, often improving their opportunities.

Therefore, this duality has to be reflected in a two-tiered strategy. One line of action should strengthen the economic and cultural autonomy of the indigenous people and traditional extractivists, through secure reserve areas where low-intensity forms of natural resource management could sustain relatively intact forests. The ecological gains are consistent with the goal of respect for the cultural heritage and the political and social rights of minorities.

Another line of action, equally important for the defense of Amazon ecosystems, would reduce the pressures on the forests, not only by creating an effective system of protected conservation units and natural reserves, but by the reform of agricultural technologies that contribute to deforestation. One of the ways is by reducing extensive ranching on poorly managed pastures, which can be altered positively by new technologies, as was described in Chapter 9. But the most pervasive and pernicious behavior that must be changed is that of shifting agriculturalists, who destroy the forest to eat, like locusts, because they don't have a sustainable method of production. Natural extractivism and stable agriculture need equal priority

It is estimated that there are 500,000 families in Amazonia who live primarily from some traditional extractive product—latex, nuts, palm hearts, coconut oil, *jaborandi* and other medicinals, fruits, building materials, and similar renewable forest products. Many of these are indigenous people whose cultural conditions are best protected on Indian reservations. Numerous others are old-time Amazon residents, the *caboclos*, who hunt, fish, and do a bit of logging and some petty farming, and if wildlife buyers appear they jump at the chance to sell them crocodile skins, leopard hides, or rare monkeys and parrots that are worth as much as $25,000 in the United States or Europe. Only in some cases are these sylvan residents full-time extractivists, like the rubbertappers, who live at specific locations in forests or along rivers. These are the potential beneficiaries of what has come to be known as "extractive reserves." Allegretti has proposed that 250,000 square kilometers of the Brazilian Amazon be converted immediately into extractive reserves with a total area that is larger than the whole state of Rondonia.

The extractive reserve proposal has been embraced by many of the international NGOs as "a local Amazon response to deforestation that [is] . . . a new way of integrating conservation and rural economic development in the Amazon."[6] Environmentally, an extractive reserve is desirable simply as a large, continuous forested area with a minimum of human alterations. But if a reserve can be managed to provide an adequate family income for those who live from the forest, the social benefits are that it will maintain the cohesion of sylvan communities that might otherwise be forced off the land and into an urban slum. To support their case for extractive reserves, advocates produce numbers that show that extractivism offers higher economic return than extensive cattle ranching or agriculture.

The economic arguments in favor of extractivism are questioned by some analysts of Amazonia's economic alternatives who are not committed to the "peoples of the forest" as a political movement with an ecological veneer, as are CNS, UNI, and the NGOs that back them. Among these technical critics of natural extractivism is Alfredo Kingo Oyama Homma, winner of Brazil's National Ecology Prize in 1989 for a doctoral thesis on Amazonian agronomic development.[7] Homma is the grandson of a Japanese colonist credited with adapting Asian jute to the Amazon after many others failed. Homma knows Amazon agriculture from the ground up. As a boy, he worked in the fields in Tomé–Açu, in Pará, where a Nisei colony got rich converting forest land into family-size black pepper plantations. His research work on agricultural economics is at the EMBRAPA's Center for the Humid Tropics (CPATU) in Belém. Homma's basic point is that natural extractivism, without improved productivity through research and technology, is economically unsustainable because it can't compete with cultivated competitors:

> Amazonia lives by myths, and the latest is the Chico Mendes myth of extractive reserves. . . . Ecological sensibility is noble, so demanding questions are not always asked. . . . But if it were true that extractive activities offer high returns, as some studies suggest, the "invisible hand" of Adam Smith's marketplace would not have led to the disappearance of dozens of extractive products. . . . So, one has to ask whether international financiers will supply the resources without which extractive reserves will not be viable, politically and economically, in the long run.[8]

As of now, Homma's question can be answered positively only on environmental grounds. There is no example, as yet, of an extractive reserve that shows that gathering and processing of wild rubber and Brazil nuts can be economically sustainable without subsidies. The most important Amazonian extractive product, wild rubber, is being challenged by a powerful surge in Brazil of natural rubber from plantations, which outproduced the traditional sources for the first time in 1990. Large plantations of Brazil nut trees in Pará and Amazonas are coming into production at locations that are strategically placed for exports. As Homma pointed out, domesticated production of wild species is the normal course of agricultural development since neolithic times because it is more productive than gathering in the wild.

The only extractive reserve experience that can shed light on this debate is a CNS pilot project on an old rubber estate at Cachoeira, two hours by truck from Xapuri, where 80 "free" *seringueiros* and their families live in 7,600 hectares of forest. This is where Mendes was born and grew up tapping rubber before he separated from his first wife and went to Xapuri, where he joined the rural union movement as an organizer. In 1987, Cachoeira became a citadel of resistance. When the rubbertappers at Cachoeira learned that their old *patron*, as owners are called, had sold the forest rights to a group of cattlemen, including Darcy Alves da Silva, they refused to leave. With help from Mendes and the Xapuri Rural Workers Union, they organized a forest sit-in that prevented a chainsaw gang from starting to cut trees. The impasse ended with the property's being expropriated by the government's agrarian reform institute and turned over to the rubbertappers as an extractive reserve. But the defense of Cachoeira led to the violent death of Mendes a year later.

By mid-1989, the international reaction to the Chico Mendes murder had inundated the CNS with offers of financial assistance. At least $125,000 in voluntary contributions was raised by the Environmental Defense Fund and others to help the CNS, which had operated from remote Xapuri, set up a professionally staffed office in Rio Branco, where "welcome" signs in the reception are in three languages. Millionaire proposals were being considered for the rights to make a movie on Mendes and the rubbertapper movement from the likes of Robert Redford, Guber Peters, and David Putnam. The movie proposals led to bad blood between the CNS leaders and Ilzamar, Chico's young widow, who sold the rights to which she was entitled for $1.4 million to a Brazilian film producer, who later sold them to Peters. But money was being offered from many other sources, and the CNS leaders and their NGO advisers decided that it was time to push ahead with an extractive reserve project, based on Cachoeira and an existing rubbertapper marketing cooperative that Mendes had founded in Xapuri.

James LaFleur, an American economic consultant living in Brazil who had helped draft Rondônia's PLANAFLORO, was called in to prepare a project proposal. LaFleur, a well-tanned beachboy from California, came to Brazil as a Peace Corps volunteer after graduating from Santa Clara College. He stayed on as an economist with the Cacao Research Center in Bahia, which sent him to Oxford for graduate studies in international commodity trading. He subsequently became a sugar buyer in Recife for an American trading company, but he also did consultancy for the World Bank on agro-environmental planning and peasant agriculture. LaFleur, an environmental activist, is the representative in Brazil of International Development Enterprises (IDE), a nonprofit, privately funded development group of wealthy businessmen in Denver that provides seed capital for small Third World entrepreneurs.

The study LaFleur prepared in 1989 presented solid prospects for an "agro-extractive cooperative" that would generate annual sales of $820,000 in processed rubber and Brazil nuts after seven years, with a small profit for 420 members. The operation was on a scale that was intended to make the cooperative the dominant buyer of rubber and nuts for 33 rubber estates in a radius of up to 45 kilometers from Xapuri with an annual rubber production of 600 tons and 700 tons of nuts. This required a large investment in a transportation system of boats, trucks, and mule teams. Cash flows and cost calculations were presented in computer printouts along with an analysis of domestic and international markets.

The prospectus looked good to NGO donors, who embarked on a Brazil nut venture at Xapuri that by early 1991 had absorbed at least $200,000 without reaching half the production target for the first year. The Ford Foundation made an initial grant of $50,000 to the Xapuri cooperative, which installed a nut processing factory in a warehouse that was donated by the federal government. Trucks and boats were provided by foreign embassies. But the major input, about $130,000, came from Cultural Survival (CS) of Cambridge, Massachusetts, where Harvard academic Jason Clay and his fellow anthropologists have become commodity brokers to promote the consumption of "green" products from the "peoples of the forest," such as Ben and Jerry's Rainforest Crunch ice cream and the Body Shop's cosmetic line using oils and essences from Amazonia.

"We are not riding a wave, we are creating it," said Clay, a kinetic redhead with promotional flair who is chief executive of Cultural Survival Enterprises (CSE). CSE sales, which totaled only $58,000 the first year of operation in 1989 rose to more than $3 million in 1991 and were projected to reach $10 million in 1992. This nonprofit operation had a $2 million budget in 1991 financed by donor contributions and commissions on sales. Sixteen companies that Clay signed up for the CSE program, including Ralston Purina, the big breakfast cereal producer, agreed to pay CSE a "green premium" of 5 percent on raw materials certified to be from sustainable forest sources and a share of the profits from the sale of consumer goods made from 21 different forest products. These payments go back through

CSE to the extractive suppliers for projects that raise their living standards. Clay made it clear that CSE is interested not only in the forests, but in the well-being and behavior of people who live in and around them. Clay says:

> Our target is the people who supply these products and their behavior is the key. Poverty is a major driving force in environmental degradation. We address that when what we do makes a difference in the way people make a living.[9]

The extractive reserve ideal of an autonomous forest-dwelling community, integrating conservation with economic and social improvement, would be attainable if the rubbertappers were competitive producers, with a sense of business management and in command of technologies for the conversion of their natural resources into marketable products. Given their background, that is not the case.

The people to whom the Xapuri project was addressed are descendants of rural workers who were brought to Amazonia a century ago from drought-stricken northeast Brazil to tap rubber. They made Amazonia the largest exporter of natural rubber to the industrial world, but when this lucrative activity collapsed, as southeast Asian plantations took over the market, the Amazonia rubbertappers were reduced to Appalachian-like poverty. Many fled the forest; others interbred with Indians and survived much like the natives through hunting, fishing and subsistence agriculture. Those who continued working as tappers kept on doing the only thing they knew how to do, totally dependent on the owners of the rubber estates. These *patrones* kept part of the smoked rubber and Brazil nuts as rent, paid for the rest at knock-down prices, and charged dearly for supplies, which were sold on credit. This is known as debt peonage, but there was a patron, or his manager, who made the economic decisions and who could be hit up for a loan in an emergency. It was a system based on dependency on an owner.

A visit to Cachoeira shows how difficult it will be to change the situation. This rubbertapper community is reached over a muddy track, passing through fenced cattle ranch pastures on both sides until reaching a forested area, where a path leads to ten wooden shacks in a clearing. Most are homes, where clothes washed in a nearby stream are hanging out to dry. There is a school which is also the community assembly hall, a store, and a small warehouse. Donkeys graze under the goalposts of a football field where games are played on Sundays. A battery-powered radio plays country music. Around the settlement there are plots of rice, beans, and mandioc, and thin cattle graze on weedy pasture, as in every hamlet in Amazonia. Cachoeira is backwoods poor, but not miserable.

The men are out in the forest. Under a high noon sun, there is cool twilight under the dense overhead canopy. The tappers trot along forest trails and cross streams on well-worn fallen logs. Markings on the massive trunks of dominant trees point the route to the seringa trees scattered in the forest maze. The tappers carry a knife that cuts at one end and scrapes on the other. With it they clean the fishbone cuts that "bleed" the seringa of latex, which drips into little tin cups fixed to the trunk. Over their shoulder, they sling a shotgun and are accompanied by a hunting dog in the hope of finding a deer, a tapir, or other game for the table. Huck Finn would have been at home with the tappers.

To maintain production, a tapper must make daily visits to up to 300 trees that are strung out along routes covering up to 5 kilometers each way. It's a half-day's work, eight months a year, except during the rainy season when it is Brazil nut time. For good rubber production, as on plantations, collections must be made before the content of the cans congeals into an elastic gum. The white latex can be kept liquid with a touch of acid, removed in jugs, and processed later into high-quality shaped blocks of solid smoked rubber that obtains a premium price. But few make the investment in molds, metal trays, and a smokehouse necessary for this treatment, which also requires cooperation. The traditional tappers are individualists who prefer to collect a foul-smelling gum—often full of sticks, dirt, and even rocks—and smoke it over their own fire into large black balls of rubber, called a *pele*, the nickname of Brazil's greatest football star. This inferior product is sold at a poor price to buyers who have often advanced some supplies and can set terms.

During the rainy months from January to May, attention shifts to the other main forest product, Brazil nuts. The towering *castanheiras* produce the nuts in hard-shelled cannisters the size of softballs that sometimes explode when they hit the forest floor. The hard ones have to be chopped open and the unshelled brown nuts are piled in sacks to be hauled out by mule or donkey. The owners of the animals are usually the buyers as well, so the deal is cash on the line in the forest, with a discount for transport. The nut collectors who don't have their own means of transport always lose.

An extractive reserve controlled by "free" rubbertappers is supposed to sever these dependencies and stimulate cooperation. About 50 of the Cachoeira rubbertappers belong to the Xapuri coop where members can sell their rubber and nuts and buy consumer goods, such as sugar and kerosene, on credit against future deliveries of rubber and nuts. The coop offers an alternative to dealing with rubber traders, middlemen, and merchants whom tappers have always pictured as their "exploiters." But it has only 200 members.

Ronaldo Olivera, an independent social worker who lives with 50 rubbertapper families in the Floresta sector of the Chico Mendes reserve, said the low level of participation was because the coop doesn't deliver supplies in the forest or provide transport for the rubber and nuts it buys, both of which private merchants do. "When the people from the union and the coop show up in the forest they talk politics when what the people need is technical help to produce food, keep their cattle from dying, and make a better living than rubber can provide. You can't survive for more than three months a year on what is earned from rubber," said Olivera, who has lived in Acre since 1978 and was a founding member of CNS with Mendes.[10]

Although Cultural Survival had agreed to buy all the nuts that Xapuri could produce at a guaranteed price, the nut-gathering campaign at Cachoeira went lame during the first year because of a lack of mules and donkeys to get the nuts out of the forest, and many tons rotted. The 150 tons that got to the coop warehouses were enough to produce 50 tons of shelled nuts, but the plant was slow in training women to use the simple nutcracker equipment. As a result, fewer than 40 tons of shelled nuts were produced when orders were open for 300 tons. The second year of operation again failed to shell more than 100 tons, losing money.

A well-organized nut operation should earn a lot more for the tappers than raw rubber, and a large export market for processed nuts could make the coop a powerful tool for the development of the Chico Mendes reserve. But this calls for management that the CNS has not been able to provide. The political heirs of Mendes are an honest, dedicated group. Julio Barboza, the CNS president, has a quick wit behind a country-boy face, and he has shown himself to be a good diplomat in negotiations with donors. Osmarino Amancio Rodriguez, the secretary, is a courageous fighter, a fiery orator with piercing black eyes, who reads the bible and carries a .38 caliber revolver because of threats on his life. Raimundo "Big Ray" Barros, who shocked a rainforest seminar in Germany by displaying the bloodstained shirt of an assassinated rubbbertapper, is a good union organizer who prefers to be in the forests talking with tappers than taking part in seminars. But no one in the leadership has had experience in administration and there is little clear thinking on how an extractive reserve should be organized on productive lines.

Assis Monteiro, the Xapuri coop president until 1991, was a rubbertapper and a strong union man. He presided at a meeting of coop members at Cachoeira one Sunday morning accompanied by Rolando Polanco, the manager, a young economist from Rio Branco who is politically active in the Workers Party but inexperienced in management. The assembly at the schoolhouse was called to discuss how to overcome delays in getting nuts out of the forest that were causing a loss of potential exports. The 25 members who were present said the coop should have sent mules, and Assis said the members had to figure out how to deliver the nuts for pickup by the coop truck. "I am not your patron to tell you what to do. You have to decide," said Assis. But after three hours of debate there was no decision.

It was only a year later, after Assis had been replaced, that the coop decided to buy and maintain a troop of transport animals for rubber and nuts, with a charge to the members. Meanwhile, many tappers had sold off rubber and nuts needed to pay their debts at the coop to traditional buyers who provided transport. Debt delinquency was decapitalizing the coop, said the manager, but it seemed that the Cachoeira rubbertappers looked to the coop as their new patron, or debt holder, whereas the coop looked to the environmental NGOs as its benefactor.

Because of problems like these, in April 1991 Clay notified the CSN/UNI alliance, more in sorrow than in anger, that the Xapuri nut project was in trouble. The first shipment produced at the factory was 14 tons of shelled nuts, in vacuum-packed aluminum bags, for which CSE had paid a $35,000 advance. Not only did the shipment arrive months behind schedule but the nuts "were rancid and could not be sold on the US market."[11] The fulfillment of the scheduled delivery of 150 tons of shelled nuts from Xapuri during 1991–92 depended on increasing the average daily output of the women cracking nuts at the factory from 7 kilos (a very low level) to 12 kilos in eight hours. An evaluation report in August 1990 warned that without this increase in worker output the "situation is unsustainable."[12] The project failed to generate enough cash during the first year to buy the 400 tons of raw nuts needed for processing in 1991, which required a $40,000 emergency loan from CSE to keep the factory going. By any regular method of accounting, the nut project was bankrupt.

But Clay and LaFleur were undeterred as the second year of operations began, with some improvement in output. "I am not discouraged, they will learn from experience," said LaFleur, who had become a consultant to CSE for various commodity projects. Clay was fully committed to a strategy that would continue to provide money to CNS and the Xapuri project, regardless of production. The money was from CSE sales of $3 million, including $1 million in Brazil nuts, which generated a "green premium" of 5 percent and profit shares for CNS/UNI—even when 90 percent of the nuts were acquired from other suppliers. "What we believe that we are doing is using the production that is currently controlled by the elites to gradually take over the supply of raw materials that the elites depend upon," wrote Clay, replying to criticisms from CNS/UNI that he was "helping our traditional enemies" by buying in the open market.[13]

Nuts were bought in Bolivia as well as Brazil because CSE's commitments to its clients far exceeded Xapuri's limping production capacity of 100 tons. There was a commercial plant at Rio Branco, owned by a supermarket chain, and several private plants just across the Bolivian border at Cobija that were producing over 1,000 tons each by the same methods used at Xapuri, and obviously at lower cost. Strong external demand for nuts guaranteed a remunerative price. Ben and Jerry's Ice Cream alone was buying twice as many Brazil nuts as Xapuri could produce, and one day's consumption by a big candybar maker was equal to Xapuri's total production for the year. CSE was moving much faster on the marketing side than the extractive reserve project was on the production side.

"I keep telling these guys that this nut thing is a business, not a social program," said LaFleur during one of his frequent trips to Acre as a troubleshooter. "The factory can be managed so that collectors receive higher prices, workers get good salaries and an overall profit can still be turned. As the coop consolidates and expands, it will force other buyers in the region to pay its price for rubber and nuts. They can make the market," said Clay.[14] But the NGOs involved in the Xapuri "model" are motivated more by a social and political strategy than by market economics.

The coop paid a raw-nut price during the 1991 season that was as much as 20 percent higher than what was paid by traditional traders, and gatherers who sold to the coop earned altogether about $35,000, a large amount of money in Acre.[15] Nut processing in Xapuri created a payroll during eight months for about 65 workers, including 50 women, who earned a total of about $80,000. A "cottage industry" system for nutcracking by housewives working part-time at their homes in the rubber estates was tried to increase production while spreading the earnings to the field. All these benefits depend, however, on the economic viability of the project.

If the Xapuri cooperative is seen as a pilot project for the Chico Mendes reserve, the costs of setting up this much larger extractive reserve, involving 1,200 families, can be calculated in millions of dollars. The reserve extends from Xapuri to Brasileia on the border with Bolivia, and to Assis Brazil, on the Peruvian border. This southeast corner of Acre is one of the most strategic places in the Amazon for the protection of the rainforest because it is the most likely route for a highway from Brazil to Peru's road system to the Pacific. The Chico Mendes Extractive Reserve stands astride that route and would be served by this trans-Amazon link.

Because of the enormous costs, the Chico Mendes reserve remained nothing more than a legal formality two years after the decree creating it was announced. IBAMA, alleging lack of money, had done nothing to demarcate the perimeter. A long list of ranchers who had land within the reserve were waiting to be expropriated and illegal loggers continued to extract wood. Rubbertapper families were abandoning the reserve; in all Amazonia the tappers were in desperate economic straits because of the collapse of wild rubber prices. Without any source of income, the reserve had no administrative directorate or any technical staff and the STR unions at Brasileia and Xapuri provided only informal monitoring of the area. No census had been taken to determine elegibility for membership in the reserve association called for in the law and no election had been held to choose a directorate. For all practical purposes, the Chico Mendes reserve was adrift and no one seemed sure what to do next, until PMACI came up with $500,000 to start demarcation in 1992.

More practical results were being obtained at the Antimari forest management project, west of Rio Branco, as was described in Chapter 5. With the support of ITTO, 80 extractivist families at Antimari an old rubber estate, were moving toward new forms of production under a management plan prepared by FUNTAC, with the participation of international consultants and CNS representatives.

As a testing ground for sustained production of not only rubber and nuts, but improved agriculture for family consumption and selected timber cutting, Antimari is a package that is important for the entire Amazon region. An ITTO report on the Antimari project phase II, which began in August 1991, said it was "crucial" for the "peoples of the forest" in Amazonia because:

> The economic future of the existing extractivist communities is surrounded by considerable uncertainty due to the low prices being received for rubber, and integrated, conservationist, community-based forest management systems need to be urgently developed if forest-based development is to succeed in halting environmental degradation and social and economic problems associated with the drift of rural people to inadequately served urban areas.[16]

The precarious state of the wild-rubber economy on which the extractivist movement is based preoccupied an international symposium to consider alternative uses of the forest organized by CNS/UNI and the usual NGOs at the Federal University of Acre in February 1991. The three-day meeting put rubbertappers and Indians face-to-face with technicians and theorists in agro-forestry, rural sociology, ethnobotany, handicrafts, natural-product marketing, and other ways out of the rubber-and-nut dependence. There were even two conventional foresters, who described recent experiences in sustained yield at the Tapajós national forest; but the suggestion that the extractive reserves consider cutting trees was indignantly rejected. "The trees are my brothers. It would be like killing my mother," Krenak told the forum. Barboza was more cautious. He said extractive reserves had to be better organized and economically secure before lumbermen could be allowed in to do selective logging. But when ordinary tappers had the floor, the tone was desperate as they described dramatic conditions of impoverishment on their rubber estates.

Juarez de Todos os Santos, from the STR at Feijo, reported that prices paid in the forest for rubber were so low that it required three times as much rubber to buy bread or sugar as the year before, and half a year's production was needed to buy a shotgun. "*Seringueiros* and their families are starving. To survive they are destroying the trees by stripping them of latex and killing all the forest animals for meat," said Juarez.

Pedro Tellez, an old-time tapper, spoke of the past as if the *patron* were the protector of the forest and of the sylvan people.

> The *patrones* were the real ecologists. They conserved the forests with care for production of better quality rubber. . . . The *seringueiros* then worked the trees more kindly. Why? Because the *patrones* took care of us. There was *charqui*, cooking oil, mandioc flour and the supplies were

> bought in large quantities, so it was cheaper. . . . Now the *seringuiero* has to mistreat the trees to live, or he has to sell his location and leave. . . there isn't time enough to collect and till the soil for food, so we are in the hands of the *mareteiros* (peddlers) who sell us food. Even if you produce 1,000 kilos they will pluck you clean.[17]

Time may have blurred the old tapper's memory, but there was no protest from his companions, who all shared his view that "to survive there has to be something different from rubber and Brazil nuts." But nobody seemed to have any good ideas on alternatives, so the major conclusion of the symposium was that the government should buy all Amazon wild rubber at a guaranteed price about twice as high as the international price. As was seen in Chapter 8, IBAMA turned down such a rescue operation because the Collor government refused to finance a special price subsidy for the rubbertappers that would exceed the already protected price paid for domestic plantation rubber.

The growth of plantation rubber production was a sword of Damocles hanging over the Amazon wild-rubber collectors. In 1990, for the first time in Brazilian history, the supply of cultivated rubber (16,200 tons) was greater than wild rubber (14,000 tons). Plantations in Sao Paulo, Bahia, Espirito Santo, and relatively cool, dry areas of Mato Grosso were coming into production with good yields and free of the fungal South American leaf blight that killed off many rubber plantations in Amazonia. Tire companies (Michelin, Pirelli, and Firestone) have plantations that allow them to satisfy quotas of domestic rubber purchases needed to obtain rubber import licenses. If the movement toward freer trade opens the Brazilian market to external competition from rubber products, this would drive a further nail into the coffin of wild-rubber production.

Allegretti's response to the crisis was a proposal that the defense of wild rubber be shifted from economic grounds—"where the competitiveness of wild rubber is fragile"—to social and ecological grounds. She argued that in the short run there was no alternative for rubbertappers, so the case for a government fund to buy wild rubber at a subsidized price, equivalent to a monthly minimum wage for each rubbertapper, had to be fought with the ecological argument that without rubbertappers in the forest there would soon be no forests in Amazonia.

> The technocrats argue that the *seringueiros* have to demonstrate the economic viability of extractive reserves before any new ones will be created. That is a trap we will not fall into because we are not talking about short-term viability. The *seringuieros* are the guarantors of the Amazon forests, and that's enough! That is the argument on which we will build our political alliances with everyone interested in preserving the forests, including workers in the tire factories and NGOs who know how to pressure the multinational owners.[18]

This comes dangerously close to casting the rubbertappers into the role of ecological tokens, like the spotted owls of the Pacific Northwest forests. But owls don't need to be subsidized to survive.

* * *

The high visibility given Chico Mendes, the rubbertappers, and extractive reserves has withdrawn attention from the main source of human alteration of the Amazonian environment—the landless, migrant peasant. Millions of these "forgotten men" are gnawing away at the forests every year for a piece of land on which to sow annual crops for two of three years until fertility gets so low that they have to clear more land. Although big ranchers have contributed dramatically to deforestation by torching large individual areas, the cumulative effect of land clearing in small bites is greater. Most students of tropical forest depletion now agree that "shifting agriculture is the single most important cause of tropical deforestation."[19] This clearly appears to be the case in Amazonia now that the major expansion of cattle ranches has slowed.

The figures on land clearing in Acre between 1987 and 1989—a period in which ranches were still expanding—show that as much forest was removed for colonization settlements and small farm lots as went into new pastures.[20] In Pará and western Maranhão, areas of continuous conflict between ranchers and landless peasants, the most visible form of deforestation is the result of land cleared in the forests that lie along streams and irregular land between the ranches. The landless migrants camp in rickety shelters, always fearful of expulsion, and farm in niches of poor soil between the larger properties. They are usually tolerated by the proprietors if they don't try to occupy cleared land.

Near Acailândia, where BR-222 (the trans-Maranhão highway) intersects the Belém–Brasília highway, Julio Motta, a pioneer rancher from Pernambuco, has a 2,000-hectare property that he began to develop in 1985, putting in pasture on which he maintains a herd of 2,000 nelore cows. But before getting down to animal husbandry, he had to deal with a human problem.

> When I bought this place there were already ninety families camped here along the highway. I needed workers to put in fences, but they weren't interested. Some were *garimpeiros* in the gold placers, others fished and hunted, and the rest planted rice and mandioc wherever there was a place to squat, which worried me. You know how many I killed? None. I just paid for some building materials and relocated them all five kilometers down the road where they now have bars, fruit stalls and shops. There are also some charcoal makers who sell to the pigiron makers in Acailândia.[21]

The number of landless peasants on the margin of the economy is an environmental problem of major dimensions. The ranchers offer no solution because they employ relatively few full-time workers. Motta operates his spread with a foreman he brought from Pernambuco and six cowboys.

Distribution of land through agrarian reform programs would help direct the peasants into more organized settlements, but experience has shown that without technologies that sustain agricultural yields, many abandon their unproductive lots. The line of action to reduce pressure on the forests requires a combination of stable tenure and farming techniques for Amazonian conditions that will make agriculture, combined with agroforestry and animal husbandry, a sustainable economic activity.

Nowhere in Amazonia is this more critical than in eastern Pará, where tens of thousands of migrant peasants have taken a foothold in the Tocantins–Araguaia valley between Imperatiz, Marabá, and Tucurui, with some advances to the west between Carajás and the Mato Grosso–Pará border. Here is where the sharpest lines of conflict have been drawn between the migrants (usually very poor northeast peasants with no capital or technical skills, but with an almost messianic thirst for land) and the entrepreneurial ranchers and colonists from the southern states of São Paulo, Goiás, and Minas Gerais who have carved out modern agricultural and cattle operations.

There is a substantial literature that focused on the land conflicts in the region in the 1960s and 1970s between squatters and the the big holdings. There were hundreds of victims, which drew the attention of human rights activists such as Amnesty International.[22] During the military governments, there was a small left-wing guerrilla insurgency that brought severe repression to the Tocantins–Araguaia region. Some land was distributed to peasants, but not enough to eliminate conflicts in which peasant leaders continued to be killed by hired gunmen and ranches that were not protected were invaded by a rural movement called "sem terra" ("the landless"), supported by Roman Catholic missionaries and local politicians, particularly of the Workers' Party. These conflicts did not appear to have a clear relationship to the environment, however, and as global ecology eclipsed human rights as a trendy international issue, the landless peasants got less attention.

But "sustainable agriculture" remains the central question for peasants, in terms of both social stability and a balanced relationship with the environment. Philip Fearnside, the sharp-eyed critic of many Amazonian government programs, has addressed the problem in relation to the CVRD reforestation program in the Carajás corridor, which includes a large agroforestry component, as described earlier. Fearnside said:

> An agricultural program linked to the Carajás development represents one of the best opportunities to return to the area some of the wealth being removed from the mining operations. However, the way in which such an agricultural program is designed is crucial to the role such a program would play; it can either be used to develop truly sustainable agroecosystems for the long-term benefit of the region's inhabitants or it can be yet another continuation of Amazonia's traditional role as a mere colony of economic centers beyond its boundaries, such as Brazil's Central-South.[23]

The development of a "sustainable agro-ecosystem" for Amazonian conditions is the great challenge for researchers now. This challenge has produced a major change in how some activists for agrarian reform see their role. Land distribution remains a social need, but the development of productive technologies to work the land has taken on equal importance.

Imperatriz, a booming commercial center of southern Maranhão, is one of the "bang-bang" places in the Carajás corridor. It is linked to Acailândia by the Brasília–Belém highway and a railroad spur off the main Carajás line. It has gone through a ten-year construction boom, fed by *garimpeiros* who struck it rich at Serra Pelada and other placer mines in Pará and by wealthy ranchers from Goiás and Minas Gerais who have bought land in the region. There are modern hotels and ten-story office buildings downtown and expensive new residences for the wealthy families on still unpaved streets on the outskirts. But the most predominant housing is shanties along the river that are filled with numerous peasants and their families, many of whom say they were driven off land they had worked for years by the new ranchers.

Beside Our Lady of Fatima Church, the Roman Catholic bishop of Imperatriz, Afonso Felipe Gregory, occupies a residence where the walls facing a side street are pock-marked with bullet holes. This is where Josimo Tavares, a young priest working with peasant organizations opposing the ranchers, was killed in an ambush in 1986 by gunmen who have been identified but never brought to trial. According to church figures, rural violence connected with land disputes caused 138 deaths in western Maranhao between 1980 and mid-1988. The same pattern of violence extended from Conceiçáo do Araguaia to Marabá on the west side of the Tocantins.

Shortly after his appointment as the first bishop of Imperatriz in 1988, Gregory organized a state-wide pilgrimage of peasants to Imperatriz in support of land reform. "This is supposed to be a frontier but there is a scandalous concentration of land in the hands of a few who are forcing families off the land, causing conflicts and killings. Without an agrarian reform that changes the unjust distribution of land there will be no peace in the countryside and no genuine democracy in the country," said the bishop. But two years later, after watching the collapse of several squatter invasions of ranches, Gregory was warning the pastoral land commission in his diocese against stirring up peasant land seizures. "Until there is technical assistance, credit support, and marketing, the peasants who occupy the land are worse off than they were before. It isn't a social solution and it isn't worth the risks," said Gregory.[24]

The Roman Catholic church has played an important role in Amazonia helping peasants organize in rural unions through a so-called Land Pastoral Commission (CPT), which maintains a network of priests and pastoral agents. In the Tocantins–Araguaia area, the work was organized initially by a group of six French priests who came into the region in the 1960s, continuing a tradition begun early in the century by a French dominican friar, Gil Vilanova, who founded Conceiçáo do Araguaia as a missionary outpost among the Kayapo Indians.

At the height of the violence in the region, landowners accused priests of fomenting land conflicts through the creation of unions and the organization of "Christian base communities," as was done in many peasant areas. The military expelled two of the French priests, arrested a Brazilian bishop, and tortured lay pastoral agents who worked with the unions. One of these was Emmanuel Vonbergue, a French agricultural technician who went to Marabá to run the CPT office. He and his Brazilian wife worked closely for a decade with peasant communities that produced a crop of effective rural union leaders, such as Almir Ferreira Matos, coordinator of STRs around Marabá. "The Christian base communities have been seedbeds for leaders. They are now heading rural unions that are more active and some have been elected to municipal office, including a mayor," said Vonbergue.

The unions led by Ferreira invaded forest areas rich in *castanheiras*, (Brazil nut trees), for which Maraba has been for many years the main commercial center. Most of the forests were controlled by the Mutran family, who ran the $50 million Brazil nut export trade from Belem. To fight off invasions, the Mutrans surrounded their forests with cattle ranches, burning down thousands of hectares of dense forest. Ferreira argued that the Mutrans had no right to the properties because they only had long-term permits to gather nuts. As the violence grew, the Sarney government reached agreement with the Mutrans and other *castanha* concessionaires on expropriation, and paid them several million dollars in compensation, leaving the properties in the hands of the peasant invaders.

The peasants then discovered that the Brazil nut business was not good enough to make a living, and they turned to agriculture, particularly rice farming, with continuous clearance of land. Vonbergue, with his training in agronomy, realized that the loss of soil fertility would lead to a collapse of the occupation. He turned to Jean Hebette, one of the original French priests, who had gone to the Federal University of Para as a researcher in Amazonian sociology and rural economy. Hebette hit on a plan to create a research center that would provide the rural unions around Marabá with technical input from the university's school of agriculture and a multidisciplinary institute on Amazon studies that he had established, called NAEA (Nucleo de Altos Estudos Amazonicos).

"We are in the presence of a consummated occupation. We would not recommend the settlement in that area, but we have to respond to the needs of the settlers," said Hebette. The result has been the creation of the Tocantins Agrarian Center on a 66-hectare property between Marabá and Itupiranga, which in 1990 began a five-year program with financial support from the European Community and a technical team led by Vanbergue and a specialist in tropical agriculture from Martinique. The liaison with the University of Pará has brought five young agronomists to the center, where they work directly with rural unions in Marabá, Jacundá, Itupiranga, and São João da Araguaia.

Vonbergue said that cooperation between technicians and grassroots producers is oriented toward combining improved agricultural technology, such as rice varieties developed for low-fertility tropical soils at EMBRAPA's tropical agriculture center in Belem, and a better understanding of how the peasant economy works. "For the first time, there is a study of small farmer decision-making before agronomic technology is passed on. We are working with a total of 40,000 families in this area, some of whom own land and others who are 'sem terras'. The first thing we have learned is that marketing is as important as production," said Vonbergue. The simple expedient of having a rice-hulling facility at the center has allowed 200 small producers to hull, sack, and store rice in the harvest season that can be sold six months later at double the price. This makes a dramatic difference for producers who are cultivating an average of 3.5 hectares annually.

Hebette and Vonbergue are not drawn to extractivism. They point out that it takes 500 hectares of forest to maintain one family collecting nuts in the Marabá, at least at the prices paid by Mutran. "The problem of Amazonia is not management of forests, but management of already deforested areas that are going back into secondary growth [*capoeira*, in Brazil]. We think agroforestry could make a contribution in *capoeira* forest land which could be planted with coffee, *cupuacu*, and *castanheira*. But there will still have to be agriculture for food, so technologies will have to be developed that maintain soil fertility," said Vonbergue.

Organized peasants, with backing from the church and political parties like the PT, extend down the rutted mud track called the Transamazonian highway, which begins in Marabá and runs 1,700 kilometers to Itaituba. At Altamira, the main town midway on the highway, the Rural Workers Union has 700 active members and claims to represent 30,000 peasants and farm workers on settlements strung out along the highway. Benedito do Prado, the union president, took office in August 1988 with a new directorate of younger farmers, several of whom had been trained by the CPT for union leadership.

The talks were tough at an assembly of farmers attended by organizers from the state federation. "There, across the street, is the headquarters of the big landowners," said one delegate, pointing at the office of the Ruralist Democratic Union (UDR), a national organization of proprietors. "They are our enemies and we run risks by organizing against them, because they hate to see us united, but that is the price for a better life," said a union delegate. Conflicts along the Transamazonia are frequent. Altamira's urban area already has 40,000 people and is a thriving commercial center. It is poised to be a boomtown if Eletronorte, the state power company, begins the construction on the Xingú, downstream from Altamira, of the huge hydroelectric project, Cararao. Despite the disastrous condition of the highway, land speculators, ranchers, and spontaneous settlers have kept moving in on forest land. INPE's satellite images showed the Transamazonia was one of the most severely burned forest areas during 1988 to 1990, indicating the intensity of land clearing.

"The dream of getting land still brings people here all the time, although the reality is different from the dream. Most new migrants have not been able to settle on a piece of land because the best land is already occupied, and many people are afraid to go deep in the forest without any help from the government in roads, as we had to do," said do Prado. He came from Rio Grande do Sul in southern Brazil and lives on 12 hectares he cleared in an official settlement at kilometer 61, toward Itaituba. Do Prado said he and his four neighbors, who work together in a community with religious inspiration, have to bring the cacao, rice, and black pepper they grow out on their backs over a 7-kilometer path because there is no feeder road. Such determined settlers are not deterred easily, and they are growing in numbers and political awareness.

"We have done a rough census from Altamira to Itaituba and there are at least 200,000 people. There are land clearings as much as 40 kilometers off the highway, including some big ranches. The Transamazonia is not dead, and people here are going to pressure more and more for land and for government assistance," said Domingos de Moraes, a pastoral agent who works with Edwin Krautler, the bishop of Altamira and the Xingu, and a former president of the church's Indian Missionary Council (CIMI).

Moraes, whose parents came as settlers from Paraná, grew up on the Transamazonia. "There is a new reality to be faced here. People have not given up. The cacao that is produced here is the best in Brazil. We don't get a fair price, but the production is worth a lot, and that justifies government investments in roads and services. We are going to fight until we get what is our right," he said. For settlers like these, a stable relationship with Amazonia's environment will come only when there are solid economic foundations for the Green Cathedral.

EPILOGUE

The unriddling of Amazonia cannot come too soon. The twenty-first century, with its prospects for population growth and demand for food and fiber, is not going to be gentle for tropical regions. Now that the isolation of the past has been broken, Amazonia must prepare for regional and global economic integration. A brief respite has been granted by an economic recession, which cut back public works investments in the Amazonia countries, but that is a lull that will end with financial recovery. Highways, railroads, dams, pipelines, mining projects, and colonization settlements will resume and ecosystems that are still virtually intact will be at risk, unless opportune measures are taken to protect the natural heritage.

This study does not offer a blueprint for sustainable development. There is no single recipe for a region where there is such diversity of resources and ecosystems. The conclusions that can be drawn from the experiences that have been described here serve only as a guide for the directions in which research, public policy, development investment, and economic enterprise should go in Amazonia. The overriding goal is simply to discover ways in which man and the other living organisms of the rainforest ecosystems can share the land, water, and energy of Amazonia in a mutually reinforcing way. This requires action at several levels.

A macro-ecological framework must be established. The principal measure is economic-ecological zoning, which involves scientific, technical, legal, and political decisions on land use, forest management, watershed conservation, and the creation of permanent ecological preservation units and indigenous reserves. Only Brazil is moving in this direction in a comprehensive way. Full zoning of Amazonia on the basis of uniform criteria should be a primary objective of the eight-nation Amazon Pact.

Within such a framework, micro-ecological actions at the local community level become the essential element of environmental protection. Amazonian settlers have to acquire a fuller understanding of the nature of the ecosystems where they live and their real potential for economic production. The goal here is to discover how man's intervention in the Amazonian ecosystems can make efficient use of natural resources while maintaining an environment that is stable and healthy. This is a social learning process that is transfered culturally within a community when patterns of economic behavior have a sustainable foundation. This process needs to be advanced by scientific and economic research and supported by technical assistance, credit, and marketing systems.

A special reference should be made to the proper role for private enterprise. Much of the predatory exploitation of Amazonian resources can be attributed to what John Maynard Keynes refered to as the "animal spirits" of economic man. These excesses can be easily recognized in the acquisitive behaviors of the frontier, where the line between legitimate gain and illegal plunder is often blurred. Yet, the vitality of the spirit of enterprise in Amazonia should be considered one of the region's greatest strengths. Amazonia needs profit centers as much as prophets of ecological doom. The economic adventurer, predatory and unbounded by law, can be converted from occasional profits to rational conduct in a sustainable economic system. The last thing Amazonians want is an economic moratorium based on a utopian preservation of the region. But this does not mean unbridled exploitation of natural resources with disregard for ecological stability.

As has been shown, local communities in Amazonia can be organized to make environmental protection and conservation one of their social values. This depends strongly on an informed local leadership, knowledge of the laws, and technical expertise, but it rests on the social needs and the cultural values of the community. Conservation self-interest makes a stronger case in Amazonia than ecological ethics. When residents of the region are asked to set their priorities, they come up with human needs, not the "needs of the forest," as such. The way to achieve a broader social understanding of the need for ecosystem protection is to address the palpable human needs in what I call a "five-Ms" approach: malaria, maternity, mercury, municipalities, and marketing.

Malaria in Amazonia is on the rise and afflicts an estimated 500,000 victims each year. In many communities in Amazonia, more than half the people are chronically debilitated by the mosquito-borne illness, which can be fatal. Health programs that provide protection against malaria in Amazonia are a step toward environmental protection because the threatened species is man. This is true of other endemic diseases in the region that could be eliminated with adequate prevention. The problem is lack of resources and rural medical services.

Maternal-infant care is where a society's sense of long-term values begins. Mothers are more likely to become receptive to conservation programs to "save the rainforests" if they have already seen that something is being done to "save the children." The infant mortality rates in Amazonia run at over 100 deaths for every 1,000 live births, and easily avoided intestinal infections are a major cause. Again, health services are woefully inadequate.

Mercury pollution of the rivers of Amazonia from wildcat gold mining is viewed as a major health hazard for populations that consume fish. In some rivers, fish show toxic levels that are far above permissable levels for human consumption. Tons of mercury are burned in crude retorts for gold recovery each year, and the volatilized mercury returns in rainfall to the rivers where it enters the food chain of fish. The introduction of closed retorts, which avoid waste of mercury, would save miners money and protect them from toxic fumes. Only a determined effort by regulatory agencies requiring *garimpo* cooperatives to operate under a licensing system, which would bring mine sites under control, would produce this reform. This is an expensive program for which international financing is necessary.

Municipalities are the basic administrative unit in most of Amazonia for the implementation of environmental controls and rural extension services. Some municipalities are as large as some European countries, and they have puny budgets that are consumed almost entirely in payrolls. Without specific outside help, mayors are unable to mobilize their communities for environmental issues. This is an area where NGOs, working through local teachers, church groups, communicators, and citizen organizations, can make a major input at low cost by helping mayors and town councils set up environmental boards. The transfer of agricultural and forestry technologies through municipal extension agents also requires NGO backing, because the towns do not have the money or organization to do the job alone.

Marketing of goods that are produced in an environmentally benign way, such as agroforestry products, is one of the weakest links in Amazonia. This is an area for private sector activity in which some NGOs, such as Cultural Survival of Cambridge, Massachusetts, have shown great skill in linking up small producers, usually organized in cooperatives, with foreign buyers. This marketing is an essential component of economic growth based not only on traditional native products, but on new products that have greater value added and therefore higher returns for the producers. Marketing is a challenge and opportunity for private entrepreneurs who have a commitment to sustainable development and long-term capital growth.

The "five Ms," or any combination of economic and social actions for sustainable development in Amazonia, involve long-term programs that require a major commitment of resources. International responsibility for sustainable development of Amazonia has been assumed politically by the leaders of the G-7 countries. But the financial support pledged for the Brazilian rainforest pilot program by the G-7 was slow to materialize. Only after President Collor complained in a letter to the G-7 leaders over foot-dragging on the program did the first $50 million materialize.[1] The question worrying Collor was the degree of commitment by the industrial world to an Amazonian program in relation to their other priorities, such as the Soviet economy or their own domestic economic problems. The

First World rhetoric over saving the rainforests, global climate change, and biodiversity calls for a major, long-term Amazonian program based on close cooperation between industrial and developing nations.

This is not an easy relationship. It involves politically sensitive issues of national sovereignty and mutual trust in areas where suspicion over real intentions is deep. Can this environmental crusade be isolated from the cross-currents of political, trade, and financial conflict in North-South relations? The Amazonian rainforest program is, by its cost and complexity, one of the greatest tests of international cooperation facing the world. Stable funding at adequate levels is essential, but so is good will and a measure of idealism, which is not evident in official bureaucracies.

Ecological activists in the industrial countries are a form of human capital that could be mobilized if the G-7 would support an international voluntary "Green Corps" for service in tropical rainforests and other critical areas. There are examples in Amazonia of privately sponsored volunteers with economic and social skills who have made useful contributions working through NGOs in the region. Many more volunteers with appropriate skills and adequate language training could be usefully employed by NGOs if they were supported by international funds. The shared experience between volunteers drawn from many countries would provide a North-South interface in a crucial area of global concern with long-term projections for international relations. The "Green Corps" could be one of the most tangible peace dividends of the post-Cold War era.

Notes

CHAPTER 1

1. *Time International*, no. 38 (Sept. 18, 1989); *The Economist* (Oct. 15, 1988), p. 25; Adrian Cowell, Central Television, London, 1988; *National Geographic*, Vol. 174, no. 6 (Dec. 1988).

2. Among these are Norman Myers, *Conversion of Tropical Moist Forest* (Washington, D.C.: National Academy of Sciences, 1980); Catherine Caufield, *In the Rainforest* (New York: Alfred J. Knopf, 1985); Robert Repetto, *The Forest for the Trees? Government Policies and the Misuse of Forest Resources* (Washington, D.C.: World Resources Institute, 1988); Suzanne Head and Robert Heinzman (eds.), (San Francisco: Sierra Club Books, 1990).

3. The World Resource Institute (Washington, D.C.) said in its 1990–91 annual report that Brazil was the world's third largest creator of greenhouse gases, after the United States and the Soviet Union. This claim has been rejected by José Goldemberg, Brazil's secretary of science and technology, who said in a letter to *The New York Times* (July 27, 1990) that Amazonia's contribution to increased CO_2 during the past decade has been only 3.7 percent, or one-third WRI's figure, which was based largely on estimated burning during one peak year, 1987 (World Resource Institute, *World Resources 1990–91*, Table 24.1, p. 347).

4. Philip Fearnside, an American ecologist at Brazil's National Amazon Research Institute in Manaus, drew international attention in 1982 when he extrapolated rates of forest clearing observed in some areas of intense colonization and predicted that the entire Brazilian Amazon would be deforested by the year 2000.

5. Government of Brazil/The World Bank/Commission of European Communities, Pilot Program for the Conservation of the Brazilian Rainforests, Brasília (May 1991).

6. Brazilian standard works include Fernando Henrique Cardoso and Geraldo Muller, *Amazonia: Expansao do Capitalismo* (São Paulo: Editora Brasiliense, 1977); Octavio Ianni, *A Luta pela Terra* (Petropolis: Vozes, 1979); and Berta K. Becker, *Geopolitica da Amazonia* (Rio de Janeiro: Zahar, 1982). Some leading examples of this literature in English include Joe Foweraker, *The Struggle for Land—A Political Economy of the Pioneer Frontier in Brazil* (Cambridge: Cambridge University Press, 1981); Stephen G. Bunker, *Underdeveloping the Amazon* (Chicago: University of

Chicago Press, 1985); and Sue Branford and Oriel Glock, *The Last Frontier* (London: Zed Books Ltd., 1985). Also, two collections of essays, Emilio F. Moran (ed.), *The Dilemma of Amazonian Development* (Boulder, Colo.: Westview Press, 1983), and Marianne Schmink and Charles H. Wood (eds.), *Frontier Expansion in Amazonia* (Gainesville: University of Florida Press, 1984). Two more recent works are Susanna Hecht and Alexander Cockburn, *The Fate of the Forest* (London: Verso, 1989) and *Developing Amazonia: Deforestation and Social Conflict in Brazil's Carajás Programme* (Manchester University Press, 1990).

7. Oity Faria Leite, personal communication, April, 1991. Internal problems emerged at Bela Vista with the defeat of the landowners. The community split into political factions, one led by Oity and the other by CPT followers of a radical Roman Catholic priest, Ricardo Resende, who preaches a policy of continuous land seizures. The dispute between the "agro-ecologists" and the "agitators" divided the ranch into two hostile camps where party politics and religious rivalries prevented unity.

8. Jean Hebette, "Reservas Indigenas hoje. Reservas camponesas amanha?" *Para Desenvolvimento*, no. 20/21. Instituto de Desenvolvimento Economico-Social do Pará (IDESP), Belém (1987), pp. 28–29.

9. Judith Lisansky, *Migrants to Amazonia* (Boulder, Colo.: Westview Press, 1990), p. 58.

10. Ibid., pp. 40–58.

11. Ibid., pp. 140–165.

12. This definition coincides with that adopted by the Amazon Pact countries in a workshop on protected areas held in Leticia, Colombia, in 1989 (Carlos Castano Uribe and Martha Rojas Urrego, "Areas Protegidas de la Cuenca del Amazonas," *Diagnostico Preliminar* (Bogotá: FAO/UNEP, 1990)).

13. Thomas E. Lovejoy and Herbert O. R. Schubart, "Land, People and Planning in Contemporary Amazonia." *Conference on the Development of Amazonia in Seven Countries*, Center for Latin American Studies, Cambridge University (Sept. 23-26, 1979). Also, Herbert Schubart, "Ecologia e Utilizaxao das Florestas," in Eneas Salati et al., *Amazonia: Desenvolvimento, Integracao, Ecologia* (São Paulo: Conselho Nacional de Pesquisa, Editora Brasiliense, 1983)

14. Pat Aufderheide and Bruce M. Rich, "Debacle in the Amazon," *Defenders Magazine* (Mar./Apr. 1985); FAO/World Bank Investment Center, "Brazil: Northwest I, II, and III," *Technical Review* (Rome, 1987); Dennis Mahar, "Government Policies and Deforestation in Brazil's Amazon Region," World Bank Environment Department Working Paper no. 7 (Washington, D.C.: World Bank, June 1988).

15. Herman E. Daly, "Sustainable Development—From Concept and Theory towards Operational Principles," *Population and Development Review* (manuscript) (Palo Alto, Calif.: Hoover Institution, 1988); Clem Tisdell, "Sustainable Development: Differing Perspectives of Ecologists and Economists," *World Development*, Vol. 16, no. 3 (Washington, D.C.: World Bank, 1988); Ignacy Sachs, "Sustainable Development: From Normative Concept to Action," *Seminar on Environment*, 30th annual meeting (manuscript) (Amsterdam: Inter-American Development Bank, 1989); Robert J. A. Goodland, "Environmental Sustainability in Economic Development—With Emphasis on Amazonia" (mimeo) (Washington, D.C.: The World Bank, 1988).

16. World Commission on Environment and Development, *Our Common Future* (Oxford University Press, 1987).

17. Business Council for Sustainable Development, "Leadership for Sustainable Development in Developing Countries: An Entrepreneurial Vision" (draft) (Geneva, June 1991).

18. Daly, op. cit., p. 3; Richard B. Norgaard, "The Development of Tropical Rainforest Economics," in Head and Heinzman (eds.), op. cit., pp. 171–183.

19. Donald O. Henry, *From Foraging to Agriculture: The Levant at the End of the Ice Age* (Philadelphia: University of Pennsylvania Press, 1989).

20. For a detailed historical account, see John Hemming, *Amazon Frontier: The Defeat of the Brazilian Indians* (Cambridge, Mass.: Harvard University Press, 1987). For a militant cultural anthropological view, see David Maybury-Lewis, "Demystifying the Second Conquest," in Schmink and Word (eds.), op. cit., pp. 127–134. An informed Brazilian perspective is provided by Darcy Ribeiro, *Os Indios e a Civilizacao* (Petropolis: Editora Vozes, 1977).

21. Among the many examples of this view see Philip Fearnside, "Agriculture in Amazonia," (Oxford: Pergamon Press, 1985), and Susanna B. Hecht, "The Environmental Question in Amazonian Development: Current Dynamics" (unpublished consultant report, Interamerican Bank, 1988).

22. Moran (ed.), op. cit.; particularly J. J. Nicolaides et al., "Crop Production Systems in the Amazon Basin," and E. F. Moran, "Government-directed Settlement in the 1970s: An assessment of Transamazon Highway Colonization," pp. 300–312.

CHAPTER 2

1. Cristobal de Acuña, "Nuevo Descubrimiento del Gran Rio del Amazonas en el ano 1639. Informes de Jesuitas en el Amazonas 1660–1684," *Monumenta Amazonica*, (Iquitos: IIAP-CETA, 1986).

2. Sir Walter Raleigh, *The Discovery of Guiana, in Voyages and Travels Ancient and Modern* (New York: Collier & Son, 1910), p. 389.

3. Alfred R. Wallace, "A Narrative of Travels on the Amazon and Rio Negro" (London: Reeve & Co., 1853), p. 335.

4. Theodore Roosevelt, *Through the Brazilian Wilderness* (New York: Scribner Sons, 1914), pp. 125, 144.

5. Warren Dean, *Brazil and the Struggle for Rubber*, (New York: Cambridge University Press, 1987), Table 1.

6. Ibid., p. 64.

7. Eneas Salati et al., *Amazonia: Desenvolvimento, Integracao, Ecologia* (São Paulo: Brasiliense/CNPQ, 1983), pp. 12–13.

8. Octávio Ianni, "A Luta pela Terra: Conceicao do Araguaia, Estudo de Caso" (São Paulo: CEBRAP, 1978), p. 151

9. See F. H. Cardoso and G. Muller, *Amazonia: Expansao do Capitalismo* (São Paulo: Brasiliense, 1977); Malori José Pompermayer, "Strategies of Private Capital in the Brazilina Amazon," in Marianne Schmink and Charles H. Wood (eds.), *Frontier Expansion in Amazonia* (Gainesville: University of Florida Press, 1984).

10. R.J.A. Goodland, "Environmental Sustainability in Economic Development—With Emphasis on Amazonia" (mimeo) (1988).

11. Joelmir Beting, *O Globo* (Rio de Janeiro, July 20, 1991), p. 30.

12. Bill McKibben, an American wildlife philosopher, describes this condition as a state of human alienation in which the environmental problems of modern society are "deeply rooted in our ways of life, in our thinking, and in our estrangement from nature." ("The Mountain Hedonist," *New York Review of Books* [April 11, 1991]).

13. Fernando Collor de Melo, speech at Carajás, state of Pará (July 16, 1990) announcing Forestry Poles program. Agencia Brasil text.

14. Quote in Harald Sioli, *Amazonia. Fundamentos da Ecologia da Maior Regiao de Florestas Tropicas* (Petropolis: Editora Vozes, 1985), p. 12.

15. Salati, op. cit., p. 25.

16. H. E. Daly, in R.J.A. Goodland, "Environmental Implications of Major Projects in Third World Development," (mimeo) (Washington, D.C.: The World Bank, 1988).

17. Salati, op. cit., p. 12.

18. Ibid, p. 31.

19. Luiz Carlos Molion, "A Amazonia e o Clima do Globo Terrestre," paper presented at seminar, University of Brasília (Nov. 1987), p. 5.

20. Roger Sedjo, "Forests to Offset the Greenhouse," *Journal of Forestry*, Vol. 12, no. 14 (July 1989).

21. Juan de Onis, "Brazil's Wild West," Travel Section, *The New York Times* (March 25, 1990).

22. Salati, op. cit., p. 30.

23. Peter Raven, "We're Killing Our World," Keynote address, American Association for the Advancement of Science, Chicago (Feb. 14, 1987); "Forests, People, and Global Sustainability," National Audubon Society Biennial Convention, Western Washington University, Bellingham (Aug. 24, 1987).

24. Edward O. Wilson, *Biophilia* (Cambridge, Mass.: Harvard University Press, 1984).

25. Les Kaufman, *The Last Extinction* (Cambridge, Mass.: The MIT Press, 1986).

26. Estudos Avancados, "Emopresas de biotecnologia em busca desolucoes." University of São Paulo (June/July 1991).

CHAPTER 3

1. Arnold Toynbee, *Between Maule and Amazon* (Oxford: Oxford University Press, 1967).

2. "Relacion del Descubrimiento del Rio Amazonas y Hoy San Francisco del Quito" (anonymous, circa 1639), *Biblioteca Amazonas*, Vol. III (Quito: Instituto Ecuatoriano de Estudios del Amazonas, 1942). This document, attributed to Father Alonso de Rojas S.J., is an account of the Amazon expedition led by Captain Pedro Texeira written in Quito. It was sent to the Spanish Council of the Indies by Martin de Saavedra y Guzman, governor and captain general of Nueva Granada.

3. Donald W. Lathrap, *The Upper Amazon, Thames and Hudson.* (Southampton, Eng.: The Camelot Press, 1970).

4. Ibid., pp. 112, 127.

5. Charles La Condamine, "A Voyage Made within the Inland Parts of South-America" (London, 1747), p.9.

6. Ibid., p. 15.

7. Ibid., p. 19.

8. Amerigo Vespucio, letter to Lorenzo de Pierfrancesco Medici (1500), in Ger-

man Arciniegas, *Latin America: A Cultural History* (New York: Alfred A. Knopf, 1967), p. 92.

9. Alfred W. Crosby, *Ecological Imperialism* (Cambridge: Cambridge University Press, 1986), p. 21.

10. Ibid., p. 19.

11. Gerardo Reichel–Dolmatoff, "Goldwork and Shamanism," (Bogotá: Gold Museum, 1987).

12. Claude Levi-Strauss, *Tristes Tropiques*. (New York: Washington Square Press Publications, Simon & Schuster, 1977). Strauss, whose early years of anthropological research were in Brazil, recognized the inherent antagonism between the primitive and European cultures."The fact is that these primitive peoples, the briefest contact with whom can sanctify the traveler . . . [are] enemies of our society, which pretends to itself that it is investing them with nobility at the very time when it is completing their destruction, whereas it viewed them with terror and disgust when they were genuine adversaries. The savages of the Amazonian forest are sensitive and powerless victims, pathetic creatures caught in the toils of mechanized civilization." p. 31.

13. John Hemming, *Red Gold: The Conquest of the Brazilian Indians* (Cambridge, Mass.: Harvard University Press, 1978), and *Amazon Frontier: The Defeat of the Brazilian Indians* (Cambridge, Mass.: Harvard University Press, 1987). Levi-Strauss, op. cit.

14. David Maybury-Lewis, "Demystifying the Second Conquest," in Marianne Schmink and Charles H. Wood (eds.), *Frontier Expansion in Amazonia*, (Gainesville: University of Florida Press, 1984), p. 127–134.

15. Arciniegas, op. cit., p. 95.

16. Cristobal de Acuña, *Informes de Jesuitas en el Amazonas* (Iquitos, Peru: Monumenta Amazonica, IIAP-CETA, 1986), pp. 37–107.

17. Ibid., p.105.

18. Robert L. Anderson, "Following Curupira: Colonization and Migration in Pará, 1758 to 1930" (Ph.D diss., University of California, 1976. University Microfilms), pp. 14–18.

19. Ibid., p. 250.

20. Manuel de Oliveira Lima, *O Imperio Brasileiro* (São Paulo: Edicoes Record, 1962), p. 2.

21. A complete analysis of the boom, the bust, and the consequences of the Amazon rubber cycle can be found in Barbara Weinstein, *The Amazon Rubber Boom, 1850–1920* (Stanford Calif.: Stanford University Press, 1983), and Warren Dean, *Brazil and the Struggle for Rubber* (New York: Cambridge University Press, 1987).

22. Montaigne F. Maury, *The Amazon and the Atlantic Slopes of South America* (Washington, D.C., 1853).

23. Ibid., p. 21.

24. Ibid., p. 49.

25. Arthur Cezar Ferreira Reis, *A Amazonia e a Cobica International* (Rio de Janeiro: Civilizacao Brasileira, 1982), pp. 55–78.

26. Ibid., pp. 76–77.

27. Placido de Castro, *O Aspecto Militar na Questao Acreana* (Brasília: Comissao Nacional do Centenario de Placido de Castro, Ministerio de Educacao e Cultura, 1973), p. 58.

28. Roberto Santos, *Historia Economica*, p. 290.

29. Fernando Jacques Magalhaes Pimenta, "The Treaty for Amazonian Cooperation: An Analysis of the Brazilian Proposal in the Light of Brazil's Regional and International Constraints" (Unpublished master's thesis. Washington, D.C.: Graduate School of Arts and Sciences, George Washington University, 1982), p. 121.

30. *Tratado de Cooperacao Amazonica* (Brasília: Ministry of Foreign Relations, 1978), p. 64.

31. Virgilio Barco, "Political del Gobierno Nacional para la defense de las derechos indigenas y la Conservacion Ecologica de la Cuenca Amazonica" (Bogotá: Republica de Colombia, 1988).

CHAPTER 4

1. For detailed accounts of settler experiences in Rondônia on the Transamazon road in Pará, in Ecuador, Peru, and Bolivia, see FAO/World Bank Investment Center, "Brazil: Northwest I, II and III," *Technical Review* (Rome, 1987); Emilio F. Moran (ed.), *The Dilemma of Amazonian Development* (Boulder, Colo.: Westview Press, 1983); and Marianne Schmink and Charles H. Wood (eds.) *Frontier Expansion in Amazonia* (Gainesville: University of Florida Press, 1984).

2. World Bank/FAO Investment Center, op. cit., p. 5.

3. Ibid., Summary, p. 1.

4. Federative Republic of Brazil, Constitution (1988), Article 25 (Brasília).

5. In Instituto de Planejamento Economicoe Social (IPEA), *Amazonia Legal, Descricao Principais Polos Estrategia de Ocupacao* (Brasília: Ministerio de Planejamento, 1972), pp. 27–28. The figures are from a survey by the Brazilian Agrarian Reform Institute (IBRA) in 1966 that covered only the states of Pará, Amazonas, and Acre and the federal territories of Amapá, Rondônia, and Roraima.

6. Maria de Piedade Morais, "A Dimensao Social da Fronteira: A Terra e o Pequeno Produtor" (monograph) (Department of Economy, Federal University of Rio de Janeiro, 1988).

7. *The New York Times* (Aug. 29, 1988), p. 18.

8. Ruralmedia, "An Institutional Marketing Strategy for the Association of Amazonian Investors" (mimeo) (Brasília, 1988).

9. IBGE, *Sinopse Preliminar do Censo Agropecuario, 1985.* Vol. 4, nos. 1 and 5 (Rio de Janeiro: Fundacao Instituto Brasileiro de Geografiae Estatistica, 1987).

10. Robert Goodland, "Environmental Ranking of Amazonian Development Projects in Brazil," *Environmental Conservation*, Vol. 7, no. 1 (1980), pp. 9–26.

11. Early basic studies include Susanna Hecht, "Cattle Ranching in the Eastern Amazon: Evaluation of a Development Strategy" (Ph.D. diss., University of California, Berkeley, 1982), and Hecht, "Environment, Development and Politics: Capital Accumulation and Livestock Raising in Eastern Amazonia," *World Development*, Vol. 13 (1985). For a more recent general treatment, see Robert Repetto, "Overview," in Robert Repetto and Malcolm Gillis (eds.), *Public Policies and Misuse of Forest Resources* (New York: Cambridge University Press, 1988); Philip Fearnside, "Deforestation in the Brazilian Amazon," in George Woodwell (ed.) *The Earth in Transition*, (New York: Cambridge University Press, New York, 1990).

12. Susanna B. Hecht, "Cattle Ranching in Amazonia: Political and Ecological Considerations," in Schmink and Wood (eds.), op. cit.

13. Clando Yokomizo, "Financial and Fiscal Incentives in the Amazon," in *Amazonia: Facts, Problems and Solutions. Symposium Annals* (University of São Paulo, 1990), pp. 58–101.

14. For a detailed account of this period, see Sue Branford and Oriel Glock, *The Last Frontier* (London: Zed Books Ltd., 1985). During the Brazilian military period (1964–85), human rights violations provided more militant issues for the political opposition than the environment did. A major target was Amazon ranching projects that evicted peasants from lands where they had lived for years. There were also organized invasions of large properties by landless peasants. During a decade of strife, there were murders on both sides, but most of the victims were rural union leaders, lawyers, politicians, and agrarian reform activists, including Roman Catholic priests and nuns who supported the peasants. When the conflicts in the so-called Parrot's Beak of northern Goiás, southern Pará, and Maranhão generated a Maoist guerrilla movement, the military imposed martial law, wiped out some 80 guerrillas, and distributed some land under a "pacification" plan. Agrarian reform remains a major issues, but rural violence has subsided since democratic government was restored.

15. *Amazonia, New Universe* (Belém: SUDAM, 1972).

16. Yokomizo, op. cit., p. 73.

17. E.A.S. Serrao and J. M. Toledo, "Sustaining Pasture-Based Production Systems for the Humid Tropics," in T. Downing, S. Hecht, and H. Pearson (eds.), *Development or Destruction of the Livestock Sector in Latin America* (Boulder, Colo.: Westview Press, 1990). Also, for regeneration of tree species in abandoned pasture, see Christopher Uhl et al., "Disturbance and Regeneration in Amazonia: Lessons for Sustainable Land Use," *The Ecologist*, Vol. 19, no. 6 (Nov./Dec. 1989).

CHAPTER 5

1. *International Herald Tribune* (Mar. 23, 1989), travel section.

2. Paul Vantomme, "Forest Extractivism in the Amazon: Is It Sustainable and Economically Viable?" First International Symposium on Environmental Studies on Tropical Rain Forests, Manaus (mimeo) (Oct. 1990). Vantomme calculated the value on the basis of production volumes reported by the Institut Brasileiro de Geografia e Estatistica (IBGE) in its 1989 *Yearbook*. The author said that his calculations based on other sources, including the Brazilian Forestry Development Institute and personal observation, indicated that log production was closer to 42 million cubic meters than the 29 million reported by IBGE.

3. Roberto Samanez Mercado and Sabina Campagnani, "Exportacoes da Floresta Amazonica," Instituto de Florestas da Universidade Federal de Rio de Janeiro (mimeo) (1989).

4. Campbell Plowden and Yuri Kasuda, "Logging in the Brazilian Rainforest," paper presented at the Workshop on the U.S. Tropical Timber Trade, Conservation Options and Impacts, The Rainforest Alliance, New York (Apr. 1989).

5. Alan Grainger, "The Future Role of the Tropical Rain Forests in the World Forest Economy" (Ph.D. diss., University of Oxford, 1986).

6. Duncan Poore, *Natural Forest Management for Sustainable Timber Production* (London: IIED, Oct. 1988).

7. Jean-Paul Lanly, *Evaluation of Tropical Forest Resources* (Rome: FAO/UNEP, 1982); Eneas Salati, *Amazonia: Desenvolvimento, Integracao, Ecologia* (São Paolo: Conselho Nacional de Pesquisa, Editora Brasiliense, 1983); José Carlos Carvalho, "A Politica de Acao do IBDF e Sua Aplicacao na Amazonia" (mimeo) (Brasília: Instituto Brasileira de Desenvolvimento Florestal–IBDF, 1988); John O. Browder, *Subsidies, Deforestation and the Forest Sector in the Brazilian Amazon* (Washington, D.C.: World Resources Institute, 1985); FAO, *Potential for Agriculture*, Annex V, Forestry (1988).

8. Grainger, op. cit., Appendix 2.2, p. A7.

9. Carvalho, op. cit.

10. For example, see Rainforest Action Network, "Rainforest Action Guide" (San Francisco, 1990). "In a generation the last [of the rainforests] could be gone. At stake is the health of our planet. Saving the rainforest is the adventure of a lifetime. And this is your invitation to join. . . ." A tax-deductible donation of $15 (minimum) for membership provides access to eco-products such as T-shirts, a book club, bumperstickers, posters, slideshows, and a global beachball. Natural Resources Defense Council of New York offers a similar package of "Rainforest Rescue Products." (*NRDC Newsline*, Vol. 5, no. 5 [Oct. 1990]).

11. Lei de Politica Agricola, Article 104, Congreso Nacional (Brasília, 1990); and Executive Legislative Decree no. 209, "Rural Land Tax Modification," Article 4 (Brasília, Dec. 18, 1990).

12. Nicholas Burch, letter to Peter Splett, counselor for economic development assistance, Embassy of the Federal Republic of Germany, Brasília (Oct. 1990). In this letter, Burch requested grant assistance of $125,000, which he said would be necessary to kept the project going. "I feel it would be a great pity to abandon a project . . . whose positive examples could be of considerable value in influencing environmental reforms."

13. Dennis Mahar, "Desenvolvimento Economica da Amazonia" (Rio de Janeiro: IPEA, 1978).

14. FAO, report on "Natural Resources for Food and Agriculture in Latin America and Caribbean" (Rome, 1987), p. 78.

15. A. T. Tardin et al., *Subprojeto Desmatamento.* Convenio IBDF/CNPQ–INPE. Relatorio INPE–RPE/103 (São José dos Campos, São Paulo: INPE, 1979).

16. Warwick, Kerr, " A Amazonia: Sua Floresta e Sua Ocupacao," (Brasília: Federal Senate, Parliamentary Investigating Committee on Amazonia, May 1979).

17. *Ciencia Hoje*, Vol. 2, no. 10, São Paulo, Brazil (Jan./Feb. 1984).

18. Alberto W. Setzer et al., *Relatorio de Atividades do Projeto IBDF-INPE "SEQE"—Ano 1987* (Brasília: Ministerio de Agricultura, 1988).

19. Dennis Mahar, "Government Policies and Deforestation in Brazil's Amazon Region," World Bank Environment Department Working Paper no. 7 (Washington, D.C.: World Bank, June 1988).

20. Philip M. Fearnside, Antonio T. Tardin, and Luiz G. Meira Filho, "Deforestation Rate in Brazilian Amazonia" (Brasília: National Secretariat of Science and Technology, Aug. 1990).

21. FAO's estimate for the annual rate of deforestation during 1981–85 in the Amazon region of Peru was 0.4 percent, the same rate assigned to Brazil, and for the Bolivian Amazon the rate was 0.2 percent. For purposes of comparison, FAO's estimate for the same period showed annual rates of deforestation of 6.5 percent in the Ivory Coast, 5 percent in Nigeria, 4.1 percent in Nepal, 4 percent in Costa Rica, 2.6 percent in Thailand, and 1.2 percent in Malaysia.

22. Hildegard O'Reilly Sternberg, *The Amazon River of Brazil* (Weisbaden: Geographische Zeitschrift 40, 1982), p. 17.

23. Vantomme, op. cit.

24. Plowden and Kasuda, op. cit.

25. Brazilian Association of Plywood Industries, *Revista Compensado*, Ano 4, no. 36 (São Paulo, 1990), pp. 19–20. Interview with Luis Favero in a special issue on Amazonia.

26. Christopher Uhl and Ima Celia Guimares Vieira, "Selecao Predatoria," *Ciencia Hoje*, Vol. 10, no. 55 (July 1989), pp. 36–41.

27. Interview (Jan. 11, 1988).

28. FAO, "Training and Investment Preparation for Forestry Development in the Amazon Region," Terminal Mission report (mimeo) (Rome, 1978).

29. Ibid., p. 1.

30. Ibid., p. 10.

31. Orlando Valverde, *O Problema Forestal da Amazonia Brasileira* (Petropolis, Brazil: Editora Vozes, 1980).

32. IBAMA, *Programa Nacional de Conservacao e Desenvolvimento Florestal* (Brasília, 1990).

33. TFAP, "Informe de la Reunion de Coordinadores Nacionales del PAFT en America Latina" (mimeo) (Santiago, May 1990), p. 17.

34. FAO, "Review of the Tropical Forestry Action Plan (TFAP)," (Rome: Committee on Forestry, Tenth session, September 1990), p. 28.

35. Environmental Defense Fund, "The Tropical Forestry Action Plan (TFAP): A Case Study of the Cameroon" (mimeo) (Washington, D.C., Apr. 1990)

36. Manuel Martin Lazaro, personal communication (Sept. 1988).

37. Robert Simeone, report on Palcazu (mimeo) (1988).

38. Simeone, an ecologist who has studied Palcazu closely, describes the system like this:

> In this model, a series of long, narrow strips are clearcut in the forest. The length of the strip is determined by topography, quality of the adjacent forest stand and harvesting logistics. All trees greater than 2 inches in diameter that lie within the 60–100 feet wide strip are felled and extracted with oxen, which reduces soil compaction. They are then left to regenerate naturally from seeds in the adjoining untouched forest. Branches, leaves, and bark from the trees that are felled are cut and left in the forest to serve as nutrients. The smaller trees are made into fence posts, the next larger into telephone poles, and finally those with a diameter greater than 12 inches are used as sawlogs and are processed into timber at a sawmill on the property. The defective material is converted into charcoal. This is supposed to make total use of the available materials. A cutting cycle of 30–40 years is projected, with intermediate commercial thinnings to regulate optimum stand density, species selection, etc.

39. For a succinct description, see Robert Buschbacher, "Natural Forest Management in the Humid Tropics: Ecological, Social and Economic Considerations," *Ambio*, Vol. 19, no. 5 (Aug. 1990), pp. 253–258.

40. Martin Lazaro, personal communication (1989).

41. ITTO, "Memorandum of Understanding, Project: PD 24/88" (Quito: Chimanes Bolivia, 1991).

42. ITTO, "Integration of Forest-Based Development in the Western Amazon—Phase I" (Project document, 1988).

43. SEPLAN/Acre, "Projecto Acre—Humanizar o Desenvolvimento," Gilberto

Siqueiros, coordinator (Rio Branco, 1989), and "Estudo de Viabilidade Tecnico-Economica de BR-364: Trecho Rio Branco–Cruzeiro do Sul–frontiera com Peru" (Rio de Janeiro: Stewart & MacDowell, Engheniaria e Consultoria Ltda., 1989).

44. Flaviano Melo, personal communication, Oct. 1989.

45. Osmarino Amancio Rodriguez, personal communication (Nov. 1990).

46. FUNTAC, "Integrated Forest-Based Development in the Western Amazon—Phase II. Technology for Sustainable Utilization of Forest Raw Materials" (Quito, Ecuador: PD 94/90 Rev.2, ITTO, Eighth session, June 1991).

47. Philip Fearnside and Judy Rankin, "Jarí and Carajás: The Uncertain Future of Large Silvicultural Plantations in the Amazon," *Interciencia*, Vol. 7, no. 6 (1982).

CHAPTER 6

1. Ministerio das Minas e Energia, Centrais Eletricas Brasileiras S.A.–Eletrobras, *Plano 2010, Relatorio Geral. Plano Nacional de Energia Eletrica, 1987–2010* (Rio de Janeiro, 1987), Table 2.1–2.2, p.62.

2. A barrel of oil contains about 6 million BTUs, or the equivalent of 1,748 kilowatt-hours. The conversion of energy wood into electricity is based on calculations made by Eletrobras at the Balbina dam in Amazonas, where a 50-megawatt thermoelectric plant was installed using native forestwood. See ibid., p. 101–103. At a conversion factor of 0.7, 50 billion cubic meters of timber become 35 billion tons of energy wood. It takes 1.5 kilo of wood to produce 1 kilowatt-hour, so each cubic meter of wood generates about 450 kilowatt-hours, and 450 kilowatt-hours times 35 billion tons of wood is 15,750 tetrawatt-hours. Therefore,the forest, as energy wood, has a potential electric power equal to 67 years of Brazil's 1990 consumption of 235 tetrawatt-hours, or 23 years at the projected level of 668 tetrawatt-hours for 2010. (From ibid., Table 1.32–3, p.44.)

3. In small thermal electric plants, it takes about 25 cubic meters of wood daily to generate 1 megawatt continuously, so a 1,000-megawatt plant would consume 25,000 cubic meters of wood each day. With each hectare providing about 120 cubic meters of wood, that would require clearing 80,000 hectares of native forest each year that the plant operated. In a 25-year cycle, which would be the time necessary for regrowth, each 1,000-megawatt plant would consume 20,000 square kilometers of forest. Large wood-fired boilers could be more efficient, but the limiting cost factor would be cutting native forest at distances ever farther from the power plant; only plantation forests would reduce costs.

4. Plano 2010, op. cit., p. 150.

5. Primerio Plano Quinquenal (1956–60), Superintendencia do Plano de Valorizacao Economica da Amazonia (SPVEA), Vol. 2 (Belém, 1955), pp. 65–70.

6. Ibid., Vol. 1, p. 25

7. Peruvian production and reserve figures are adapted from *The Andean Report* (Lima) (Dec. 1988), p. 307, (Oct. 1990), p. 178, and (Jan. 1991), p. 9; and Doreen Gillespies, personal communication.

8. *The Andean Report* (Jan. 1991), p. 2.

9. Nelson de Figueiredo Ribeiro, "A Insercao da Amazonia no Processo de

Desenvolvimento Nacional" (mimeo) (Belém, State of Pará: Secretario de Industria, Comercio e Mineracao, 1988).

10. José Roberto Moreira, "Energy Alternatives" (mimeo) (Institute of Electrotechnology and Energy, University of São Paulo, Brazil). Prepared for International Symposium on Amazonia: Facts, Problems and Solutions, University of São Paulo, 1989.

11. Robert Goodland, "Environmental Assessment of the Tucurui Hydroelectric Project, Rio Tocantins, Amazonia" (mimeo) (Brasília: Eletronorte, 1978).

12. "Aguas de Tucurui Encobrem Denuncia Graves," *Ciencia Hoje* Vol. 3, no. 14 (São Paulo) (Sept. 1984).

13. Andrew D. Johns, "Effects of Habitat Disturbance on Rainforest Wildlife in Brazilian Amazonia" (Washington, D.C.: World Wildlife Fund, 1986).

14. Environmental Policy Institute, letter to Barber Conable, president of World Bank (Sept. 24, 1987).

15. *Jornal do Brasil* (Rio de Janeiro) (July 22, 1991).

16. "Indians Unite to Preserve Forest," *Miami Herald* (Feb. 24, 1989).

17. Senator Robert W. Kasten, Jr., to James A. Baker (June 17, 1986).

18. "Citing Environment, Wis. Senator Assails Loan Plan for Brazil," *Boston Globe* (Jan. 10, 1989), p. 7.

19. *Gazeta Mercantil* (São Paulo) (Nov. 23, 1988).

20. *Folha da São Paulo* (Nov. 12, 1990), Section E-1.

21. *Andean Report* (Lima) (Oct. 1990), p. 178.

22. James D. Nations, "Road Construction and Oil Production in Ecuador's Yasuni National Park" (mimeo) (Antigua, Guatemala: Center for Human Ecology, 1988).

23. Ecuador briefing document, IUCN Tropical Forest Program, Gland, Switzerland, 1985.

24. David Neill, "Oil Wells, Indians and Rainforests of the Upper Amazon," *Missouri Botanical Garden Bulletin*, Vol. 76, no. 3, (May 1988).

25. *World Rainforest Report*, Vol. 6, no. 4 (San Francisco: Rainforest Action Network, 1990).

26. The World Bank, *Ecuador: Development Issues and Options for the Amazon Region.* Report no. IDP-0054 (Washington, D.C., Dec. 1989).

27. Theodore Macdonald, Jr., *Processes of Change in Amazonian Ecuador: Quijos Quichas Indians Become Cattlemen* (Chicago: University of Illinois Press, 1979).

28. Seminario Diagnostico de la Provincia de Naps, "Potencialidades y Limitantes," Tomo IV (Quito: Comision Ecuatoriana Permanente de Cooperacion Amazonica, Ministerio de Relaciones Exteriores, 1989), pp. 255–279.

29. "Crise no Golfo faz a Petrobras rever investimento na Amazonia," *Jornal do Brasil* (Rio de Janeiro) (Oct. 7, 1990), p. 29.

CHAPTER 7

1. *Anuario Mineral Brasileiro, 1987* (Brasília: Departamento Nacional de Producao Mineral, Ministerio de Minas e Energia).

2. Herbert Otto Schubart, "Ecology and Utilization of the Forests," in *Amazonia: Desenvolvimento, Integracao e Ecologia* (São Paulo: Brasiliense, 1983), pp. 104–105.

3. For an authoritative analysis of the Brazilian gold rush and comparison with previous similar events, see Norman Gall, "The Last Gold Rush," *Harper's* (Dec. 1988), pp. 59–65.

4. "Os garimpos fizeram a reforma mineral na Amazonia," *Jornal do Brasil* (Rio de Janeiro) (Nov. 15, 1987), p. 18.

5. O Estado de São Paulo, "Mineracao: Retrato do momento brasileiro" (São Paulo) (Sept. 24, 1987).

6. Gazeta Mercantil, "Brasil contribui para queda" (São Paulo) (July 3, 1990).

7. Wilson de Figueiredo Jardim, "Contaminacaao por Mercurio: Fatos e Fantasias," *Ciencia Hoje*, Vol. 7, no. 41 (Apr. 1988), pp. 78–79.

8. Marc Dourojeanni, "Environmental Management of Placer Gold Mining in the Brazilian Amazon" (unpublished manuscript) (1990).

9. Personal communication, Feb. 1991.

10. João Orestes do Santos, "Os recursos minerais da Amazonia Ocidental" (Manus: Sociedad Brasileira de Geologia, 1987).

11. *Revista Amazonia* (Jan.–Feb. 1990), p. 21. Associacao dos Empresarios da Amazonia, São Paulo.

12. Breno Augusto dos Santos, "Amazonia: Potencial Mineral e Perspectivas de Desenvolvimento" (São Paulo: Editora de Universidade de São Paulo, 1981), p. 23.

13. World Bank, "Brazil: Carajás Iron Ore Project." Project Performance Audit Report (mimeo) (Washington, D.C.: Operations Evaluation Department, Apr. 1990).

14. Iara Ferraz and Eduardo Viveiros de Castro, "Projeto Carajás e os Povos Indigenas: Expectativas e realidade," *Para Desenvolvimento*, no. 21 (Belém: IDESP, 1987).

15. Instituto de Desenvolvimento Economico e Social do Pará (IDESP), *Estatisticas Demograficas do Estado do Pará* (Belém, 1987).

16. World Bank. Carajás Iron Ore Project, op. cit., p. vii.

17. Ibid., p. 24.

18. "Forest Management Centers in the Eastern Amazon," *CVRD* (Rio de Janeiro) (Dec. 1989).

CHAPTER 8

1. "Sarney acusa Bird de lesar o Brasil," *O Estado de São Paulo* (Mar. 7, 1989), p. 47.

2. "Presidente ataca ecologistas," *O Estado de São Paulo* (Apr. 7, 1989), p. 13.

3. "Mesquita Admite Falha no Programa Nossa Natureza," *O Estado de São Paulo* (May 16, 1989).

4. "As criticas ao Brasil sao hipocritas," *O Estado de São Paulo* (May 2, 1989), p. 17.

5. *A Amazonia Brasileira em Foco*, no. 16 (Rio de Janeiro, 1982), pp. 86–95.

6. A similar proposal had been made 40 years earlier, at the height of the rubber crisis, by Efigenio Ferrera de Sales, governor of Amazonas, as a way to "save the region." His plan was to raise a dam 32 meters high at Obidos, where the Ama-

zon narrows, that would have formed an interior "Amazon sea." The high-water mark was to have been "ten meters below the entrance to the Amazon Theatre in Manaus." The dam was to have provided a highway bridge across the Amazon, linking southern Brazil to the northern Amazon regions. The proposal never got anywhere, but it was not considered an assault on Brazil's sovereignty. See *Revista Brasileira de Politica Internacional,* Ano XI, no. 41–42 (Brasília: Ministerio das Relacoes Exteriores, 1968), pp. 95–100.

7. Houston Economic Declaration, G-7. July 12, 1990, para. 66.

8. Government of Brasil, "Pilot Program for the Conservation of the Brazilian Rain Forests," (Brasília, Apr. 1991).

9. FUNAI, "Indian Area Work Plan, 1990–1994" (mimeo) (Brasília, Oct. 1990).

10. Personal communication (Brasília, Be. 21, 1991); "Pesquisa mostra desinteresse em demarcar terras indigenas," (São Paulo, Dec. 26, 1990).

11. Adalberto Ferreira da Silva, "Ocupacao Recente das terras do Acre" (mimeo) (Ph.D. diss., Centro de Planejamentyo Regional, Universidade Federal de Minas Gerais, Belo Horizone, 1982).

12. Jeronimo Santana, letter to President José Sarney (Brasília, Sept. 14, 1988).

13. Nucleo de Diretos Indigenas, "A Atual Situacao dos Indios Uru-eu-wau-wau" (mimeo) (Brasília, 1990). Folha de São Paulo, "Decreto de Sarney mina Plano Ambiental" (Feb. 9, 1990).

14. Personal communication (Feb. 1991).

15. "O indio precisa de tempo para mudar," *Jornal do Brasil* (Rio de Janeiro) (Jan. 28, 1991), p. 13.

16. Dennis Mahar, "Government Policies and Deforestation in Brazil's Amazon Region," World Bank Environment Department Working Paper no. 7 (Washington, D.C.: World Bank, June 1988), pp. 26–27.

17. FAO/World Bank Investment Center, "Brazil: Northwest I, II, and III," *Technical Review* (Rome, 1987), p.1.

18. Programa de Meio Ambiente e das Comunidades Indigenas (PMACI)–Plano de Acao Definitiva. Secretaria de Planejamento (SEPLAN) (Brasília, 1988).

19. "Pilot Program for the Conservation of the Brazilian Expanded Outline." Annex 1 p. 16. (mimeo) (Brasília, Jan, 1991).

20. Comissao Especial de Politica Agricola, Lei 4.086. A/89. (Brasília: Congresso Nacional, Dec. 6, 1990).

21. *Official Gazette* (Apr. 18, 1991). Decree regulating Law 8367 of January 1991 regulating fiscal incentives through regional development funds, Art. 15, item 2.

22. "Politica ambiental permanece indefinida," *Folha da Tarde* (São Paulo) (Dec. 26, 1990).

23. José Lutzenberger, interview, in *O Estado de São Paulo* (May 14, 1991), p. 10.

24. "Plano Agropecuario e Florestal de Rondônia (PLANAFLORO)" (Porto Velho: Secretaria de Planejamento, Governo do Estado de Rondônia, 1989).

25. IBGE, "PMACI I, Diagnostico Geoambiental e Socio-Economico" (Rio de Janeiro: Fundacao Instituteo Brasileiro de Geografia e Estatistica, 1990).

26. Personal communication (Feb.14, 1991).

27. Salim Jordy Filho, director, CVRD Forestry Center, Acailândia, Personal communication. (June 16, 1990).

28. Ibid.

29. Renato Moraes de Jesus, "The Need for Reforestation" (mimeo), U.S. Environmental Protection Agency Symposium (Corvallis: University of Oregon, May 1990).

CHAPTER 9

1. Personal communication (April 1990).

2. First National Rubbertapper Congress, Final Resolution, Article II (Agrarian Reform), item 4. "The areas occupied by rubbertappers are defined as extractive reserves, with their use assured to them" (Brasília, Oct. 17, 1985).

3. Stephan Schwartzman, "Extractive Reserves: Distribution of Wealth and the Social Costs of Frontier Development in the Amazon," Symposium on Extractive Economies in Tropical Forests. National Wildlife Federation/ World Wildlife Fund/ Conservation Foundation (Washington, D.C., Nov. 1989). For another view, see Philip M. Fearnside, "Extractive Reserves in Brazilian Amazonia," *Bioscience* (June 1989), p. 387–393.

4. The best biography of Chico Mendes available in English is Andrew Rivkin's *The Burning Season: The Murder of Chico Mendes and the Fight for the Amazon Rain Forest* (New York: Houghton Mifflin, 1990).

5. Judson F. Valentim, "Research Indicates technology to Avoid Deforestation and Burning for Cattle Ranching in the State of Acre" (mimeo). Seminar (Gainesville: University of Florida, Agronomy Department, Oct. 12, 1989).

6. Instituto de Meioambiente do Acre (IMAC), Environmental Impact Hearing, Bela Alianca project (transcript mimeo) (Rio Branco, Acre, May 11, 1990).

7. For a description of the NGO strategy toward MDBs, see Pat Aufderheide and Bruce M. Rich, "Environmental Reform and the Multilateral Banks," *World Policy Journal* (Spring 1988); Bruce Rich, "The Emperor's New Clothes: The World Bank and Environmental Reform," *World Policy Journal* (Spring 1990).

8. Bruce Rich, "The Multilateral Development Banks, Environmental Policy and the United States," *Ecology Law Quarterly* (1985).

9. EDF, Bruce Rich, and Stephan Schwartzman, Letter to E. Patrick Coady (Jan. 9, 1990); and Memorandum, "Issues That Need to Be Addressed and Resolved in the Rondônia Natural Resources Management Project" (Washington, D.C.).

10. Shahid Husain, letter to Frank E. Loy, chairman of the board, Environmental Defense Fund, World Bank (Washington, D.C., March 5, 1990).

11. "Mobil Accepts Offers for Ecosystem Impact Study" *The Andean Report* (Lima, March 1991).

CHAPTER 10

1. Personal communication (São Paulo, 1990).

2. *The Economist* (Oct. 20, 1990), p. 89.

3. Krenak's prestige received a blow in June 1991 when he was disqualified as

spokesman for the indigenous movement by 15 Indian leaders from the states of Amazonas and Acre, including chiefs from the major Tikuna, Tucano, and Kashinawa people. During a meeting in Manaus, these Indians issued a declaration saying that Krenak was not their legitimate representative because he had never been elected in an assembly, and they denounced "those who use the movement for their own interest." This group was closely liked to the Roman Catholic Church's Indian Missionary Council (CIMI), which is not on the best terms with Krenak. Krenak made a tactical retreat and turned over the titular presidency of UNI to Mario Terena, another acculturated Indian who worked in Brasília for IBAMA. It should be noted that the "friends of the forest" community is rife with politics.

4. Ailton Krenak, Forum of the Forest Peoples Alliance, "Avaliacao da Amazonia no Mundo" (São Paulo, 1991).

5. COICA, "La Poblacion Indigena y el Medio Ambiente en la Amazonia," presentation by Evaristo Nagkuag Ikanen, Seminar on Global Development and the environment" (Montreal, Mar. 30, 1990). Also, COICA, "Our Agenda for the Bilateral and Multilateral Funders of Amazon Development" (mimeo statement) (Montreal, 1990).

6. Stephan Schwartzman, "The Rubber Tappers Strategy for Sustainable Use of the Amazon Rain Forest" (mimeo) (1988).

7. Alfredo Homma, "A Extracao de Recursos Naturais Renovaveis: O Caso do Extrativismo Vegetal na Amazonia" (Ph.D. diss. Universidad Federal de Vicosa, Minas Gerais, 1989).

8. Alfredo Homma, "O Futuro da Economic Extrativa na Amazonia" (mimeo) (Belém: CPATU, 1989). See also "Ecosistema e desenvolvimento, a harmonia possivel," *Amazonia Rural*, no. 11 (1989), pp. 9–10.

9. Personal communication (Rio Branco, Feb. 26, 1991).

10. Personal communication (Rio Branco, Feb. 28, 1991).

11. Jason Clay, facsimile letter to Alianca dos Povos da Floresta (Apr. 5, 1991).

12. James LaFleur and Willem Groeneveld, "Avaliacao da Usina de Beneficiamento de Castanha do Pará da Cooperative Agroextrativista de Xapuri-Acre" (mimeo) (Xapuri, Aug. 1990).

13. Clay, op. cit.

14. Jason Clay, Personal communication (Rio Branco, Feb. 1991).

15. Rolando Polanco (Xapuri coop manager), Personal communication (Rio Branco, Feb. 1991).

16. ITTO, "Integration of Forest-Based Development in the Western Amazon—Phase I. Forest Management to Promote Policies for Sustainable Management." Report on PD24/88. Permanent Committee on Reforestation and Forest Management, Eighth session (Quito, Ecuador, June 1991).

17. CNS/Instituto de Estudos Amazonicos, Seminar: "Alternativas Economicas para as Reservas Extrativistas" debate in Working Group on Rubber (Rio Branco, Feb. 26, 1991).

18. Ibid. Plenary debate (Feb. 25, 1991).

19. Judith Gradwohl and Russell Greenberg, *Saving the Tropical Forests* (Washington, D.C.: Island Press, 1988).

20. FUNTAC, Remote Sensing Laboratory. The comparative figures for the categories of land clearing between 1987 and 1989 showed:

	1987	*1989*
Total area (ha)	628,642	736,123
Pasture	340,169	390,975
Colonization	96,000	123,117
Small lots	72,230	90,374
Fallow lots	60,565	69,700

21. Personal communication (June 17, 1990).

22. Land conflicts in eastern Amazonia have been documented in Sue Branford and Oriel Glock, *The Last Frontier.* (London: Zed Books Ltd., 1985). Useful insights are provided in José Souza Martins, "The State and Militarization of the Agrarian Question in Brazil," in Marianne Schmink and Charles H. Wood (eds.), *Frontier Expansion in Amazonia* (Gainesville: University of Florida Press, 1984), pp. 463–487. Numerous Brazilian studies include Octavio Ianni, "Conceicao do Araguaia: A Luta Pela Terra" (São Paulo: CEBRAP, 1976); Otavio Guilherme Velho, "Frentes de Expansao e Estrutura Agraria" (Rio de Janeiro: Zahar, 1972); and Jean Hebette et al., "Area de Fronteira em Conflitos—O Leste do Medio Tocantins" (Belém, Universidade Federal do Pará, Nucleo de Altos Estudos Amazonicos, 1983).

23. Philip Fearnside, "Agricultural Plans for Brazil's Grande Carajás Program," *World Development,* Vol. 14, no. 3 (London: Pergamon Press, 1986).

24. Personal communication (Imperatriz, Aug. 1990).

EPILOGUE

1. Fernando Collor, letter to Chancelor Helmut Kohl (Oct. 21, 1991). Text distributed by Brazilian Foreign Ministry.

Index